Friedel Hartmann

Introduction to Boundary Elements

Theory and Applications

With 194 Figures

Springer-Verlag
Berlin Heidelberg New York
London Paris Tokyo Hong Kong

Dr.-Ing. Friedel Hartmann

University of Dortmund
Department of Civil Engineering
4600 Dortmund 50
FRG

ISBN 978-3-642-48875-7   ISBN 978-3-642-48873-3 (eBook)
DOI 10.1007/978-3-642-48873-3

Library of Congress Cataloging-in-Publication Data
Hartmann, F. (Friedel)
Introduction to boundary elements: theory and applications/Friedel Hartmann.
ISBN 978-3-642-48875-7
1. Boundary value problems. I. Title.
TA347.B69H37 1989
515.3'5--dc19   89-4160

© Springer-Verlag Berlin Heidelberg 1989
Softcover reprint of the hardcover 1st edition 1989

2161/3020 543210

Cherchez la FEM, c'est mon plaisir
tandis que la BEM, c'est mon délir.

Preface

This book is an introduction into the boundary element method. It offers both an elementary and advanced exposition of the method. The topic of the book is the application of the boundary element method to elastostatics, elastodynamics, plasticity, acoustics and heat conduction. Many elementary examples from mechanics illustrate the underlying principle and help to understand the differences between finite elements and boundary elements. Additional exercises illustrate the technique. The numerical details are fully worked out. The book contains the complete influence matrices for plate-bending problems (linear elements), for membrane and plate problems (quadratic elements) and for problems in three-dimensional elasticity.

As a supplement to the book three boundary element programs for the solution of the Laplace equation, for two-dimensional elasticity problems and plate-bending problems are offered. The programs run on the IBM-PC, PS/2 or on any compatible computer.

Dortmund, Spring 1989   Friedel Hartmann

Acknowledgements

I wish to thank my colleagues T. P. Akyol, R. Dallmann, K. Kremer, H. Kroener, K. Latz, W. Möhrmann (Daimler-Benz AG), P. Schoepp, U. Weber (Siemens AG), M. Ottenstreuer, H. Sippel (BMW AG) and R. Zotemantel (MBB AG) who contributed in various ways to make this book possible.

Contents

Introduction

If you hold a ruler to the end points of a linear function you can draw then the function with a pencil, see Fig. 1.

This is the simplest application of the principle behind the boundary element method: the boundary values of a function determine a function uniquely. In the language of mechanics this principle can be expressed as:

The displacements and the forces on the surface of a body and the exterior load determine the displacements and the stresses within the body uniquely.

In the case of a rod, as in Fig. 2, the corresponding boundary terms are the displacements and the end actions,

$$u(0) = u_1, \qquad u(l) = u_2, \qquad N(0) = -f_1, \qquad N(l) = f_2,$$

and the influence function, which connects the displacement $u(x)$ in the interior with the boundary values, reads,

$$u(x) = (1 - x)u_1 + xu_2 + 1/EA\{f_1 + [(1 - l)x + 1]f_2$$

$$+ \int_0^x p(y)[(1 - x)y + 1]dy + \int_x^l p(y)[(1 - y)x + 1]\,dy\} \tag{1}$$

This influence function equals a (curved) ruler which, if we hold it to the correct marks at the ends of the interval, allows us to trace the shape of the displaced rod with a pencil. The function (1) does not only describe the displacement of the rod in Fig. 3 but the displacement of *any* rod be it a truss element of the Eiffel tower, a tie rod of a drawing bridge, or any other similar structural element. The reader might now ask: if things are so simple, if (1) is the solution to all second order differential equations of the type $-EAu'' = p$ (this is the differential equation of the rod), why then do we need finite elements or boundary elements to solve such equations? To see this check function $u(x)$ more closely: to obtain the displacement $u(x)$ at an interior point x we must know the end displacements and the end actions of the rod

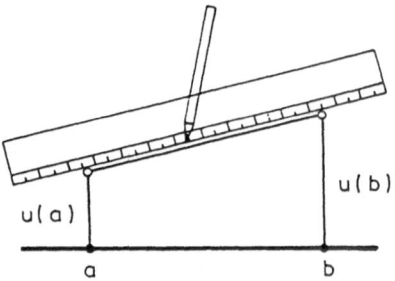

Figure 1 A linear function is determined by its boundary values

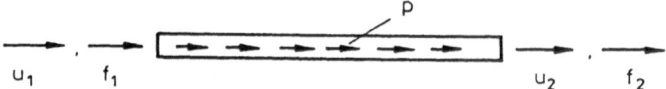

Figure 2 Displacements and end actions of a bar and the longitudinal forces p

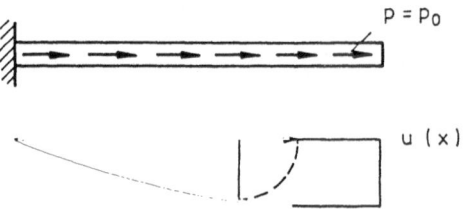

Figure 3 Longitudinal forces pull at a rod

$$u_1 = 0, \qquad u_2 = ?,$$
$$f_1 = ?, \qquad f_2 = 0$$

and these four terms are two terms too many. It is true that the support conditions imply that the displacement at the left end is zero, $u_1 = 0$, and the action at the right end as well, $f_2 = 0$, but these conditions say nothing about the conjugated terms, namely the displacement at the free end, $u_2 = ?$, and the support reaction, $f_1 = ?$. This is why we need boundary elements. In one dimension the boundary consists of just two points and so the unknowns are two numbers. In higher dimensions however the support reactions and the boundary displacements would be functions and so we would have to approximate the unknown functions by, say, linear functions. Boundary elements are just a tool to construct these piecewise linear functions. They are not beam or plate elements; they have no mechanical meaning. They represent a piece of the boundary along which we model the displacements and the force distribution by simple piecewise polynomials.

In one dimension the boundary consists, as we mentioned before, of the endpoints of the interval $[0, l]$ so we need no boundary elements. But we face the same problem as in higher dimensions: we must find the unknown boundary values and the technique used to find these terms is virtually equivalent to the technique we employ in higher dimensions: we utilize a *coupling condition* between the boundary terms u_i and f_i. In the case of a rod this coupling condition reads

$$\frac{EA}{l} \begin{bmatrix} 1 & -1 \\ -1 & 1 \end{bmatrix} \begin{bmatrix} u_1 \\ u_2 \end{bmatrix} = \begin{bmatrix} f_1 \\ f_2 \end{bmatrix} + \begin{bmatrix} p_1 \\ p_2 \end{bmatrix} \tag{2}$$

where the matrix is the stiffness matrix of the rod and the components p_i the negative *end fixing forces*. The end fixing forces are the support reactions f_i when both ends of the rod are fixed ($u_i = 0$ implies $f_i = -p_i$).

This coupling condition is the key to our problem: Of the four boundary terms u_1, u_2 and f_1, f_2 in Eq.(2) two are prescribed while the conjugated terms are unknown

$$u_1 = 0, \qquad u_2 = ?,$$
$$f_1 = ?, \qquad f_2 = 0,$$

so that the two equations in Eq.(2) (the distributed load $p_0 = 100$ kN/m and the length $l = 4$m)

$$\frac{EA}{l} \begin{bmatrix} 1 & -1 \\ -1 & 1 \end{bmatrix} \begin{bmatrix} 0 \\ u_2 \end{bmatrix} = \begin{bmatrix} f_1 \\ 0 \end{bmatrix} + \begin{bmatrix} 200 \\ 200 \end{bmatrix}$$

can be solved for the two unknowns,

$$u_2 = 0.8 \text{ m}, \qquad f_1 = -400 \text{ kN}, \qquad (\frac{EA}{l} = 250)$$

If we substitute these values and the given boundary data $u_1 = 0$, $f_2 = 0$ into the influence function (1) then we obtain the displacement of the rod

$$u(x) = 0.4\,x - 0.05\,x^2 \quad \text{m}$$

The problem is solved. This solution procedure is applicable to all kinds of boundary conditions and all kinds of loads. What we need is the stiffness matrix of the rod, a list of the negative end fixing forces p_i and an equation solver (for a system 2×2).

In a matrix notation the coupling condition (2) can be written as

$$\boldsymbol{K u} = \boldsymbol{f} + \boldsymbol{p} \tag{3}$$

and in this form it is probably familiar to the reader, though it usually appears in the literature only in the "homogeneous" form

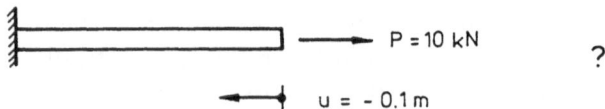

Figure 4 The end action and the end displacement cannot be prescribed independently of each other

$$Ku = f$$

that is when the end fixing forces are zero because distributed forces are absent. But this loading condition is only a special case.

Every engineer knows that he cannot pull at the end of a rod with a specific force P and require at the same time that the end displacement takes a certain value u, chosen independently of P, see Fig. 4. If P is given, then u is uniquely determined,

$$u = \frac{Pl}{EA}$$

The mathematical expression of this coupling condition between the boundary values of a rod is the system (3).

If we substitute arbitrary boundary values u_i, f_i into the influence function (1) then neither the end displacements nor the end actions of the function so constructed would coincide with the boundary values u_i and f_i. But the opposite holds true: If a set u_i, f_i of boundary values satisfies the coupling condition (3) then these values are also the boundary values of the influence function.

This solution procedure applies now to higher dimensional problems, to plates, elastic bodies, etc. as well. It is only in 2-D and 3-D that the boundary values are functions and therefore the coupling conditions become *integral equations*. Integral equations are equations $Ku = f$ with infinitely many rows and columns. Infinitely many because the boundary consists of infinitely many points and so the boundary functions u and f are symbols with infinitely many degrees-of-freedom. To solve these integral equations approximately we interpolate the given and the unknown boundary functions by n piecewise linear or quadratic functions and determine the n nodal values of the unknown functions by satisfying the integral equation only at n collocation points. By this technique the infinite system of equations becomes a system of n equations. This system is much smaller than a comparable FE-stiffness matrix.

If you are familiar with finite elements then you know that we can eliminate from any stiffness matrix the internal nodal degrees of freedom and represent the elastic properties of a domain by the displacements and actions on the boundary alone. This condensed stiffness matrix formulates a similar coupling condition between the nodal values of displacements and the equivalent nodal forces as the integral equations which we use in the boundary element method.

Figure 5a FE-discretization of a floor plate

Figure 5b BE-discretization of the same plate

In the finite element method we construct a solution by patching together a set of piecewise linear or quadratic domain functions, see Fig. 5a, and we control the nodal values of this approximation by shaking the structure n-times. Each time the virtual external work of the true forces and the FE-forces must be the same (*principle of virtual displacements.*)

Instead in the boundary element method there is no need to build a solution from scratch. We already have a beautiful solution, the influence function of the governing equation which is an analytical, smooth function in the interior and which we must "only" feed with the correct boundary functions. These boundary functions are, so to speak, the marks at the ends of the interval which we need to put the ruler, the influence function, in place. To find these functions we approximate them by piecewise linear or quadratic functions ("the boundary elements"), see Fig. 5b, and we control their nodal values by applying *Betti's principle*: n-times we let a concentrated force act on the boundary and n-times the reciprocal external work of the real elastic state and the auxiliary singular state must be the same.

1 Fundamentals

This chapter is a concise summary of the fundamentals of the boundary element method. The experienced reader may prefer to look over this material rather casually and then refer to it again when need arises. The novice reader is advised to begin with chapter 2 where the method is explained in detail by applying it to one-dimensional problems.

1.1 Notation

If we use indicial notation then summation over all indices which appear twice is implied. Hence, the trace a_{ii} of a matrix A is the expression

$$\operatorname{tr}(A) = a_{ii} = a_{11} + a_{22} + \cdots + a_{nn}$$

The scalar product of two vectors $a = \{a_i\}$ and $b = \{b_i\}$ reads,

$$a \cdot b = a_i \, b_i \, ,$$

and the notation for the scalar product of two matrices $A = [a_{ij}]$ and $B = [b_{ij}]$ is

$$A \cdot B = \operatorname{tr}(A^T B) = a_{ij} b_{ij} = a_{11} \, b_{11} + a_{12} \, b_{12} + \cdots + a_{nn} \, b_{nn}$$

The matrix product $AB = C$ of two matrices A and B is a matrix C with the elements

$$c_{ij} = a_{ik} b_{kj} \qquad \text{(sum over } k)$$

If we consider vectors to be $(n \times 1)$ matrices then the scalar product of two vectors can also be written as

$$a \cdot b = a^T b$$

If the T is attached to the second vector then we mean the matrix

$$C = a \, b^T = a \otimes b$$

with the entries

$$c_{ij} = a_i b_j$$

The concept of the scalar product

$$\boldsymbol{a} \cdot \boldsymbol{b} = a_1 b_1 + a_2 b_2 + \cdots + a_n b_n =: (\boldsymbol{a}, \boldsymbol{b})$$

can also be extended to functions. The integral

$$\int_0^l f(x)g(x)\,dx =: (f, g)$$

is called the L_2-*scalar product* of two functions. All work integrals in mechanics are L_2-scalar products between tensors of rank 0 (scalar-valued functions)

$$\delta W_e = \int_0^l p(x)\hat{u}(x)\,dx \,,$$

(*virtual external work of a rod*),

tensors of rank 1 (vector-valued functions),

$$\delta W_e = \int_\Omega \boldsymbol{p}(\boldsymbol{x}) \cdot \hat{\boldsymbol{u}}(\boldsymbol{x})\,d\Omega = \int_\Omega p_i(\boldsymbol{x})\hat{u}_i(\boldsymbol{x})\,d\Omega \,,$$

(*virtual external work of an elastic plate*),

or rank 2 (matrix-valued functions), as e.g.

$$\delta W_i = \int_\Omega \boldsymbol{E} \cdot \hat{\boldsymbol{S}}\,d\Omega = \int_\Omega \varepsilon_{ij}\hat{\sigma}_{ij}\,d\Omega \,,$$

(*energy product of an elastic body*)

Closely connected with scalar products is the concept of the transposed operator. The *transposed matrix* $\boldsymbol{A}^T$ is defined by

$$(\boldsymbol{A}\boldsymbol{x}, \hat{\boldsymbol{x}}) = (\boldsymbol{x}, \boldsymbol{A}^T\hat{\boldsymbol{x}})$$

Hence, the *transposed operator* of an integral operator

$$Ku = \int_0^l k(y, x)u(y)\,dy$$

is the operator K^T which satisfies

$$(Ku, \hat{u}) = (u, K^T \hat{u})$$

and because

$$(Ku, \hat{u}) = \int_0^l [\int_0^l k(y, x) u(y) \, dy] \, \hat{u}(x) \, dx$$

$$= \int_0^l u(y) [\int_0^l k(y, x) \hat{u}(x) \, dx] \, dy = (u, K^T \hat{u})$$

we conclude that the operator transposed to K,

$$K^T u = \int_0^l k(x, y) u(y) \, dy,$$

has the same kernel as K only that the places of x and y are interchanged. If the kernel is symmetric then we have, as in the case of a symmetric matrix, $K = K^T$.

The kernels of influence functions are functions of two variables: the source point and the integration point. The first point we call $x = \{x_i\}$ and the second one $y = \{y_i\}$. Hence, from the origin o of the coordinate system two parallel sets of axis of coordinates emanate. On the one set of axis we mark the coordinates x_i of the source point and on the other system the coordinates y_i of the integration point.

The distance between the two points, see Fig. 1.1,

$$r = [(y_1 - x_1)^2 + (y_2 - x_2)^2]^{\frac{1}{2}},$$

is a function of both coordinates and it therefore possesses derivatives with respect to y_i and x_i:

$$r_{,y_i} = \frac{y_i - x_i}{r}, \qquad r_{,x_i} = \frac{x_i - y_i}{r} = -r_{,y_i}$$

The notation

$$r_{,i} = r_{,y_i}$$

will henceforth denote the derivative with respect to the coordinates y_i.

The single derivatives of the distance r constitute two unit vectors,

$$\nabla_x r = \{r_{,x_i}\}, \qquad \nabla_y r = \{r_{,y_i}\}$$

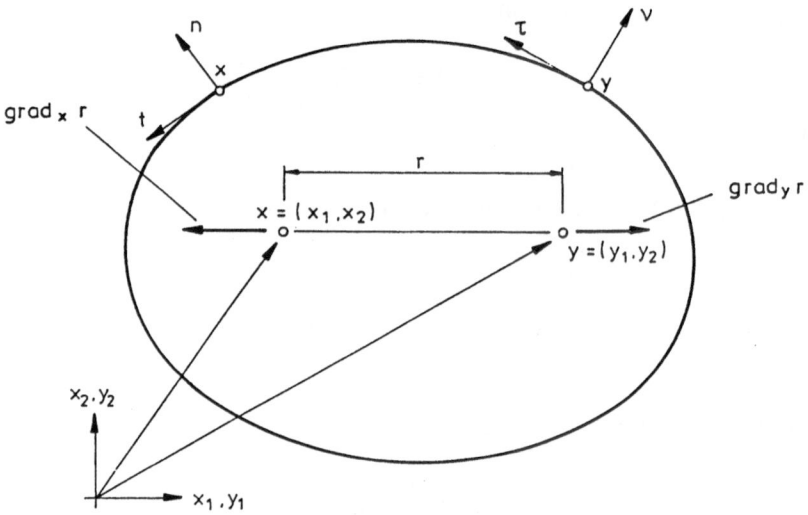

Figure 1.1 The source point x and the integration point y

These two gradients lie on the straight line that connects the two points x and y but they point in opposite directions. For to increase the distance between the two points x and y at the maximum possible rate you have to push the points in these directions, see Fig. 1.1.

If the source point x lies on the boundary then we denote the normal vector and the tangent vector at x by n and t,

$$n = \{n_1, n_2\}^T, \qquad t = \{t_1, t_2\}^T = \{-n_2, n_1\}^T,$$

and at an integration point y by ν und τ

$$\nu = \{\nu_1, \nu_2\}^T, \qquad \tau = \{\tau_1, \tau_2\}^T = \{-\nu_2, \nu_1\}^T$$

The directional derivatives

$$r_n = \frac{\partial r}{\partial n} = r_{,x_1} n_1 + r_{,x_2} n_2 = -r_{,1} n_1 - r_{,2} n_2,$$

$$r_t = \frac{\partial r}{\partial t} = r_{,x_1} t_1 + r_{,x_2} t_2 = +r_{,1} n_2 - r_{,2} n_1,$$

$$r_\nu = \frac{\partial r}{\partial \nu} = r_{,y_1} \nu_1 + r_{,y_2} \nu_2 = +r_{,1} \nu_1 + r_{,2} \nu_2,$$

$$r_\tau = \frac{\partial r}{\partial \tau} = r_{,y_1} \tau_1 + r_{,y_2} \tau_2 = -r_{,1} \nu_2 + r_{,2} \nu_1,$$

are a measure of the rate of change of the distance r if the point $\boldsymbol{x}$ or $\boldsymbol{y}$ respectively is pushed into the direction of the normal or tangent vector. These directional derivatives are the scalar product between the gradient ∇r and the pertinent direction. As all the vectors involved are unit vectors the values of these directional derivatives range from -1 to $+1$. A rate of $+1$ means that the vector points away from the opposite pole (the point $\boldsymbol{x}$ or $\boldsymbol{y}$). A rate of -1 means that the vector points directly to the opposite pole.

The higher directional derivatives are obtained as follows. Let

$$\boldsymbol{m} = \{m_1, m_2\}^T, \qquad \boldsymbol{p} = \{p_1, p_2\}^T = \{-m_2, m_1\}^T$$

be a further pair of orthogonal unit vectors, which we can interpret as a normal vector $\boldsymbol{m}$ and its associated tangent vector $\boldsymbol{p}$. Employing the notation

$$r_p = r_{,x_1}\, p_1 + r_{,x_2}\, p_2 = r_{,1}\, m_2 - r_{,2}\, m_1$$

we obtain then for the second order directional derivatives the expressions

$$r_{\nu,m} = \frac{\partial r_\nu}{\partial m_x} = \frac{1}{r} r_\tau r_p, \qquad r_{\tau,m} = \frac{\partial r_\tau}{\partial m_x} = -\frac{1}{r} r_\nu r_p,$$

$$r_{n,m} = \frac{\partial r_n}{\partial m_x} = \frac{1}{r} r_t r_p, \qquad r_{t,m} = \frac{\partial r_t}{\partial m_x} = -\frac{1}{r} r_n r_p$$

With the help of these formulas, the chain-rule

$$(r_\nu r_n)_{,m} = r_{\nu,m}\, r_n + r_\nu r_{n,m} \qquad \text{(etc.)}$$

and the rule

$$\left(\frac{1}{r}\right)_{,m} = -\frac{1}{r^2} r_m$$

we can easily calculate directional derivatives of any order.

The ε-*neighborhood* of a point $\boldsymbol{x}$ with respect to a domain Ω is the set of all those points $\boldsymbol{y}$ in Ω, whose distance from $\boldsymbol{x}$ is equal to or less than ε, see Fig. 1.2.

$$N_\varepsilon(\boldsymbol{x}) = \{\boldsymbol{y} \in \Omega \mid |\boldsymbol{y} - \boldsymbol{x}| \le \varepsilon\}$$

To isolate a singularity we often remove a small neighborhood $N_\varepsilon(\boldsymbol{x})$ of the singular point from Ω. We denote this punctured domain by

$$\Omega_\varepsilon = \Omega - N_\varepsilon(\boldsymbol{x}),$$

and its boundary by

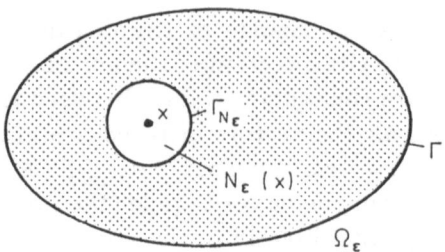

Figure 1.2 A punctured domain

$$\Gamma_{N_\varepsilon}(\boldsymbol{x}) = \{\boldsymbol{y} \in \Omega \mid |\boldsymbol{y} - \boldsymbol{x}| = \varepsilon\}$$

We often choose the point $\boldsymbol{x}$ as the origin of the system of coordinates and we then switch to spherical coordinates

$$
\begin{aligned}
y_1 &= r\cos\varphi, & y_1 &= r\cos\varphi\sin\vartheta, \\
y_2 &= r\sin\varphi, & y_2 &= r\sin\varphi\sin\vartheta, \\
& & y_3 &= r\cos\vartheta,
\end{aligned}
$$

so that

$$
\begin{aligned}
r_{,1} &= \cos\varphi, & r_{,1} &= \cos\varphi\sin\vartheta, \\
r_{,2} &= \sin\varphi, & r_{,2} &= \sin\varphi\sin\vartheta, \\
& & r_{,3} &= \cos\vartheta
\end{aligned}
$$

The surface element dS_1 of the n-dimensional unit sphere S_1 is

$$
dS_1 = \begin{cases} d\varphi, & \text{arc-length of the unit circle (2-D)}, \\ \sin^2\vartheta\, d\vartheta\, d\varphi, & \text{surface element of the unit sphere (3-D)} \end{cases}
$$

The tangent vectors which emanate from a point $\boldsymbol{x}$ on the surface of a body form a cone. This cone splits the unit sphere S_1, which surrounds the point $\boldsymbol{x}$ and is centred at $\boldsymbol{x}$, into two parts. That part of S_1, which lies inside the cone we denote by $S_1(\boldsymbol{x}, \Omega)$. Its size determines the angle of the boundary point

$$
\Delta\varphi(\boldsymbol{x}) = \int\limits_{S_1(\boldsymbol{x},\Omega)} dS_1
$$

Because the surface of the unit sphere has the size 4π and the perimeter of the unit circle the size 2π the angle of a smooth point on a surface is $\Delta\varphi = 2\pi$ and the angle of a smooth point on the edge of a plate is $\Delta\varphi = \pi$.

Influence functions are scalar products (in the L_2-sense) between a kernel $k(y, x)$ and a layer $f(y)$

$$u(x) = \int_\Omega k(y, x) f(y)\, d\Omega_y$$

The kernel usually has the form

$$k(y, x) = \frac{k(\varphi, \vartheta)}{r^m}\,,$$

that is, it is of the type

$$(distance)^{-m} \times geometry$$

If the source point x is not contained in the domain of integration, Ω, then the influence is always finite, the integral exists. If the source point lies in Ω, then the distance r becomes zero at $y = x$ and so the kernel has an infinite peak. This must not necessarily mean that the influence, the integral over Ω, is infinite. A simple way to check for this is to assume that Ω is the unit ball $N_1(x)$ and the source point x the centre of the ball. Because the volume element of the n-dimensional unit ball is

$$d\Omega = r^{n-1} dr\, dS_1\,,$$

the integral

$$\int_\Omega \frac{k(\varphi, \vartheta)}{r^m}\, d\Omega_y = \int_0^1 r^{n-1-m}\, dr \int_{S_1} k(\varphi, \vartheta)\, dS_1 \qquad (1.1)$$

exists if and only if $m < n$. Such an integral is called *weakly singular*. In case $m \geq n$, the integral is called *strongly singular*.

If the domain Ω is not a ball, then we can always split Ω into the unit ball with centre at x (the scale is relative) and a punctured domain where the integral is regular.

Tricomi calls the function $k(\varphi, \vartheta)$, which only depends on the geometry of the domain the *characteristic*, [1]. This name should not to be confused with the characteristic function, which we shall introduce later.

An integral such as Eq.(1.1) exists also for exponents $m \geq n$ in the sense of a *Cauchy principal value* if the integral of the characteristic over the unit sphere S_1 vanishes

$$\int_{S_1} k(\varphi, \vartheta)\, dS_1 = 0\,,$$

because we then have

$$\lim_{\varepsilon \to 0} \int_{\Omega_\varepsilon} \frac{k(\varphi, \vartheta)}{r^m} \, d\Omega_y = \lim_{\varepsilon \to 0} \int_\varepsilon^1 r^{n-1-m} \, dr \int_{S_1} k(\varphi, \vartheta) \, dS_1 = 0$$

Hence, the principal value of an integral is the limit of the integral obtained by, first, removing the singularity and then letting the radius ε of the cavity tend to zero. Regular integrals, as e.g. the integral of the function $\sin x$ over the unit interval $[0,1]$, do not depend on the summation process. Strongly singular integrals converge (if at all) only if the singularity is given a chance to annihilate itself in this particular way.

1.2 The basic idea

In the following we will formulate the basic idea behind the boundary element method once in the language of mathematics and once in the language of mechanics.

1.2.1 The basic idea from a mathematical point of view

A matrix A and two vectors x, $\hat{x}$, arranged in the order

$$\hat{x}^T A x$$

form a scalar. Because a scalar is invariant to transpositions this expression is equivalent to the expression

$$x^T A^T \hat{x}$$

Thus any pair of vectors x and $\hat{x}$ satisfies Betti's principle

$$B(\hat{x}, x) = \hat{x}^T A x - x^T A^T \hat{x} = 0$$

The work done by the vector of nodal forces $A\,x$ on acting through the vector of nodal displacements $\hat{x}$ is the same as the work done by the nodal forces $A^T \hat{x}$ on acting through the nodal displacements x.

Consider the solution $x = x_L$ of the equation

$$A x = b$$

and let g^i be the solution of the adjoint equation in case the right-hand side is a unit vector e_i (a "point load")

$$A^T g^i = e_i \,,$$

It then follows easily that

$$B(g^i, x_L) = g^{iT} b - x^T e_i = g^{iT} b - x_i = 0 \,,$$

or

$$x_i = g^{iT} b$$

Hence, the "fundamental solution" g^i of the adjoint operator A^T furnishes an "integral representation" of the component x_i of the solution vector, that is, it follows that the rows of the inverse of a matrix

$$x = A^{-1} b$$

are the fundamental solutions of the transposed matrix, the adjoint operator and *to solve an equation means to multiply the right-hand side with the fundamental solutions of the adjoint operator.* This is the basic principle of the boundary element method and, we dare say, of mathematics in general.

1.2.2 The basic idea from an engineering point of view

An engineer will explain the same concept as follows:

The relation between the nodal displacements u_i and nodal forces f_i of a truss as in Fig. 1.3 is expressed by a symmetric matrix, the stiffness matrix K

$$K u = f$$

According to Betti's principle (note that $K = K^T$)

$$B(\hat{u}, u) = \hat{u}^T K u - u^T K \hat{u} = \hat{u}^T f - u^T \hat{f} = W_{1,2} - W_{2,1} = 0 \,,$$

the reciprocal external work of the nodal forces of two loadcases, f_i and $\hat{f}_i$, is the same. Hence we can determine, say, the vertical displacement u_1 of node 1 by letting a point force $\hat{P} = 1$ act in the direction of the unknown displacement.

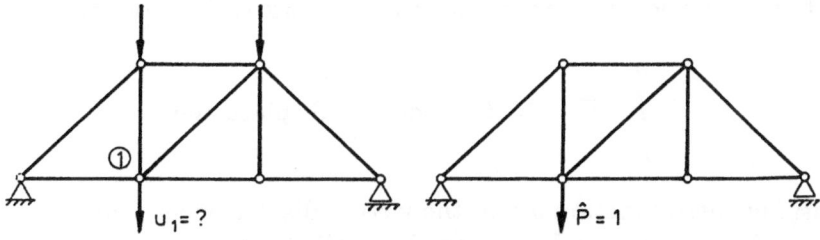

Figure 1.3 Calculation of the nodal displacement u_1 with Betti's principle

In this case the vector $\hat{f}$ of the nodal forces is the unit vector e_1 so that we have

$$K g^1 = e_1 , \qquad g^1 = \text{vector of associated nodal displacements} ,$$

and hence Betti's principle renders

$$B(g^1, u) = g^{1T} f - u^T e_1 = g^{1T} f - u_1 = 0 ,$$

or, solved for u_1,

$$u_1 = g^{1T} f$$

1.3 Influence functions

The state variables of continua are functions and we shall, therefore, in the following replace the vectors u and $\hat{u}$ again by functions u and $\hat{u}$ and the matrix K by a differential operator

vector	=	function,
matrix	=	differential operator,
scalar product	=	integral

But the basic idea remains the same: the pairing of a singular elastic state with a regular elastic state renders an influence function for the displacement u of a structural element or, more generally speaking, for the solution of a differential equation.

1.3.1 Betti's principle

The longitudinal displacement u of a rod satisfies the differential equation

$$-EAu'' = p$$

so that the expression of the virtual external work is the integral

$$\int_0^l -EAu'' \hat{u} \, dx = \text{force} \times \text{displacement}$$

If the displacement function u and the virtual displacement $\hat{u}$ are both smooth enough to allow integration by parts to be applied then we can split this integral into a "boundary integral" and a domain integral

$$\int\limits_0^l -EAu''\hat{u}\,dx = -[N\hat{u}]_0^l + \int\limits_0^l EAu'\hat{u}'\,dx\,, \qquad (N = EAu')$$

This is *Green's first identity*. Putting all terms on one side and listing the necessary requirements for the functions u and $\hat{u}$ explicitly, we may summarize our results as follows,

p: $u \in C^2[0,l]$, $\hat{u} \in C^1[0,l]$,

$$\text{q: } G(u,\hat{u}) = \int\limits_0^l -EAu''\hat{u}\,dx - [N\hat{u}]_0^l - \int\limits_0^l \frac{N\hat{N}}{EA}\,dx = \delta W_e - \delta W_i = 0$$

The letters p, q are short for *if, then*. If u and $\hat{u}$ are sufficiently smooth then the sum of the first two terms, the virtual external work, is equal to the domain integral, the virtual internal work. This is the *principle of virtual displacements*.

If we interchange the places of u and $\hat{u}$, then we obtain *the principle of virtual forces*

p: $\hat{u} \in C^2[0,l]$, $u \in C^1[0,l]$,

$$\text{q: } G(\hat{u},u) = \int\limits_0^l -EA\hat{u}''u\,dx - [\hat{N}u]_0^l - \int\limits_0^l \frac{\hat{N}N}{EA}\,dx = \delta W_e^c - \delta W_i^c = 0\,,$$

and if we subtract the two identities

$$B(\hat{u},u) = G(\hat{u},u) - G(u,\hat{u}) = 0 - 0 = 0\,,$$

then we obtain *Betti's principle*, Green's second identity.

p: $\hat{u},\, u\, \in C^2[0,l]$,

$$\text{q: } B(\hat{u},u) = \int\limits_0^l -EA\hat{u}''u\,dx + [\hat{N}u - \hat{u}N]_0^l - \int\limits_0^l \hat{u}(-EAu'')\,dx$$

$$= W_{1,2} - W_{2,1} = 0$$

These simple mathematical structures form the basis of mechanics. To allow us to focus in the following better on the main points and to avoid unnecessary repetitions we introduce a somewhat abstract but unifying notation for the different operators. We shall denote the differential operators in the domain by the capital letter D and the boundary operators by ∂^i, so that we write in the case of a rod as follows:

$$Du = -EAu'', \qquad \partial^0 u = u, \qquad \partial^1 u = EAu', \qquad \text{normal force}$$

In the case of a beam

$$Dw = (EIw'')'', \quad \partial^0 w = w, \qquad \partial^1 w = w', \qquad \partial^2 w = -EIw'', \quad \text{moment},$$
$$\partial^3 w = -(EIw'')', \quad \text{shear force}$$

In the case of a stretched membrane (N = uniform tension)

$$Du = -N\Delta u, \qquad \partial^0 u = u, \qquad \partial^1 u = N\frac{\partial u}{\partial n}, \qquad \text{traction}$$

In the case of a Kirchhoff plate

$$Dw = K\Delta\Delta w, \qquad \partial^0 w = w, \qquad \partial^1 w = \frac{\partial w}{\partial n}, \qquad \partial^2 w = M_n(w), \qquad \text{moment}$$

$$\partial^3 w = V_n(w), \quad \text{Kirchhoff-shear}$$

and in the case of an elastic plate or body

$$\boldsymbol{Du} = -\boldsymbol{Lu}, \qquad \partial^0\boldsymbol{u} = \boldsymbol{u}, \qquad \partial^1\boldsymbol{u} = \tau(\boldsymbol{u}), \qquad \text{traction vector}$$

(For an explicit representation of the system $-\boldsymbol{L}$ see chapter 4.)

To such a self-adjoint operator of order $2m$,

$$\begin{aligned}\text{rod, membrane, elastic plate and body,} &\qquad 2m = 2,\\ \text{beam, Kirchhoff-plate,} &\qquad 2m = 4,\end{aligned}$$

belong two integral identities

$$\text{p:} \; u, \hat{u} \; \in C^{2m}(\bar{\Omega}) \times C^m(\bar{\Omega}),$$

$$\text{q:} \; G(u, \hat{u}) = \int\limits_{\Omega} Du\,\hat{u}\,d\Omega - \sum_{i=1}^{m}(-1)^i \int\limits_{\Gamma} \partial^{2m-i}u\,\partial^{i-1}\hat{u}\,ds - E(u,\hat{u})$$
$$= \delta W_i - \delta W_e = 0,$$

and

p: $\hat{u}, u \in C^{2m}(\bar{\Omega})$,

q: $B(\hat{u}, u) = G(\hat{u}, u) - G(u, \hat{u})$

$$= \int_{\Omega} D\hat{u}\, u\, d\Omega - \sum_{i=1}^{2m}(-1)^i \int_{\Gamma} \partial^{2m-i}\hat{u}\ \partial^{i-1}u\, ds - \int_{\Omega} \hat{u}\, Du\, d\Omega$$

$$= W_{1,2} - W_{2,1} = 0$$

In one dimension the boundary integrals, naturally, are to be read as

$$[N\hat{u}]_a^b = N(b)\hat{u}(b) - N(a)\hat{u}(a)\,, \qquad \text{etc.}$$

The integral

$$E(u, \hat{u}) = \int_0^l EAu'\hat{u}'\, dx \qquad (\text{Rod})\,,$$

$$E(u, \hat{u}) = \int_{\Omega} N(u_{,1}\,\hat{u}_{,1} + u_{,2}\,\hat{u}_{,2})\, d\Omega \qquad (\text{Stretched membrane})\,,$$

$$E(\boldsymbol{u}, \boldsymbol{\hat{u}}) = \int_{\Omega} \sigma_{ij}\hat{\varepsilon}_{ij}\, d\Omega \qquad (\text{Elastic plates and bodies})\,,$$

$$E(w, \hat{w}) = \int_0^l EIw''\hat{w}''\, dx\,, \qquad (\text{Beam})$$

$$E(w, \hat{w}) = \int_{\Omega} [w_{,11}\,(\hat{w}_{,11} + \nu\hat{w}_{22}) + 2(1-\nu)w_{,12}\,\hat{w}_{,12}$$

$$+ w_{,22}\,(\hat{w}_{,22} + \nu\hat{w}_{,11})]\, d\Omega \qquad (\text{Kirchhoff-plate})\,,$$

is the *energy product* (= virtual strain energy δW_i) between two displacements u and $\hat{u}$. On the diagonal, $u = \hat{u}$, the integral is, if we multiply it with $1/2$, the

$$\text{strain energy} = \frac{1}{2}\, E(u, u)$$

The first identity formulates the *principle of virtual displacements*, see [2],

$$G(u, \hat{u}) = \delta W_e - \delta W_i = 0, \qquad \text{for all } \hat{u},$$

and the *principle of virtual forces*,

$$G(\hat{u}, u) = \delta W_e^c - \delta W_i^c = 0, \qquad \text{for all } \hat{u}$$

The second identity formulates *Betti's principle*,

$$B(\hat{u}, u) = W_{1,2} - W_{2,1} = 0$$

1.3.2 Betti data

We call the $2m$ boundary terms in Betti's principle *Betti data*. These $2m$ terms form m pairs of conjugated boundary terms. Under regular conditions of two such terms one term is prescribed and the other is unknown. (Two boundary terms are conjugated if the sum of their indices is $2m - 1$ where $2m$ is the order of the differential operator.) The lower terms

$$\partial^0 u, \ \partial^1 u, \dots, \ \partial^{m-1} u, \qquad \text{(essential boundary conditions)}$$

are the displacements, and the higher terms

$$\partial^m u, \ \partial^{m+1} u, \dots, \ \partial^{2m-1} u, \qquad \text{(natural boundary conditions)}$$

the forces. With a view towards the boundary conditions we speak of *essential* and *natural* boundary conditions.

1.3.3 Fundamental solutions

The equilibrium position of a beam is the function $\hat{w}$, that satisfies the differential equation

$$EI\hat{w}^{IV} = \hat{p}, \qquad (EI = c)$$

and the proper boundary conditions. A differential equation only makes sense at points where the right-hand side is smooth. If the load consists of a concentrated force $\hat{P}$ we must, therefore, define the equilibrium differently: we require that the shear force of the beam suffers a jump discontinuity of size $\hat{P}$ at the source point x

$$\lim_{\varepsilon \to 0} \left\{ \hat{Q}(x+\varepsilon) - \hat{Q}(x-\varepsilon) \right\} = \hat{P}, \qquad \hat{Q} = -EI\hat{w}'''$$

Equivalent to this requirement is that the limit of the work which is done by two shear forces at two cuts $x+\varepsilon$ and $x-\varepsilon$ on acting through a virtual deflection w tends to $\hat{P}w(x)$ if the two cuts close in on x

$$\lim_{\varepsilon \to 0} \left\{ \hat{Q}(x+\varepsilon)w(x+\varepsilon) - \hat{Q}(x-\varepsilon)w(x-\varepsilon) \right\} = \hat{P}w(x)$$

What in the case of a beam is a set of two points, is in the case of a plate a circle. If a concentrated force acts at a point x then this means that the integral of the Kirchhoff shear taken over smaller and smaller circles $\Gamma_{N\varepsilon}(x)$ centred at x must tend to $\hat{P} = 1$, see Fig. 1.4. If we let, during the same process, the Kirchhoff shear do work by acting through a virtual deflection $w(y)$ then the limit of the work is just $1 \times w(x) = w(x)$.

   Summarizing this we may state: the deflection $\hat{w} = g_0$ is the deflection of the plate under the attack of a concentrated force $\hat{P} = 1$ at x, if and only if a) g_0 is at all points $y \neq x$ a homogeneous solution of the plate equation $K \Delta \Delta g_0 = 0$, and if b) the limit of the work done by its Kirchhoff shear $\hat{V}_n$ on acting through any possible virtual deflection w is equal to $w(x)$.

   Extending this definition now to all our differential equations, we define: the function $g_0(y, x)$ is a fundamental solution of the operator D, if

a) $D_y g_0(y, x) = 0,$ at all points $y \neq x$,

b) $\displaystyle\lim_{\varepsilon \to 0} \int\limits_{\Gamma_{N\varepsilon}(x)} \partial^{2m-1} g_0(y, x)\, u(y)\, ds_y = u(x),$ for all $u \in C^{2m}(\bar{\Omega})$,

that is, if the limit of the work done by the force $\partial^{2m-1} g_0$ on acting through any displacement u is $u(x)$. If we choose as virtual displacement u the function

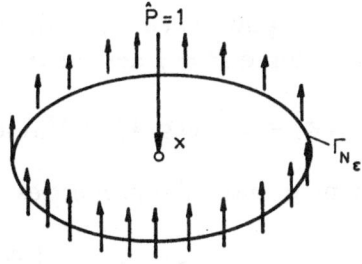

Figure 1.4 The balance between the concentrated force and the edge stresses

$u = 1$, then the limit is 1 and we again have the result that the integral of the Kirchhoff shear tends to 1.

The fundamental solutions which belong to the higher singularites, that is to a concentrated couple, to a concentrated bend or to a point dislocation, are obtained by applying the operators ∂^i to the zero-order fundamental solution and differentiating with respect to the coordinates x_i.

$$g_i(\boldsymbol{y}, \boldsymbol{x}) = \partial_{\boldsymbol{x}}^i g_0(\boldsymbol{y}, \boldsymbol{x}) \qquad i = 0, \ldots, 2m - 1$$

To the Kirchhoff plate ($2m = 4$) belong four operators ∂^i and so we obtain four fundamental solutions while a plate or a membrane ($2m = 2$) possess two fundamental solutions

$$2m = 2 \qquad\qquad 2m = 4$$

$g_0(\boldsymbol{y}, \boldsymbol{x})$,	concentrated force,	$g_0(\boldsymbol{y}, \boldsymbol{x})$,	concentrated force,
$g_1(\boldsymbol{y}, \boldsymbol{x})$,	point dislocation,	$g_1(\boldsymbol{y}, \boldsymbol{x})$,	concentrated couple,
		$g_2(\boldsymbol{y}, \boldsymbol{x})$,	concentrated bend,
		$g_3(\boldsymbol{y}, \boldsymbol{x})$,	point dislocation

In mechanical terms this means that the higher fundamental solutions are the internal actions (slope, moments, shear forces) of the zero-order solution $g_0(\boldsymbol{y}, \boldsymbol{x})$.

To explain this in more detail let us consider the deflection g_1 of a plate when a concentrated couple at $\boldsymbol{x}$ bends the plate into the direction of the vector $\boldsymbol{n}$. The deflection is identical with the slope (in the same direction) of the same plate when a concentrated force $\hat{P} = 1$ acts at $\boldsymbol{x}$,

$$g_1(\boldsymbol{y}, \boldsymbol{x}) = \partial_{\boldsymbol{x}}^1 g_0(\boldsymbol{y}, \boldsymbol{x}) = \frac{\partial}{\partial n_{\boldsymbol{x}}} g_0(\boldsymbol{y}, \boldsymbol{x})$$

Or: the deflection of the plate when a concentrated bend, which is oriented in the direction $\boldsymbol{n}$, distorts the plate is equal to the bending moment M_n of the plate when a concentrated force $\hat{P} = 1$ acts at $\boldsymbol{x}$.

$$g_2(\boldsymbol{y}, \boldsymbol{x}) = \partial_{\boldsymbol{x}}^2 g_0(\boldsymbol{y}, \boldsymbol{x}) = M_n(g_0)(\boldsymbol{y}, \boldsymbol{x})$$

The fundamental solutions g_i have the properties

$$\lim_{\varepsilon \to 0} \int_{\Gamma_{N_\varepsilon}(\boldsymbol{x})} \partial^{2m-1-k} g_i(\boldsymbol{y}, \boldsymbol{x}) \partial^k u \, ds_{\boldsymbol{y}} = \begin{cases} c_i(\boldsymbol{x}) \partial^i u(\boldsymbol{x}) & i = k, \\ 0 & i \neq k \end{cases}$$

that is, the limit of the work done by the Kirchhoff shear of the zero-order fundamental solution $g_0(\boldsymbol{y}, \boldsymbol{x})$ is 1, but the limit of the work done by the bending moment M_n of the same solution on acting through $\partial \hat{w}/\partial n$ is zero. Only that action $\partial^{2m-1-k} g_i(\boldsymbol{y}, \boldsymbol{x})$ will contribute work to the limit which has the characteristic $1/r$ singularity (in 2-D) and in each fundamental solution there is only one such term, namely that term whose index $2m - 1 - k$ adds up with the order i of the fundamental solution to $2m - 1$. The higher the order of the fundamental solution, the lower this term.

The characteristic function which accompanies $\partial^i u$, is in simple cases the integral of the characteristic of the kernel

$$\partial^{2m-1-i} g_i(\boldsymbol{y}, \boldsymbol{x}) = \frac{k_i(\varphi, \vartheta)}{r^{n-1}}$$

over the unit sphere S_1, or if $\boldsymbol{x}$ is a boundary point over the section $S_1(\boldsymbol{x}, \Omega)$ of the unit sphere

$$c_i(\boldsymbol{x}) = \int\limits_{S_1(\boldsymbol{x}, \Omega)} k_i(\varphi, \vartheta) \, dS_1$$

The name *characteristic function* is adopted from topology. In topology the characteristic function of a domain Ω is the function that has the value 1 inside the domain and the value 0 outside it. In the boundary element method these functions also have boundary values.

The characteristic function of the Laplacian operator is the function

$$c_0(\boldsymbol{x}) = \begin{cases} 1, & \boldsymbol{x} \in \Omega, \\ \Delta\varphi/2\pi, & \boldsymbol{x} \in \Gamma, \\ 0, & \boldsymbol{x} \in \Omega^c \quad \text{(complement)} \end{cases}$$

Its mechanical meaning is the following: If we load a cutout Ω of an infinite membrane with a concentrated force $\hat{P} = 1$ and displace the cutout by 1 unit of deflection then the concentrated force $\hat{P} = 1$ contributes the work $\delta W = 1$, see Fig. 1.5. If the concentrated force $\hat{P} = 1$ is located outside Ω, then its contribution is zero, $\delta W = 0$. If it is sitting on the boundary Γ of Ω then its work will be $\delta W = 1/2$ or if it is sitting at a 90°-corner then its work will be $\delta W = 1/4$. So the function $c(\boldsymbol{x})$ is the work done by a point force of magnitude 1.

Fundamental solutions are often characterized by stating that they are the solutions of the differential equations

$$Dg_i(\boldsymbol{y}, \boldsymbol{x}) = \delta_i(\boldsymbol{y} - \boldsymbol{x}) \tag{1.2}$$

where

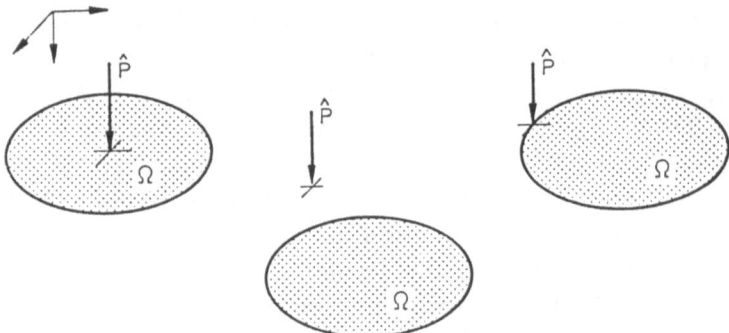

Figure 1.5 a-c. The three possible positions of $\hat{P}$

$$\int_\Omega \delta_i(y - x)u(y)\, d\Omega_y = c_i(x)\partial^i u(x)$$

Note that this differential equation has only a symbolic meaning. You cannot verify it by differentiating the left side and comparing the result with the right side, but rather you must verify this equation indirectly by testing the *action* of the solution on a certain class of functions, the virtual displacements.

Outside of the source point the fundamental solutions are analytical functions while they are singular at the source point x. Therefore integral theorems as, e.g., the principle of virtual forces or Betti's principle cannot be applied directly to loadcases with concentrated forces. To circumvent this restriction we first formulate the integral identity on the punctured domain Ω_ε (this is the domain Ω minus a small neighborhood of the source point) and we then let the radius ε tend to zero

$$\lim_{\varepsilon \to 0} G(g_0, u)_{\Omega_\varepsilon} =: G(g_0, u) = 0\,,$$

$$\lim_{\varepsilon \to 0} B(g_0, u)_{\Omega_\varepsilon} =: B(g_0, u) = 0$$

In both these equations appears an integral-free term, $c_0(x)u(x)$, which represents the work of the point load. If we place this integral-free term on the left side then we obtain two integral representations of the function u,

$$c_0(x)u(x) = \sum_{i=1}^{m} (-1)^i \int_\Gamma \partial^{2m-i} g_0\, \partial^{i-1} u\, ds_y + E(g_0, u)\,,$$

$1 \times$ disp. $= -$ work on the boundary $(W_{1,2})$ + internal virt. energy (δW_i)

$$c_0(x)u(x) = \sum_{i=1}^{2m} (-1)^i \int_\Gamma \partial^{2m-i} g_0\, \partial^{i-1} u\, ds_y + \int_\Omega g_0\, Du\, d\Omega_y\,,$$

$1 \times$ disp. $= -$ work on the boundary $(W_{1,2})$ + work on the boundary $(W_{2,1})$

+ work done in the domain $(W_{2,1})$

The first integral representation is the basis of the so-called unit-dummy-load method of structural mechanics. With the minor modification that in structural mechanics the fundamental solution g_0 is replaced by Green's function G_0, so that all the boundary integrals vanish (if there are no elastic supports), hence the deflection $w(x)$ of a beam can be expressed by the energy product of the two bending moments,

$$ w(x) = \int\limits_0^l \frac{M_0 M}{EI}\, dy = E(G_0, w)\,, $$

alone. The first integral representation will also resurface in chapter 5, when we treat nonlinear problems. But essentially it is the second equation which forms the basis of the boundary element method.

1.4 Coupling on the boundary

In the following section we will concentrate on the Laplacian operator because it is the simplest operator of mathematical physics, but the results easily extend to the other differential equations as well.

The mechanical background will provide a membrane which is stretched over a domain Ω by a uniform tension N. We assume that along the part Γ_1 of the boundary deflections $\bar{u}$ are prescribed and along the part Γ_2 of the boundary tractions $\bar{t}$

$$ -N\Delta u = p\,, \qquad u = \bar{u} \quad \text{on } \Gamma_1\,, \qquad t = \bar{t} \quad \text{on } \Gamma_2 $$

The letter t denotes the N-fold normal derivative

$$ t = N\frac{\partial u}{\partial n}\,, $$

which is the traction across a cut with the normal vector $\boldsymbol{n}$.

If we formulate Betti's principle with the zero-order fundamental solution g_0 and the deflection u, then we obtain an influence function for u

$$c_0(\boldsymbol{x})u(\boldsymbol{x}) = \int\limits_{\Gamma} \left[g_0(\boldsymbol{y},\boldsymbol{x})\,t(\boldsymbol{y}) - N\frac{\partial}{\partial\nu}g_0(\boldsymbol{y},\boldsymbol{x})\,u(\boldsymbol{y}) \right] ds\boldsymbol{y}$$

$$+ \int\limits_{\Omega} g_0(\boldsymbol{y},\boldsymbol{x})\,p(\boldsymbol{y})\,d\Omega\boldsymbol{y}\,, \tag{1.3}$$

and if we formulate the same principle with the first-order fundamental solution g_1 then we obtain an influence function for the traction $t = N\partial u/\partial n$

$$c_1(\boldsymbol{x})t(\boldsymbol{x}) = \int\limits_{\Gamma} \left[g_1(\boldsymbol{y},\boldsymbol{x})\,t(\boldsymbol{y}) - N\frac{\partial}{\partial\nu}g_1(\boldsymbol{y},\boldsymbol{x})u(\boldsymbol{y}) \right] ds\boldsymbol{y}$$

$$+ \int\limits_{\Omega} g_1(\boldsymbol{y},\boldsymbol{x})\,p(\boldsymbol{y})\,d\Omega\boldsymbol{y} \tag{1.4}$$

If the source point $\boldsymbol{x}$ lies on the boundary then the functions u and t on the left side are the same functions as on the right side, under the integral sign. They are then simultaneously the dependent variables as the independent variables so that the two equations

$$\frac{1}{2}\begin{bmatrix} u \\ t \end{bmatrix} = \int\limits_{\Gamma} \begin{bmatrix} g_0 & -N\frac{\partial}{\partial\nu}g_0 \\ g_1 & -N\frac{\partial}{\partial\nu}g_1 \end{bmatrix} \begin{bmatrix} t \\ u \end{bmatrix} ds\boldsymbol{y} + \int\limits_{\Omega} \begin{bmatrix} g_0 \\ g_1 \end{bmatrix} p\,d\Omega\boldsymbol{y} \tag{1.5}$$

($c_i(\boldsymbol{x}) = 1/2$ at smooth points)

formulate two *coupling conditions* between u and t: Two boundary functions u and t are the boundary values of the membrane if and only if they satisfy these two integral equations. Because in any situation one of the two boundary values is prescribed we may conclude that this system has the defect 1, i.e. if the first integral equation is satisfied then also the second and vice versa.

One of these two integral equations, therefore, suffices to determine the unknown boundary value. Afterwards, the influence function (1.3) will render the deflection $u(\boldsymbol{x})$ at any interior point.

1.5 Boundary elements

The boundary functions u and t are interpolated by piecewise polynomials φ_i and ψ_i

$$u = u_i\varphi_i(\boldsymbol{x})\,, \qquad t = t_i\psi_i(\boldsymbol{x}) \tag{1.6}$$

The idea is, as in finite element methods, that these basis functions can be defined piecewise over subregions of the boundary called boundary elements and that over any boundary element the basis functions can be chosen to be simple functions such as polynomials of low degree.

The unknown nodal values u_i and t_i of these piecewise polynomials are determined by collocating the first integral equation (1.3) at K points x^k on the boundary. These K equations constitute the system

$$H_{ki} u_i = G_{ki} t_i + p_i$$

with the coefficients

$$H_{ki} = \frac{\Delta\varphi}{2\pi}\varphi_i(x^k) + \int\limits_{\Gamma} N \frac{\partial}{\partial\nu} g_0(y, x^k)\,\varphi_i(y)\,ds_y\,,$$

$$G_{ki} = \int\limits_{\Gamma} g_0(y, x^k)\,\psi_i(y)\,ds_y\,,$$

$$p_i = \int\limits_{\Omega} g_0(y, x^k)p(y)\,d\Omega_y$$

However, this is not the only choice we have. It would also be possible to use the second integral equation (1.4) instead or to formulate on Γ_1 the first integral equation and on Γ_2 the second.

The coefficient H_{ki} is the work done by the traction

$$T = N\frac{\partial}{\partial\nu} g_0(y, x^k)$$

on the boundary of the cutout Ω of the infinite membrane acting through the deflection φ_i, if the concentrated force $\hat{P} = 1$ is located at the node x^k (the position of $\hat{P}$ determines the distribution of T). The coefficient G_{ki} is the work done by the unit traction ψ_i on the edge of the membrane acting through the deflection

$$U = g_0(y, x^k)$$

of the boundary of the subregion Ω, if the concentrated force $\hat{P} = 1$ is located at the node x^k. (The deflection U depends on the position of P.) In more general terms we may state: the coefficient G_{ki} or H_{ki} represents the influence which the boundary layer ψ_i or φ_i, respectively, exerts on the collocation point x^k. The influence depends on the two kernels $g_0(y, x^k)$ and $N\partial g_0(y, x)/\partial\nu$, respectively.

Because *each* boundary layer (wherever it is located) influences *each* collocation point each row of the matrices G and H is fully populated. In addition the matrices G and H are unsymmetric. This is easily understood if we consider the contour Γ to be a material wire, see Fig. 1.6. The attractive force which the two elements $\Gamma_l \cup \Gamma_{l+1}$ with "mass" φ_l exert on the collocation point $\boldsymbol{x}^k$ is, in general, not equal to the force the two elements with "mass" $\Gamma_k \cup \Gamma_{k+1}$ exert on the collocation point $\boldsymbol{x}^l$ because the form, the length and, hence, also the "mass" of the two elements differ.

Only in the limit, if the elements shrink to mere points, does the matrix G become symmetric, because the kernel $g_0(\boldsymbol{y}, \boldsymbol{x})$ is symmetric

$$g_0(\boldsymbol{x}^k, \boldsymbol{x}^l) = g_0(\boldsymbol{x}^l, \boldsymbol{x}^k)$$

But the matrix H remains unsymmetric because the kernel $N\,\partial g_0(\boldsymbol{y}, \boldsymbol{x})/\partial\nu$ depends on the normal vector at the integration point (= source point if $K = \infty$),

$$N\frac{\partial}{\partial\nu}g_0(\boldsymbol{x}^k, \boldsymbol{x}^l) = -\frac{1}{2\pi r}\nabla_y r \cdot \nu(\boldsymbol{x}^k)$$

$$\neq -\frac{1}{2\pi r}\nabla_y r \cdot \nu(\boldsymbol{x}^l) = N\frac{\partial}{\partial\nu}g_0(\boldsymbol{x}^l, \boldsymbol{x}^k),$$

and because the two normal vectors at the two collocation points $\boldsymbol{x}^l$ and $\boldsymbol{x}^k$ are, in general, not parallel the matrix is not symmetric.

The right-hand sides of a stiffness matrix $\boldsymbol{K}\boldsymbol{u} = \boldsymbol{f}$ are (equivalent) nodal forces. The right-hand sides of the influence matrices G and H are displacements. The matrix H maps displacements onto displacements

$$\boldsymbol{H}\boldsymbol{u} = \text{displacements}\,,$$

and the matrix G maps forces onto displacements

$$\boldsymbol{G}\boldsymbol{t} = \text{displacements}$$

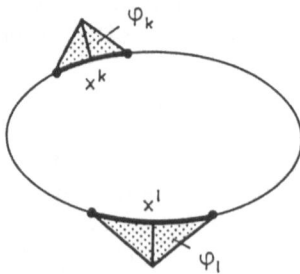

Figure 1.6 The mutual attraction of two material wires with different 'mass'

so that the entries of the matrix G contain the factor $(stiffness)^{-1}$. As the system matrix A is, in general, a mixture of both matrices G and H the columns of H are usually multiplied with the inverse stiffness so that all the entries of A have the same order of magnitude.

1.6 Conforming and non-conforming solutions

In FE-analysis we distinguish between conforming and non-conforming elements or, as we should say more precisely, functions. A basis function is conforming with regard to an operator of order $2m$ if the derivatives up to the order $m-1$ (these are the displacement terms) are continuous. Equivalent with this statement is that the strain energy of the function is finite.

This classification now carries over to boundary elements. The BE-solution

$$u_h(x) = \int\limits_\Gamma [g_0(y,x)\,t_{\mathrm{BE}}(y) - N\frac{\partial}{\partial\nu}g_0(y,x)\,u_{\mathrm{BE}}(y)]\,ds_y + \int\limits_\Omega g_0(y,x)\,p(y)\,d\Omega_y$$

has finite energy if and only if the boundary displacement $u_{\mathrm{BE}}(y)$ is continuous and the traction $t_{\mathrm{BE}}(y)$ is piecewise continuous or better. The functions u_{BE} and t_{BE} are the piecewise polynomials which satisfy the discrete coupling conditions. (Normally we denote these polynomials by the same letters u and t as the true Betti data because, normally, it is clear which functions we refer to.)

In more general terms we may state: a BE-solution is conforming if the approximation of the "highest" displacement, the term $\partial^{m-1}u$, is at least continuous and if the quality of the approximation of the "lower" displacements increases accordingly. The approximation of the force terms, on the other hand, allows no stronger singularities than admissible if the energy is to remain bounded. This excludes concentrated forces if the equation is of second order and concentrated couples or higher singularities if the equation is of fourth order.

If the basis functions are not conforming at the collocation points then the singular integrals do not exist at these points. The approximation of displacements with constant elements is only possible because the collocation points are placed at the centre of the boundary elements where the basis functions are (locally) conforming.

1.7 The interpretation of the solution

The BE-solution

$$u_h(x) = \int\limits_\Gamma [g_0(y,x)\,t_{\mathrm{BE}}(y) - N\frac{\partial}{\partial\nu}g_0(y,x)\,u_{\mathrm{BE}}(y)]\,ds_y + \int\limits_\Omega g_0(y,x)\,p(y)\,d\Omega_y$$

is based on the influence function for $u(\boldsymbol{x})$ and so we might assume that the boundary functions $u_{\text{BE}}(\boldsymbol{x})$ and $t_{\text{BE}}(\boldsymbol{x})$ under the integral sign (these are the piecewise polynomials which satisfy the discrete coupling condition) are also the boundary values of the BE-solution

$$\lim_{\boldsymbol{x}\to\Gamma} u_h(\boldsymbol{x}) = u_{\text{BE}}(\boldsymbol{x}) \qquad ?, \qquad \lim_{\boldsymbol{x}\to\Gamma} t_h(\boldsymbol{x}) = t_{\text{BE}}(\boldsymbol{x}) \qquad ?$$

But this is not true because the piecewise polynomials u_{BE} and t_{BE} satisfy the coupling condition only approximately, namely only at the collocation points. The deflection u and the traction t of the BE-solution differ by error terms ε and η from the functions u_{BE} and t_{BE},

$$\lim_{\boldsymbol{x}\to\Gamma} u_h(\boldsymbol{x}) = u_{\text{BE}}(\boldsymbol{x}) + \varepsilon(\boldsymbol{x}), \qquad \lim_{\boldsymbol{x}\to\Gamma} t_h(\boldsymbol{x}) = t_{\text{BE}}(\boldsymbol{x}) + \eta(\boldsymbol{x})$$

Only at the collocation points $\boldsymbol{x}^k$, where the coupling conditions are satisfied, the error $\varepsilon(\boldsymbol{x})$ is zero, but not the error $\eta(\boldsymbol{x})$. If the elements are non-conforming piecewise constant elements, then the error $\eta(\boldsymbol{x})$ even becomes infinite at the transition points between the elements, because dislocations must be attributed to the action of infinite forces,

$$\lim_{\boldsymbol{x}\to\Gamma} t_h(\boldsymbol{x}) = \infty$$

Only infinite forces can tear a material apart. Because of the energy balance *external work = internal energy* the energy is also infinite or, to make a less dramatic statement, the square of the first derivatives of the BE-solution can no longer be measured; the integral

$$\frac{1}{2}E(u_h, u_h) = \frac{1}{2}\int_{\Omega} (u_{h,1}^2 + u_{h,2}^2)\, d\Omega$$

does not exist.

1.8 Symmetric formulations

Symmetric and positive definite matrices are obtained if we combine the two integral equations (1.5) appropriately and solve them with Galerkin's method.

On Γ_1, where t is unknown, we formulate the first integral equation and on Γ_2, where u is unknown, the second integral equation. We always choose on the boundary Γ that integral equation whose free term is conjugated to the unknown term. We therefore obtain the following system of integral equations:

$$\int\limits_{\Gamma_1} g_0 t \, ds_y - \int\limits_{\Gamma_2} N \frac{\partial}{\partial \nu} g_0 u \, ds_y = \frac{1}{2} \bar{u} - \int\limits_{\Gamma_2} g_0 \bar{t} \, ds_y + \int\limits_{\Gamma_1} N \frac{\partial}{\partial \nu} g_0 \bar{u} \, ds_y \quad (\Gamma_1),$$

$$\int\limits_{\Gamma_1} g_1 t \, ds_y - \int\limits_{\Gamma_2} N \frac{\partial}{\partial \nu} g_1 u \, ds_y = \frac{1}{2} \bar{t} - \int\limits_{\Gamma_2} g_1 \bar{t} \, ds_y + \int\limits_{\Gamma_1} N \frac{\partial}{\partial \nu} g_1 \bar{u} \, ds_y \quad (\Gamma_2)$$

If we use the standard approximation (1.6) and if we solve this system by Galerkin's method then the resulting system of equations is symmetric,

$$\begin{bmatrix} A & B \\ B^T & C \end{bmatrix} \begin{bmatrix} t \\ u \end{bmatrix} = \begin{bmatrix} r_1 \\ r_2 \end{bmatrix}$$

Its coefficients are

$$a_{ij} = \int\limits_{\Gamma_1} \int\limits_{\Gamma_1} g_0 \, \psi_i \, \psi_j \, ds_y \, ds_x = a_{ji},$$

$$b_{ij} = -\int\limits_{\Gamma_1} \int\limits_{\Gamma_2} N \frac{\partial}{\partial \nu} g_0 \, \varphi_i \psi_j \, ds_y \, ds_x = \int\limits_{\Gamma_2} \int\limits_{\Gamma_1} N \frac{\partial}{\partial n} g_0 \, \psi_j \varphi_i \, ds_y \, ds_x,$$

$$c_{ij} = \int\limits_{\Gamma_2} \int\limits_{\Gamma_2} N \frac{\partial}{\partial \nu} g_1 \, \varphi_i \varphi_j \, ds_y \, ds_x = c_{ji}$$

To prove that the system is positive definite

$$[t, u] \begin{bmatrix} A & B \\ B^T & C \end{bmatrix} \begin{bmatrix} t \\ u \end{bmatrix} > 0 \qquad \text{for all } u, t \neq o,$$

we have to show that for arbitrary, non-vanishing functions φ und ψ the integral

$$\int\limits_{\Gamma_1} \Phi[\psi, \varphi](x) \, \psi(x) \, ds + \int\limits_{\Gamma_2} \Phi'[\psi, \varphi](x) \, \varphi(x) \, ds$$

is positive, where

$$\Phi[\psi, \varphi](x) = \Phi_1[\psi](x) - \Phi_2[\varphi](x) = \int\limits_{\Gamma_1} g_0 \, \psi \, ds_y - \int\limits_{\Gamma_2} N \frac{\partial}{\partial \nu} g_0 \, \varphi \, ds_y,$$

$$\Phi'[\psi, \varphi](x) = \Phi'_1[\psi](x) - \Phi'_2[\varphi](x) = \int\limits_{\Gamma_1} g_1 \, \psi \, ds_y - \int\limits_{\Gamma_2} N \frac{\partial}{\partial \nu} g_1 \, \varphi \, ds_y$$

The potential Φ shows the following behavior, see section 1.11,

$$\lim_{\boldsymbol{x}_i \to \Gamma_1} \varPhi = \varPhi = \lim_{\boldsymbol{x}_e \to \Gamma_1} \varPhi, \qquad \lim_{\boldsymbol{x}_i \to \Gamma_2} \varPhi = \frac{1}{2}\varphi + \varPhi,$$

$$\lim_{\boldsymbol{x}_e \to \Gamma_2} \varPhi = -\frac{1}{2}\varphi + \varPhi, \qquad \lim_{\boldsymbol{x}_i \to \Gamma_1} N\frac{\partial \varPhi}{\partial n_e} = \frac{1}{2}\psi + N\frac{\partial \varPhi}{\partial n_e},$$

$$\lim_{\boldsymbol{x}_i \to \Gamma_2} N\frac{\partial \varPhi}{\partial n_e} = N\frac{\partial \varPhi}{\partial n_e}, \qquad \lim_{\boldsymbol{x}_e \to \Gamma_1} N\frac{\partial \varPhi}{\partial n_e} = -\frac{1}{2}\psi + N\frac{\partial \varPhi}{\partial n_e},$$

$$\lim_{\boldsymbol{x}_e \to \Gamma_2} N\frac{\partial \varPhi}{\partial n_e} = N\frac{\partial \varPhi}{\partial n_e}, \qquad \varDelta \varPhi = 0 \qquad \boldsymbol{x} \notin \Gamma,$$

where $\boldsymbol{x}_i$ and $\boldsymbol{x}_e$ respectively means that the point $\boldsymbol{x}$ approaches the boundary Γ from the interior, Ω_i, or the exterior, Ω_e. The subscript e on n is to indicate that the normal vector points into the exterior.

The first identity for the potential $\varPhi$ in the interior domain (the normal vector points then into the exterior) is

$$G(\varPhi, \varPhi)_{\Omega_i} = \int_{\Gamma_1} (\frac{1}{2}\psi + N\frac{\partial \varPhi}{\partial n_e})\varPhi \, ds\boldsymbol{y} + \int_{\Gamma_2} N\frac{\partial \varPhi}{\partial n_e}(\frac{1}{2}\varphi + \varPhi) \, ds\boldsymbol{y}$$

$$- E(\varPhi, \varPhi)_{\Omega_i} = 0$$

and in the exterior domain (now the normal vector points into the interior)

$$G(\varPhi, \varPhi)_{\Omega_e} = \int_{\Gamma_1} (\frac{1}{2}\psi - N\frac{\partial \varPhi}{\partial n_e})\varPhi \, ds\boldsymbol{y} + \int_{\Gamma_2} -N\frac{\partial \varPhi}{\partial n_e}(-\frac{1}{2}\varphi + \varPhi) \, ds\boldsymbol{y}$$

$$- E(\varPhi, \varPhi)_{\Omega_e} = 0$$

The sum of these two equations is

$$G(\varPhi, \varPhi)_{\Omega_i} + G(\varPhi, \varPhi)_{\Omega_e} = \int_{\Gamma_1} \varPhi \psi \, ds\boldsymbol{y} + \int_{\Gamma_2} N\frac{\partial \varPhi}{\partial n_e} \varphi \, ds\boldsymbol{y} - E(\varPhi, \varPhi)_{R^n} = 0,$$

or

$$\int_{\Gamma_1} \varPhi \psi \, ds\boldsymbol{y} + \int_{\Gamma_2} \varPhi' \varphi \, ds\boldsymbol{y} = E(\varPhi, \varPhi)_{R^n} > 0$$

The potential Φ is zero if and only if the layers φ and ψ are zero and because the energy is positive definite the proof is complete. All this holds true, naturally, for other equations as well and so it is possible to obtain symmetric matrices; but they remain fully populated.

1.9 The integral operators and their shifts

In more abstract terms the two coupling conditions in Eq.(1.5) between the Betti data of a membrane can be formulated as

$$
\frac{1}{2}\begin{bmatrix} \partial^0 u \\ \partial^1 u \end{bmatrix} = \int_\Gamma \begin{bmatrix} \partial_y^0 \partial_x^0 g_0 & \partial_y^1 \partial_x^0 g_0 \\ \partial_y^0 \partial_x^1 g_0 & \partial_y^1 \partial_x^1 g_0 \end{bmatrix} \begin{bmatrix} +\partial^1 u \\ -\partial^0 u \end{bmatrix} ds_y + \begin{bmatrix} \partial_x^0 g_0 \\ \partial_x^1 g_0 \end{bmatrix} p\, d\Omega_y
$$

The kernels of the single integral operators are "products" of the boundary operators ∂_y^i and ∂_x^j and the zero-order fundamental solution g_0.

The integral operators can be considered as operators which map boundary functions from a Sobolev space $H^r(\Gamma)$ into a Sobolev space $H^{r-2\alpha}(\Gamma)$, see [3]. If the shift 2α is positive then the integral operator differentiates the function, its regularity decreases by the order of 2α. If the shift 2α is negative then the operator integrates the function, the regularity of the function increases by the order 2α. If we replace each operator by its shift 2α and each boundary value by the order i of its highest derivative, then the coupling conditions become

$$
\begin{bmatrix} 0 \\ 1 \end{bmatrix} = \begin{bmatrix} -1 & 0 \\ 0 & 1 \end{bmatrix} \begin{bmatrix} 1 \\ 0 \end{bmatrix} + \begin{bmatrix} -2 \\ -1 \end{bmatrix} [2]
$$

In the case of a Kirchhoff plate the four coupling conditions between the four boundary functions read

$$
\frac{1}{2}\begin{bmatrix} \partial^0 w \\ \partial^1 w \\ \partial^2 w \\ \partial^3 w \end{bmatrix} = \int_\Gamma \begin{bmatrix} \partial_y^0 \partial_x^0 g_0 & \partial_y^1 \partial_x^0 g_0 & \partial_y^2 \partial_x^0 g_0 & \partial_y^3 \partial_x^0 g_0 \\ \partial_y^0 \partial_x^1 g_0 & \partial_y^1 \partial_x^1 g_0 & \partial_y^2 \partial_x^1 g_0 & \partial_y^3 \partial_x^1 g_0 \\ \partial_y^0 \partial_x^2 g_0 & \partial_y^1 \partial_x^2 g_0 & \partial_y^2 \partial_x^2 g_0 & \partial_y^3 \partial_x^2 g_0 \\ \partial_y^0 \partial_x^3 g_0 & \partial_y^1 \partial_x^3 g_0 & \partial_y^2 \partial_x^3 g_0 & \partial_y^3 \partial_x^3 g_0 \end{bmatrix} \begin{bmatrix} +\partial^3 w \\ -\partial^2 w \\ +\partial^1 w \\ -\partial^0 w \end{bmatrix} ds_y
$$

$$
+ \int_\Omega \begin{bmatrix} \partial_x^0 g_0 \\ \partial_x^1 g_0 \\ \partial_x^2 g_0 \\ \partial_x^3 g_0 \end{bmatrix} p\, d\Omega_y \, ,
$$

or, if we concentrate on the shifts alone,

$$\begin{bmatrix} 0 \\ 1 \\ 2 \\ 3 \end{bmatrix} = \begin{bmatrix} -3 & -2 & -1 & 0 \\ -2 & -1 & 0 & 1 \\ -1 & 0 & 1 & 2 \\ 0 & 1 & 2 & 3 \end{bmatrix} \begin{bmatrix} 3 \\ 2 \\ 1 \\ 0 \end{bmatrix} + \begin{bmatrix} -4 \\ -3 \\ -2 \\ -1 \end{bmatrix} \quad [4]$$

The higher the index 2α, the greater the singularity of the integral operator. The smoothest operator is the one in the upper left corner, $2\alpha = -3$, and the most singular operator is the one in the lower right corner, $2\alpha = 3$. Its kernel is of the order $O(r^{-4})$.

Galerkin's method will render symmetric and positive definite matrices, if we choose among these four integral equations those two integral equations whose integral-free terms are conjugated to the two unknown terms on the edge of the plate.

boundary condition	integral equations
clamped	$1 + 2$
hinged	$1 + 3$
free	$3 + 4$

Along a free edge the deflection $w = \partial^0 w$ and the normal derivative $\partial w / \partial n = \partial^1 w$ are the unknowns and so the two relevant integral equations are the 3rd and 4th equation, that is the two integral equations with the worst kernels.

To avoid these hyper-singular kernels one can shift by integration by parts some of the derivatives from the kernel onto the boundary layers. Nedelec [4] makes extensive use of this technique to solve hyper-singular integral equations with Galerkin's method. For a first course on how to integrate by parts on manifolds, i.e. curves and surfaces, we recommend Gunther's book [5].

In one dimension the integral operators can be depicted graphically, see Fig. 1.7. The kernel $2\alpha = -2$ can be imagined to be the horizontal displacement $\hat{u}(y, x)$ of a rod that is under the influence of a concentrated force $\hat{P} = 1$. The kernel $2\alpha = -1$ is the associated normal force and the kernel $2\alpha = 0$ is the single force itself. Now recall the laws of mechanics: according to Betti's principle we must have

$$W_{1,2} = \int_0^l \hat{u}(y, x)\, p(y)\, dy = \int_0^l \hat{u}(y, x)\, (-EAu''(y))\, dy = 1 \times u(x) = W_{2,1} ,$$

and according to the principle of virtual forces

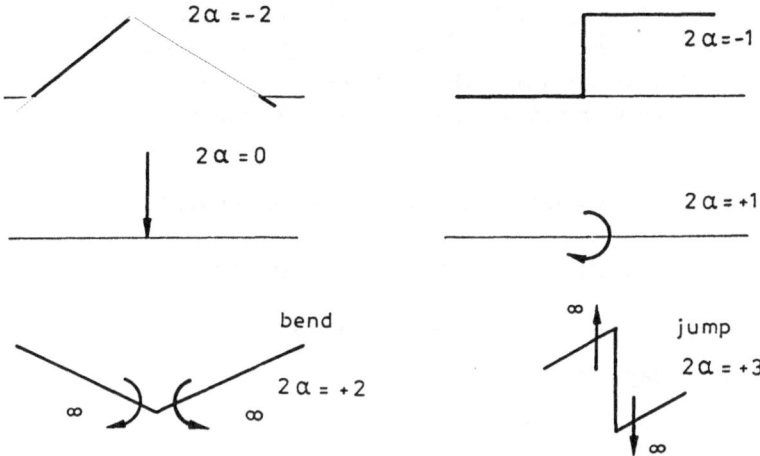

Figure 1.7 Graphical representation of integral operators

$$\delta W_i^c = \int\limits_0^l \hat{N}(y,x)\,N(y)\,dy = \int\limits_0^l \hat{N}(y,x)\,(EAu'(y))\,dy = 1 \times u(x) = \delta W_e^c,$$

and the work done by the concentrated force $\hat{P}$ on acting through the virtual displacement u should be

$$\int\limits_0^l \hat{P}\,u(y)\,dy \equiv \int\limits_0^l \delta_0(y-x)\,u(x)\,dy = 1 \times u(x)$$

Hence, the function u is integrated consecutively *two-times, one-times, zero-times*.

The positive operators can only be depicted in a symbolical way. While negative operators have the classical form

$$\int\limits_0^l k(y,x)u(y)\,dy$$

the positive operators, the operators which differentiate, consist of *point operators* and classical (negative) operators. The integral representation of the first derivative of a function u is

$$u'(x) = u(l) - u(0) + (1-l)u'(l) + \int\limits_0^l g_1(y,x)u''(y)\,dy$$

The classical integral operator in the background has a negative shift, it integrates the second derivative. The point operator in the foreground has a positive shift, it differentiates. If the function is linear then $u'' = 0$ and then only the point operator acts on the data and forms the difference quotient.

1.10 Galerkin, collocation and least square

The results in this section apply if the approximation

$$f(y) = f_j \, \varphi_j(y)$$

for the unknown layer f,

$$\int_0^l g(y, x) f(y) \, ds_y = r(x) \,,$$

consists of splines of degree $k-1$ which belong to $C^{m-1} \subset H^s(\Gamma), s \leq m+1/2 \leq k$. In the case of higher dimensional integral equations we assume that the basis functions φ_j constitute a $S_h^{k,m}$-*family* of regular functions, see [6]. The hat-functions, e.g., are splines of degree $k - 1 = 1$, but piecewise quadratic polynomials are not. Because this would also require that the first derivatives are continuous at the nodes.

In the collocation method the n nodal values f_j are determined by satisfying the integral equation at n points x^i

$$\int_0^l g(y, x^i) \varphi_j(y) \, ds_y \, f_j = r(x^i) \,, \qquad i = 1, 2 \ldots, n$$

In Galerkin's method the nodal values f_j are subject to the condition that the error is orthogonal in the L_2-sense to the n basis functions

$$\int_\Gamma [\int_\Gamma g(y, x) \varphi_j(y) \, ds_y \, f_j - r(x)] \varphi_i(x) \, ds_x = 0 \,, \qquad i = 1, 2 \ldots, n$$

Finally, in the least-square method the basis functions have to render the square of the error a minimum.

$$\int_\Gamma [\int_\Gamma g(y, x) \, \varphi_j(y) \, ds_y \, f_j - r(x)]^2 \, ds_x \qquad \longrightarrow \qquad \text{minimum.}$$

Wendland [3] showed that usually the error satisfies an estimate as

$$\|u - u_h\|_\sigma \leq ch^{s-\sigma} \|u\|_s$$

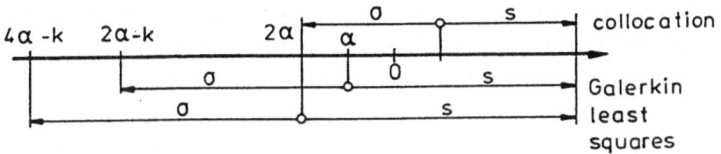

Figure 1.8 Admissible Sobolev-indices

Here u_h denotes the approximate solution. In which norms these estimates are valid depends on the dimension of the contour (is Γ a curve or a surface ?) and the indices α, k and m.

For the case of a one-dimensional integral equation and the case that $k = m + 1, \alpha \leq 0$ (2α is the shift of the integral operator) Fig. 1.8 shows the range of admissible indices r and s.

From this diagram we conclude that the highest possible order of convergence that can be achieved is

$$
\begin{array}{lll}
\text{collocation} & O(h^{k-2\alpha}) & \text{in } H^{2\alpha}, \\
\text{Galerkin} & O(h^{2k-2\alpha}) & \text{in } H^{2\alpha-k}, \\
\text{least square} & O(h^{2k-4\alpha}) & \text{in } H^{4\alpha-k}
\end{array}
$$

The interesting point is, that these rates also determine the order of the error in the domain because the error in the interior

$$
|u - u_h| = |\int_{\Gamma} g(y,x)(f(y) - f_h(y))\, ds y|
$$

$$
\leq c\, \| g[x] \|_{-2\alpha} \, \| f - f_h \|_{2\alpha}
$$

(collocation, note that $2\alpha < 0$)

can be estimated by the norm of the error on the boundary and the dual norm of the kernel. Because the kernel is infinitely smooth the norm which measures the kernel can be arbitrarily strong and therefore the norm of the error arbitrarily weak. Hence, the further the tip of the arrow moves to the left the better the convergence of the method and this means that the least-square method and Galerkin's method are superior to the collocation method.

To achieve the same rate of convergence the methods need basis functions of different order, namely

$$
m_c = 2m_G + 1 = 2m_{ls} + 1 - 2\alpha
$$

The collocation method requires basis functions with more than double the order of the other two methods. The highest rate of convergence shows the least-square method but its system matrix also has the worst condition number.

The FE-matrix of a second-order differential equation has the condition number $O(h^{-2})$, a fourth-order equation has the condition number $O(h^{-4})$, etc. In the case of integral equations the condition number depends primarily on the shift 2α. The matrix of the collocation method and Galerkin's method have the condition number $O(h^{-2|\alpha|})$, but the least-square matrix has a condition number which is twice as bad, namely $O(h^{-4|\alpha|})$, see [3].

For Galerkin's method we need to integrate twice. But if we use a symmetric formulation then the computational effort splits into half. In certain instances special quadrature formulas, see [7], offer a further reduction in computing time. From a mathematical point of view Galerkin's method and the collocation method are very close. It is well known that collocation can be seen as a Galerkin method with Dirac-functions as test functions. However, if we use Dirac-functions as boundary layers and piecewise polynomials as test functions then the system matrix is just the transposed of the collocation matrix, [8]. We can also consider Galerkin's method as a "weighted collocation method" if we interpret the Gaussian points as collocation points, [9]. In this perspective a Galerkin method means that we formulate the collocation equations at, say, $4 \times n$ quadrature points (4 Gaussian points and n linear elements) and to reduce this overdetermined system of equations to a square matrix we multiply each group of 4 rows with the pertinent Gaussian weights and store them in one single row, see Fig. 1.9.

Both methods in turn can be seen as special instances of the more general Petrov Galerkin method which provides a unifying frame for the analysis of both methods, [10]. The mathematical theory of integral equations is embedded into the modern Fourier analysis for pseudo differential operators. (Recall that strongly singular and hyper-singular integral operators differentiate boundary

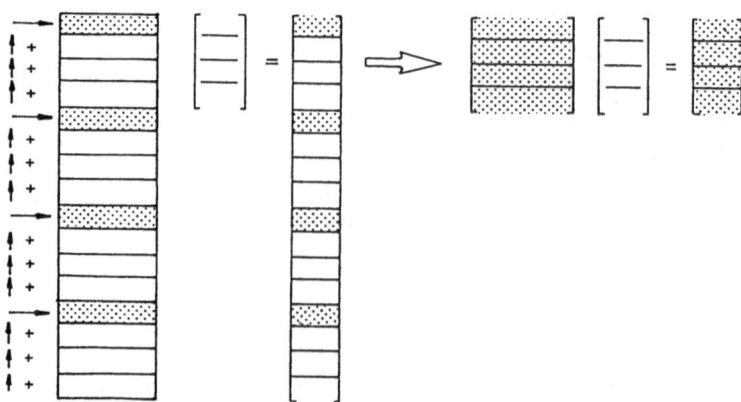

Figure 1.9 Galerkin as a weighted collocation method

functions, so they have similar local effects as differential operators.) The decisive concept in this analysis is the concept of the *strong ellipticity* of an integral operator, [10]. If Gårding's inequality is satisfied then this is equivalent to the positive definitness of the principal symbol of the pseudo differential operator, and since the Gårding inequality implies stability and convergence of every Galerkin approximation this is particularly true for the boundary element Galerkin methods. For two-dimensional boundary integral equations and spline Galerkin or spline collocation the strong ellipticity (albeit somewhat relaxed) is not only sufficient but also necessary for stability and convergence [10].

The convergence and asymptotic accuracy of collocation methods for integral equations on curves was investigated by many mathematicians, [11], [12], [13], [14], [15], [16], [17], [18], [19], and the investigation can now be considered nearing completion. An important result of these investigations is that the naive collocation is not always the best. The optimal position of the collocation point depends on the principal part of the pseudo differential operator. A first kind integral equation in the plane with a Cauchy singular kernel gives poor results if we use naive collocation but the solution improves dramatically if we place the collocation points in between the boundary element nodes. Vice versa: a second kind integral equation with a smooth kernel gives good results with naive collocation and not so good results if we place the collocation points in between the nodes, [20].

1.11 Potentials

Two point charges q_1 and q_2 act on each other with a force which is directly proportional to the product of the charges and inversely proportional to the distance between them. If r is the vector distance between the two charges then this force F in empty space is given by

$$F = K \frac{q_1 q_2}{r^3} r \,,$$

where K depends on the units employed. The electric field E is defined such that the force on a point test charge q is given by

$$F = q E$$

in the limit as q goes to zero. The electric field E is the gradient of the potential Φ

$$E = \frac{q\,r}{r^3} = -\nabla \Phi \qquad \Phi = \frac{q}{r}$$

produced by the point charge q in empty space. If we consider electrical charges which are distributed over a wire or a sphere, see Fig. 1.10, then the potential Φ becomes

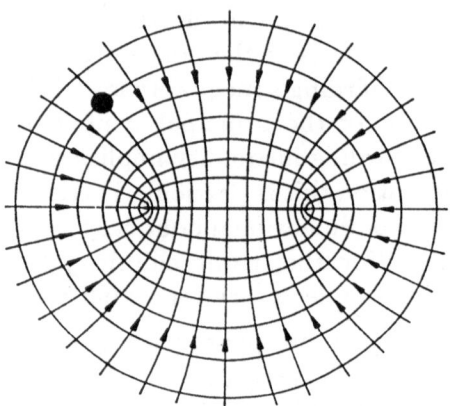

Figure 1.10 An electrical charge in the potential field of a wire

$$\Phi(x) = \frac{1}{4\pi} \int_\Gamma \frac{1}{r} f(y) \, ds_y \tag{1.7}$$

Such potentials are homogeneous solutions of the Laplace equation at any point not on Γ because their kernel, the function $(4\pi r)^{-1}$, is the fundamental solution of the three-dimensional Laplacian operator

$$\Delta\Phi(x) = \frac{1}{4\pi} \int_\Gamma \Delta x \frac{1}{r} f(y) \, ds_y = \int_\Gamma 0 f(y) \, ds_y = 0, \qquad x \notin \Gamma$$

This property makes potentials interesting. A three-dimensional boundary value problem for the Laplacian

$$\Delta u = 0 \quad \text{in } \Omega, \qquad u = \bar{u} \quad \text{on } \Gamma$$

becomes an integral equation problem on the boundary, a two-dimensional manifold, if we assume that u is the potential field generated by a certain (unknown) distribution f of electrical charges on the boundary of the domain

$$u(x) = \int_\Gamma \frac{1}{r} f(y) \, ds_y$$

To determine the distribution $f(y)$ approximately we interpolate it by piecewise polynomials and determine the nodal values by satisfying the boundary condition at K collocation points,

$$\int_\Gamma \frac{1}{r} f(y) \, ds_y = u(x^k), \qquad r = |y - x^k|, \qquad k = 1, 2 \ldots K$$

This transformation of a boundary value problem into an integral equation on the boundary can be done with any of our operators. With each differential operator D of order $2m$ we can associate $2m$ kernel functions. These functions are obtained on applying the operators ∂^i to the zero-order fundamental solution and differentiating with respect to the coordinates y_i.

$$\partial_y^0 g_0 = g_0\,, \qquad \partial_y^1 g_0\,, \qquad \partial_y^2 g_0\,, \qquad \cdots \qquad \partial_y^{2m-1} g_0$$

The scalar products of these $2m$ kernel functions and a boundary layer f generate $2m$ potentials,

$$u_i(\boldsymbol{x}) = \int_\Gamma \partial_y^i g_0(\boldsymbol{y}, \boldsymbol{x}) f(\boldsymbol{y})\, ds_{\boldsymbol{y}} \qquad i = 0, 1, \ldots, 2m-1,$$

which are homogeneous solutions of the governing differential equation.

The attentive reader will have noticed that we differentiated the fundamental solutions here with respect to $\boldsymbol{y}$ while we differentiated the same functions in section 1.3.3 with respect to $\boldsymbol{x}$. But this is no real difference because the fundamental solutions are symmetric. We only interchanged the names of the variables when integrating; otherwise the name of the free variable would have been $\boldsymbol{y}$.

Essentially there are two approaches to boundary element methods: Betti's principle or potential theory. With Betti's principle we derive the influence function by letting a unit point force act on the domain. The point force is located at $\boldsymbol{x}$ and so we obtain an expression as

$$1 \times u(\boldsymbol{x}) = \ldots$$

However in potential theory we consider distributions of sources $f(\boldsymbol{y})$ and we evaluate their influence on a point $\boldsymbol{x}$ by integrating over $\boldsymbol{y}$. The source points are now the integration points $\boldsymbol{y}$ and the observation point (or field point) becomes the point $\boldsymbol{x}$, so that the interpretation is just reversed. But, as we said before, fundamental solutions are symmetric and so it is simply a matter of taste, not of mathematics, whether you consider the point $\boldsymbol{x}$ or the point $\boldsymbol{y}$ to be the source point.

More important is that each of the $2m$ potentials is a homogeneous solution of the governing differential equation

$$Du = \int_\Gamma D_{\boldsymbol{x}} \partial_y^i g_0(\boldsymbol{y}, \boldsymbol{x}) f(\boldsymbol{y})\, ds_{\boldsymbol{y}} = \int_\Gamma \partial_y^i D_{\boldsymbol{x}} g_0(\boldsymbol{y}, \boldsymbol{x}) f(\boldsymbol{y})\, ds_{\boldsymbol{y}}$$

$$= \int_\Gamma 0 f(\boldsymbol{y})\, ds_{\boldsymbol{y}} = 0$$

The boundary layers $f(y)$ which in physics represent electrical charges or mass distributions are interpreted in mechanics as displacements $\partial^0, \partial^1, \cdots, \partial^{m-1}$ or forces $\partial^m, \partial^{m+1}, \cdots, \partial^{2m-1}$.

If a concentrated force $\hat{P} = 1$ acts at a point y of an infinite membrane then the deflection at some distant point x is

$$g_0(y, x) = -\frac{1}{2\pi N} \ln r, \qquad r = |y - x|, \qquad (-\ln r = \ln \frac{1}{r})$$

At the source point itself the deflection is infinite. This is what we must expect if we press an infinitely thin needle into a membrane. The term N in the denominator is the uniform tension in the membrane. The greater this force, the smaller the overall deflection of the membrane will be. In the following we set the force N equal to 1.

If we replace the point force by a series of forces $f(y)$ which are aligned along a curve Γ then the deflection at a point x will be

$$u(x) = -\frac{1}{2\pi} \int_{\Gamma} \ln r \, f(y) \, ds_y$$

and because the integral of the logarithm is bounded, the deflection directly under the line forces is also now bounded.

The integral $u(x)$, too, is a potential and because its kernel is the fundamental solution of the Laplacian operator, the potential satisfies at all points x, which do not lie on Γ the differential equation $\Delta u = 0$.

If the same curve Γ is the locus of dislocations $d(y)$, that is if the deflection on both sides of the curve is different, then these dislocations $d(y)$ will cause at distant points the deflection

$$u(x) = \frac{1}{2\pi} \int_{\Gamma} \partial_y^1 g_0(y, x) d(y) \, ds_y$$

The kernel of this influence function, this potential

$$\partial_y^1 g_0(y, x) = \frac{\partial}{\partial \nu} g_0(y, x) = -\frac{r_\nu}{2\pi r} = -\frac{1}{2\pi r} [r_{,1} \nu_1(y) + r_{,2} \nu_2(y)] \,,$$

is the normal derivative of the fundamental solution g_0 with regard to the normal vector $\nu(y)$ at the integration point y.

On crossing the curve Γ at x we expect that the deflection u suffers a jump of size $d(x)$, the magnitude of the dislocation at $y = x$,

$$\lim_{\varepsilon \to 0} \{u(x + \varepsilon n) - u(x - \varepsilon n)\} = d(x)$$

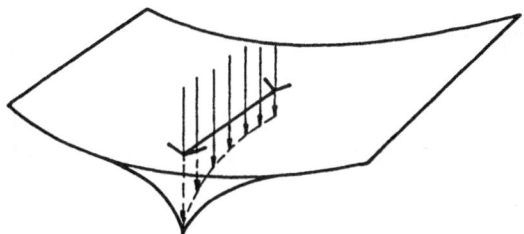

Figure 1.11 Line forces act on a membrane.

The vector n is the normal vector at x parallel to which the two points approach the curve from the left and from the right.

If forces $f(y)$ were acting on Γ, then we would expect that the traction $(=$ slope $\times N)$ suffers a jump discontinuity on Γ,

$$\lim_{\varepsilon \to 0} \left\{ N \frac{\partial u}{\partial n}(x + \varepsilon n) + N \frac{\partial u}{\partial n}(x - \varepsilon n) \right\} = f(x),$$

and that the magnitude of this jump is equal to the value of f at x. This slope discontinuity is visible as a bend in the surface of the membrane, see Fig. 1.11.

In more general terms potentials exhibit such jump discontinuities according to the following rule: The "derivative" ∂^j of a potential u_i

$$\partial^j \int_\Gamma \partial_y^i g_0(y, x) f(y) \, ds_y, \qquad (\partial^j \text{ acts on } x),$$

will be discontinuous on passing through Γ if and only if the index j is conjugated to the order of the potential, that is if $i + j = 2m - 1$.

The number $2m$ is the order of the differential operator

$$2m = 2 \qquad \text{membrane, elastic plates and bodies}$$
$$2m = 4 \qquad \text{Kirchhoff plate}$$

According to this rule the zero-th derivative of the *single layer potential* $(i = 0)$ of the membrane,

$$u_0(x) = \int_\Gamma g_0(y, x) f(y) \, ds_y = -\frac{1}{2\pi} \int_\Gamma \ln r \, f(y) \, ds_y,$$

is continuous, $j + i = 0 + 0 \neq 1$ but the normal derivative ∂^1 jumps because $j + i = 1 + 0 = 1 = 2m - 1$, see Fig. 1.12. The limit of the normal derivative

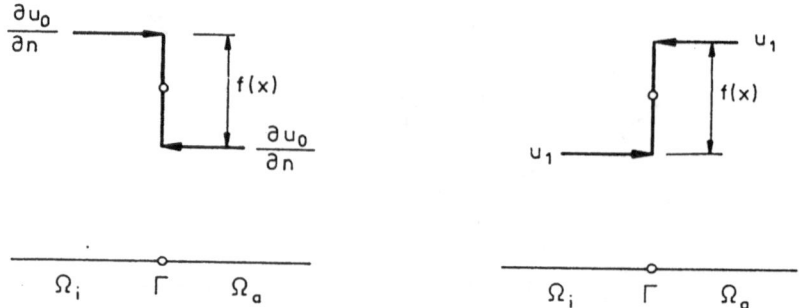

Figure 1.12 The jump discontinuity of the potentials on passing through the boundary Γ

on approaching a smooth boundary point (only at such a point the normal derivative is defined) from within Ω_i is

$$\lim_{\boldsymbol{x}_i \to \Gamma} \frac{\partial}{\partial n} u_0(\boldsymbol{x}) = \frac{1}{2} f(\boldsymbol{x}) - \frac{1}{2\pi} \int_{\Gamma} \frac{\partial}{\partial n} \ln r \, f(\boldsymbol{y}) \, ds_{\boldsymbol{y}} \, ,$$

and on approaching it from the outside, from Ω_e, is

$$\lim_{\boldsymbol{x}_e \to \Gamma} \frac{\partial}{\partial n} u_0(\boldsymbol{x}) = -\frac{1}{2} f(\boldsymbol{x}) - \frac{1}{2\pi} \int_{\Gamma} \frac{\partial}{\partial n} \ln r \, f(\boldsymbol{y}) \, ds_{\boldsymbol{y}}$$

The zero-th derivative of the *double layer potential*, $(i = 1)$,

$$u_1(\boldsymbol{x}) = \int_{\Gamma} \partial_{\boldsymbol{y}}^1 g_0(\boldsymbol{y}, \boldsymbol{x}) f(\boldsymbol{y}) \, ds_{\boldsymbol{y}} = -\frac{1}{2\pi} \int_{\Gamma} \frac{r_\nu}{r} f(\boldsymbol{y}) \, ds_{\boldsymbol{y}} \, ,$$

shows the opposite behavior. It is discontinuous

$$\lim_{\boldsymbol{x}_i \to \Gamma} u_1(\boldsymbol{x}) = -\frac{\Delta \varphi_e}{2\pi} f(\boldsymbol{x}) - \frac{1}{2\pi} \int_{\Gamma} \frac{r_\nu}{r} f(\boldsymbol{y}) \, ds_{\boldsymbol{y}} \, ,$$

$$\lim_{\boldsymbol{x}_e \to \Gamma} u_1(\boldsymbol{x}) = +\frac{\Delta \varphi_i}{2\pi} f(\boldsymbol{x}) - \frac{1}{2\pi} \int_{\Gamma} \frac{r_\nu}{r} f(\boldsymbol{y}) \, ds_{\boldsymbol{y}} \, ,$$

while the normal derivative ∂^1 is continuous. The terms $\Delta \varphi_e$ and $\Delta \varphi_i$ denote the exterior and interior angles of the boundary point, see Fig. 1.13.

Note also that the normal derivative of a single-layer potential does not exist at non-smooth points on the boundary. This holds true for other poten-

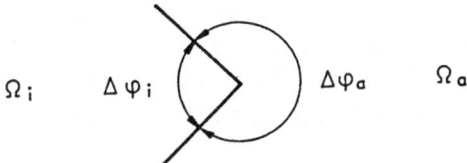

Figure 1.13 The interior and exterior angle of a boundary point

tials as well. If you calculate the directional derivative of a potential at some point inside the domain and if you then pass on to a non-smooth point on the boundary, no limit exists.

Let us summarize: The two kernels

$$g_0 = -\frac{1}{2\pi}\ln r\,, \qquad \partial_y^1 g_0 = -\frac{r_\nu}{2\pi r}$$

of a membrane represent the deflection due to a concentrated force and a point dislocation respectively so that the layers f in the two associated potentials

$$u_0 = -\frac{1}{2\pi}\int_\Gamma \ln r\, f(y)\,ds_y\,, \qquad u_1 = -\frac{1}{2\pi}\int_\Gamma \frac{r_\nu}{r} f(y)\,ds_y$$

can be considered as *forces* and *dislocations*. On passing through Γ the potential u_0, consequently, suffers a jump in the normal derivative (= traction) and the potential u_1 a jump in the zero-th derivative (= deflection).

These classical results of potential theory now easily carry over to elastic plates and bodies. It is only their kernels that are matrix-valued functions and so the layers and the potentials themselves vector-valued functions.

The *single-layer potential* u_0 has the components

$$u_{0i}(x) = \int_\Gamma U_{ij}(y,x)f_j(y)\,ds_y\,, \qquad i = 1,2,3$$

and the *double-layer potential* u_1 the components

$$u_{1i}(x) = \int_\Gamma T_{ij}(y,x)f_j(y)\,ds_y\,, \qquad i = 1,2,3$$

The kernels U_{ij} are the entries of the *Somigliana-matrix* and the kernel T_{ij} are the entries of the matrix of the associated traction vectors, see chapter 4. The single-layer potential represents the displacement field of the elastic continuum under the action of forces $f_j(y)$, which are distributed over a surface Γ and the double-layer potential the displacement field caused by dislocations f_j.

The slope operator $\partial^1 = N\partial/\partial n$ of a membrane corresponds to the traction operator $\partial^1 = \tau(\ )$, which renders the tractions across a section with the normal vector n,

$$\tau(u) = t = S\,n$$

By $\tau(u)_i$ we denote the i-th component of this traction vector.

The jump relations for the traction vector of the potential u_0 read

$$\lim_{x_i \to \Gamma} \tau(u_0)_i = +\frac{1}{2} f_i(x) + \int_\Gamma T_{ij}^*(y, x) f_j(y)\, ds_y,$$

$$\lim_{x_e \to \Gamma} \tau(u_0)_i = -\frac{1}{2} f_i(x) + \int_\Gamma T_{ij}^*(y, x) f_j(y)\, ds_y$$

The matrix T_{ij}^* is obtained on interchanging in the matrix T_{ij} the points y and x and on replacing the normal vector ν by the normal vector n, that is, the kernels T_{ij}^* are the kernels of the adjoint integral operator.

The jump relation for the potential u_1 reads, s. [21],

$$\lim_{x_i \to \Gamma} u_{1i}(x) = -C_{ij}^e(x) f_j(x) + \int_\Gamma U_{ij}(y, x) f_j(y)\, ds_y,$$

$$\lim_{x_e \to \Gamma} u_{1i}(x) = +C_{ij}^i(x) f_j(x) + \int_\Gamma U_{ij}(y, x) f_j(y)\, ds_y$$

The matrix C^i of the interior angle and the matrix C^e of the exterior angle, s. (4.9), add up to the unit matrix

$$C^i(x) + C^e(x) = I, \qquad \text{(unit matrix)}.$$

Up to now we considered only second-order differential equations but potential theory also applies to fourth-order equations. To the biharmonic equation belong four kernels, see chapter 6,

$$g_0, \qquad \partial_y^1 g_0 = \frac{\partial}{\partial \nu} g_0, \qquad \partial_y^2 g_0 = M_\nu(g_0), \qquad \partial_y^3 g_0 = V_\nu(g_0),$$

which represent the deflection of an infinite plate under the action of a

$$\textit{concentrated force} \qquad g_0(y, x),$$

$$\textit{concentrated couple} \qquad \partial_y^1 g_0(y, x),$$

$$\textit{concentrated bend} \qquad \partial_y^2 g_0(y, x),$$

$$\textit{point dislocation} \qquad \partial_y^3 g_0(y, x)$$

The layers f in the associated potentials

$$u_0 = \int_\Gamma g_0 f \, ds_y , \qquad u_1 = \int_\Gamma \partial_y^1 g_0 f \, ds_y ,$$

$$u_2 = \int_\Gamma \partial_y^2 g_0 f \, ds_y , \qquad u_3 = \int_\Gamma \partial_y^3 g_0 f \, ds_y ,$$

can be identified as *forces, moments, bends* and *dislocations* and the behavior of these potentials conforms to this interpretation. The potential u_0 suffers a jump in the "third derivative", the Kirchhoff shear, the potential u_1 in the "second derivative", the bending moment, the potential u_2 in the normal derivative and the potential u_3 in the deflection.

If we integrate the kernel functions over the domain Ω then we obtain *volume potentials*,

$$v_i(x) = \int_\Omega \partial_y^i g_0(y, x) p(y) \, d\Omega_y$$

These potentials are not homogeneous solutions, instead the action of the differential operator D will *reproduce* the "derivatives" $\partial^i p$ of the volume distribution p,

$$Dv_i(x) = D \int_\Omega \partial_y^i g_0(y, x) p(y) \, d\Omega_y = \partial^i p(x)$$

To explain this behavior consider an infinite membrane and assume we blow up a patch Ω of this membrane with a pressure p. The membrane will then show the following deflection:

$$u(x) = -\frac{1}{2\pi} \int_\Omega \ln r \, p(y) \, d\Omega_y ,$$

and because the equilibrium condition requires that the deflection satisfies the differential equation

$$-\Delta u = p \qquad \text{in the domain of the load} ,$$

we conclude that the kernel $g_0(y, x) = -(2\pi)^{-1} \ln r$ must be such a reproducing kernel.

Potential theory forms the backbone of the boundary element method because each influence function, as e.g. the influence function of a membrane $(N = 1)$

$$u(\boldsymbol{x}) = -\frac{1}{2\pi} \int\limits_\Gamma \ln r \, \frac{\partial u}{\partial \nu}(\boldsymbol{y}) \, ds\boldsymbol{y} + \frac{1}{2\pi} \int\limits_\Gamma \frac{r_\nu}{r} u(\boldsymbol{y}) \, ds\boldsymbol{y}$$

$$-\frac{1}{2\pi} \int\limits_\Omega \ln r \, p(\boldsymbol{y}) \, d\Omega\boldsymbol{y} \,, \tag{1.8}$$

is the sum of a single-layer potential, a double-layer potential and a volume potential.

The *coupling condition* between the Betti data of a membrane which we derived before with Betti's principle can be derived in the framework of potential theory as well. To this end we let the point $\boldsymbol{x}$ in Eq.(1.8) simply tend to the boundary. The limit of the left-hand side is $u(\boldsymbol{x})$. The limit of the right-hand side is the sum of the limits of the single potentials so that

$$u(\boldsymbol{x}) = -\frac{1}{2\pi} \int\limits_\Gamma \ln r \frac{\partial u}{\partial \nu}(\boldsymbol{y}) \, ds\boldsymbol{y} + \frac{\Delta\varphi_e}{2\pi} u(\boldsymbol{x}) + \frac{1}{2\pi} \int\limits_\Gamma \frac{r_\nu}{r} u(\boldsymbol{y}) \, ds\boldsymbol{y}$$

$$-\frac{1}{2\pi} \int\limits_\Omega \ln r \, p(\boldsymbol{y}) \, d\Omega\boldsymbol{y}$$

If we put the term $\Delta\varphi_e/2\pi \, u(\boldsymbol{x})$ on the left-hand side and consider that

$$\frac{\Delta\varphi_e}{2\pi} + \frac{\Delta\varphi_i}{2\pi} = 1\,, \qquad\qquad 1 - \frac{\Delta\varphi_e}{2\pi} = \frac{\Delta\varphi_i}{2\pi}\,,$$

then we obtain

$$\frac{\Delta\varphi_i}{2\pi} u(\boldsymbol{x}) = -\frac{1}{2\pi} \int\limits_\Gamma \ln r \frac{\partial u}{\partial \nu}(\boldsymbol{y}) \, ds\boldsymbol{y} + \frac{1}{2\pi} \int\limits_\Gamma \frac{r_\nu}{r} u(\boldsymbol{y}) \, ds\boldsymbol{y} - \frac{1}{2\pi} \int\limits_\Omega \ln r \, p(\boldsymbol{y}) \, d\Omega\boldsymbol{y} \,,$$

which is exactly the coupling condition (1.3) between the Betti data of the membrane.

1.12 The indirect method

Point charges or charges distributed over wires or surfaces generate fields which satisfy the Laplace equation. In mechanics the point charges are single forces and the dipoles are dislocations but as in electrostatic we can think of a stress field as a field being generated by forces or dislocations distributed over manifolds (curves or surfaces) embedded into the elastic medium. This is the approach of the *indirect method* which either uses a single-layer potential or a double-layer potential. Because each influence function, as the influence function for a membrane ($N = 1$),

$$u(x) = -\frac{1}{2\pi} \int\limits_{\Gamma} \ln r \frac{\partial u}{\partial \nu}(y)\, ds_y + \frac{1}{2\pi} \int\limits_{\Gamma} \frac{r_\nu}{r} u(y)\, ds_y$$

$$-\frac{1}{2\pi} \int\limits_{\Omega} \ln r\, p(y)\, d\Omega_y\,, \tag{1.9}$$

is a sum of a single-layer potential, a double-layer potential and a volume potential the indirect method is, so to speak, a shortened variant of the *direct method*, the method we applied up to now.

The difference between the two methods becomes manifest in the interpretation of what is or what are the source points. In the direct method the free point x is the source point while in the indirect method the integration points y are the source points.

To compare the approach of the two methods let us consider a boundary-value problem as

$$\Delta u = 0 \quad \text{in } \Omega\,, \qquad u = \bar{u} \quad \text{on } \Gamma$$

The direct method starts with the influence function (1.9) and so it must solve the integral equation

$$-\frac{1}{2\pi} \int\limits_{\Gamma} \ln r\, t(y)\, ds_y = \frac{\Delta\varphi}{2\pi} \bar{u}(x) - \frac{1}{2\pi} \int\limits_{\Gamma} \frac{r_\nu}{r} \bar{u}(y)\, ds_y$$

for the unknown traction $t = \partial u/\partial \nu$.

The indirect method either uses a single-layer potential

$$u_0(x) = -\frac{1}{2\pi} \int\limits_{\Gamma} \ln r\, f(y)\, ds_y$$

this leads to the integral equation

$$-\frac{1}{2\pi} \int\limits_{\Gamma} \ln r\, f(y)\, ds_y = \bar{u}(x)\,,$$

or a double-layer potential,

$$u_1(\boldsymbol{x}) = -\frac{1}{2\pi} \int_{\Gamma} \frac{r_\nu}{r} f(\boldsymbol{y}) \, ds\boldsymbol{y} \, ,$$

this leads to the integral equation

$$-\frac{\Delta\varphi_a}{2\pi} f(\boldsymbol{x}) - \frac{1}{2\pi} \int_{\Gamma} \frac{r_\nu}{r} f(\boldsymbol{y}) \, ds\boldsymbol{y} = \bar{u}(\boldsymbol{x})$$

While the layers u and t of the direct method are the boundary values of the solution, the layers $f(\boldsymbol{y})$ of the indirect method are auxiliary quantities which have no relation to the original problem. This is why the method got its name. But these fictitious layers have a physical meaning for the so-called *complementary problem*, see [22].

To explain this connection we consider a stretched rectangular plate, see Fig. 1.14. The displacement field $\boldsymbol{u}$ of the plate satisfies the equations

$$-L_{ij}u_j = 0 \quad \text{in } \Omega, \qquad \tau(u)_i = \bar{t}_i \qquad \text{on } \Gamma$$

where $\bar{t}_i$ denotes the prescribed tractions on the edge. These tractions are, up to $\bar{t}_1$, which takes on the values $\bar{t}_1 = \pm p$ at the two vertical edges, zero on Γ.

If we represent the displacement field by a potential of the first kind (single-layer potential)

$$u_{0i}(\boldsymbol{x}) = \int_{\Gamma} U_{ij}(\boldsymbol{y}, \boldsymbol{x}) f_j(\boldsymbol{y}) \, ds\boldsymbol{y} \, ,$$

then the traction boundary condition

$$\lim_{\boldsymbol{x} \to \Gamma} \tau(\boldsymbol{u}_0)_i = \bar{t}_i$$

leads to the integral equation

$$\frac{1}{2} f_i(\boldsymbol{x}) + \int_{\Gamma} T_{ij}^*(\boldsymbol{y}, \boldsymbol{x}) f_j(\boldsymbol{y}) \, ds\boldsymbol{y} = \bar{t}_i(\boldsymbol{x}), \qquad i = 1, 2$$

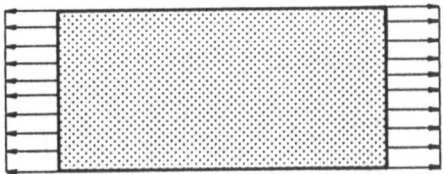

Figure 1.14 A stretched plate

for the unknown layer $f_i(y)$. With regard to these layers holds: the prescribed tractions $\bar{t}_i$ minus the solutions $f_i(y)$ of this integral equation are the components

$$s_i = \bar{t}_i - f_i$$

of the traction vector which we must apply to the edge of the complement if it is to undergo the same deformations as the edge of the stretched plate, if, so to speak, the stretched plate fits like the piece of a puzzle into the opening in the infinite elastic plate, see Fig. 1.15.

Now, to force upon the edge of the infinite plate this "rectangular" displacement pattern is certainly a formidable task and so we must expect that the tractions s_i which accomplish this become singular at the four corner points. But this means, because the tractions s_i are the sum of the components $\bar{t}_i$ (which are fixed and bounded) and the components f_i, that the components f_i must provide the necessary action, that is these components must become singular at the corners.

If we would calculate the displacement field of the stretched plate with the direct method then the unknown layers would be the components u_i of the displacement field and because displacements, unlike stresses, are always bounded, the solution would be well behaved.

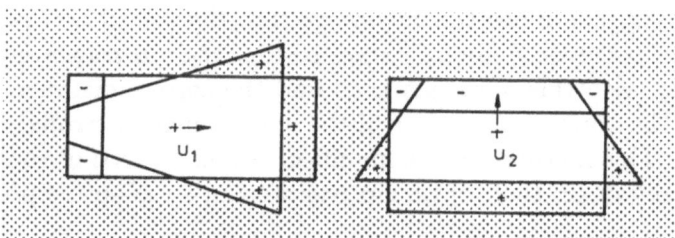

Figure 1.15 The horizontal and vertical displacement on the edge of the complement

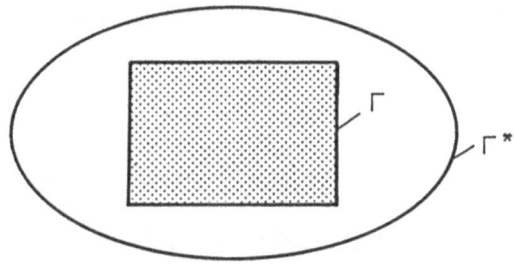

Figure 1.16 An auxiliary curve Γ^* outside of the problem domain

Closely related with the indirect method is the idea to place the fictitious layers on an auxiliary curve outside of the original domain, see Fig. 1.16.

To solve the boundary-value problem of a membrane

$$\Delta u = 0 \quad \text{in } \Omega, \qquad u = \bar{u} \quad \text{on } \Gamma,$$

with a single-layer potential u_0 essentially means to assume that the membrane Ω is embedded into an infinite membrane and to distribute forces $f(\boldsymbol{y})$ on the boundary curve Γ in such a way that the deflection of the points of Γ equals the prescribed deflection $\bar{u}$. But then we could also imagine that the forces are distributed on an auxiliary curve Γ^* outside the problem domain. These forces cause the deflection

$$u_0(\boldsymbol{x}) = -\frac{1}{2\pi} \int\limits_{\Gamma^*} \ln r f(\boldsymbol{y}) \, ds_{\boldsymbol{y}}$$

and it only remains to tune the forces $f(\boldsymbol{y})$ in such a way that the points on the real boundary Γ take on the prescribed deflection

$$-\frac{1}{2\pi} \int\limits_{\Gamma^*} \ln r f(\boldsymbol{y}) \, ds_{\boldsymbol{y}} = \bar{u}(\boldsymbol{x}) \qquad \text{on } \Gamma$$

Such an auxiliary curve has the advantage that the collocation points and the quadrature points lie on different curves so that the integrals are regular. The main difficulty with this approach is to find the correct distance between the two curves. The condition number of the system matrix strongly depends on this distance. This method can be modified further by letting the singularities move freely and attain their optimum positions in an adaptive process, [23].

1.13 Weighted residuals

Many numerical methods can be interpreted as a weighted residual technique. The same holds true for the boundary element method. To solve a boundary-value problem as

$$\Delta u = p \quad \text{on } \Omega$$

$$u = 0 \quad \text{on } \Gamma$$

we can start with a weighted residual formulation as

$$\int\limits_{\Omega} (\Delta u - p) u^* \, d\Omega_{\boldsymbol{y}} = 0$$

where $u^* = g_0$ is the fundamental solution and we can then apply integration by parts. But because the fundamental solution is not smooth enough for integration by parts to apply we must first exclude from the domain a small neighborhood of the source point

$$\int\limits_{\Omega_\epsilon} (\Delta u - p)u^* d\Omega_y = 0$$

After integration by parts we then let the radius ϵ shrink to zero. The result of this weighted residual approach is naturally the same result as if we had directly applied Green's first identity

$$\lim_{\epsilon \to 0} G(g_0, u)_{\Omega_\epsilon} =: G(g_0, u) = 0,$$

or Betti's principle

$$\lim_{\epsilon \to 0} B(g_0, u)_{\Omega_\epsilon} =: B(g_0, u) = 0$$

so there is no difference between the two approaches.

1.14 Influence functions and finite elements

If we replace a distributed load that acts on a beam, see Fig. 1.17, by a concentrated force and if we argue with a colleague whether this is admissible, then we really are arguing about the properties of an influence function: what difference will it make if in the influence function for the bending moment

$$M(x) = -f_2[x]^T u + u_2[x]^T f + \int\limits_0^l g_2(y, x) p(y)\, dy$$

the domain integral is replaced by the term

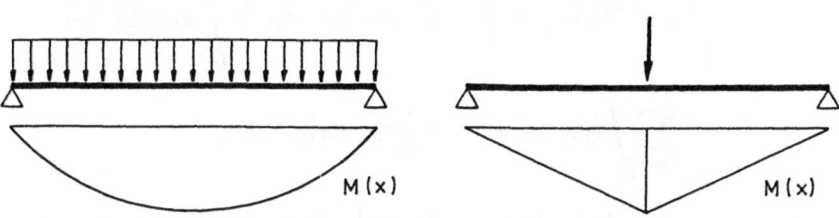

Figure 1.17 The distributed load is replaced by a statically equivalent concentrated force

$$g_2(x^P, x)P, \qquad (x^P = \text{centre of gravity of } p), \qquad P = \int\limits_0^l p(y)\, dy,$$

the influence of the statically equivalent concentrated force P? (This is equivalent to an approximate calculation of the domain integral). Similar questions arise when we ask ourselves: what effect will incorrect assumptions about the stiffness EI have on the deflection w of a concrete beam? Or how will errors in the boundary values u, f influence the internal actions? All these questions concern the properties of the influence function.

When we turn to approximate methods then we do so because we cannot solve the original problem. Instead we replace the original problem by a neighboring problem and this we solve *exactly*. This is why influence functions play a *central role*.

To calculate the stresses in a structure with finite elements means replacing the original load by a load which is equivalent in the sense of the principle of virtual displacements and solving this approximate loadcase exactly.

When we try to assess the error in the internal actions of the FE-solution then we are essentially trying to estimate the properties of the influence function: how do the internal actions change if the distributed forces are replaced by forces which, say, are concentrated on the element boundaries? How sensitive is the influence function to such modifications?

Let us elaborate on this close connection between finite elements and influence functions in more detail by considering a stretched membrane.

The deflection of a membrane

$$u(x) = \int\limits_\Gamma g_0\, t(y)\, ds_y - \int\limits_\Gamma N \frac{\partial}{\partial \nu} g_0 u(y)\, ds_y + \int\limits_\Omega g_0 p\, d\Omega_y,$$

depends on three terms: the traction $t(y)$ and the deflection $u(y)$ on the boundary and the lateral pressure p that acts on the membrane.

If along interior lines l_m additional line forces t_Δ are distributed or if dislocations u_Δ occur, then the influence function must be supplemented correspondingly:

$$u(x) = \int\limits_\Gamma g_0\, t(y)\, ds_y - \int\limits_\Gamma N \frac{\partial}{\partial \nu} g_0 u(y)\, ds_y + \int\limits_\Omega g_0 p(y)\, d\Omega_y$$

$$+ \sum_m \int\limits_{l_m} \left(g_0\, t_\Delta(y) - N \frac{\partial}{\partial \nu} g_0 u_\Delta(y) \right) ds_y$$

This function is, as we shall show in the following, the general representation of an FE-solution.

To approximate the deflection u of a membrane with finite elements means seeking a solution u_h in terms of n basis functions

$$u_h = u_i \varphi_i(\boldsymbol{x})$$

Elementwise, the basis functions φ_i are smooth but on the element boundaries their slope is discontinuous or they might even suffer a jump discontinuity. Now recall that the normal derivative of the deflection u_h is, up to the factor N, the traction t within the membrane. Hence such slope discontinuities mean that line forces t_Δ act along the element boundaries. Even if the deflection is discontinuous then this means that dislocations u_Δ occur. Hence, the influence function given above is precisely the integral representation of all possible FE-solutions. *Whether we consider the FE-solution to be the sum of the n basis functions $\varphi_i(\boldsymbol{x})$ or to be the response of the membrane to the FE-forces (the right-hand sides of the basis functions φ_i) and to the FE-line-forces and FE-dislocations is the same.*

$$u_h(\boldsymbol{x}) = u_i \varphi_i(\boldsymbol{x}) = \int_\Gamma \left(g_0\, t_h - N \frac{\partial}{\partial \nu} g_0 u_h \right) ds_y + \sum_e \int_{\Omega_e} g_0 p_h^e \, d\Omega_y$$

$$+ \sum_m \int_{l_m} \left(g_0\, t_\Delta - N \frac{\partial}{\partial \nu} g_0 u_\Delta \right) ds_y \qquad (1.10)$$

The terms

$$t_\Delta = N \frac{\partial u_h}{\partial n}\bigg|_{\text{left}} + N \frac{\partial u_h}{\partial n}\bigg|_{\text{right}}, \qquad u_\Delta = u_h|_{\text{left}} - u_h|_{\text{right}},$$

are the jumps that occur between neighboring elements on the element boundaries l_m. If the FE-solution is conforming (C^0-basis functions) then the discontinuities u_Δ are zero. The term

$$p_h^e = -N \Delta u_h^e$$

represents the lateral pressure that acts on the single element Ω_e. Its distribution is obtained by substituting the local representation u_h^e of the FE-solution into the membrane equation.

For an application of these ideas consider the membrane in Fig. 1.18a whose deflection we approximated with linear elements. Such an approximation corresponds to a loadcase where line forces on the element boundaries replace the distributed load. The size of these forces is equal to the N-fold bend (the slope discontinuity) in the deflection surface. If the basis functions were piecewise quadratic then the elements themselves would deflect (and not just perform rigid-body motions as do the linear elements) because the non-zero pressure

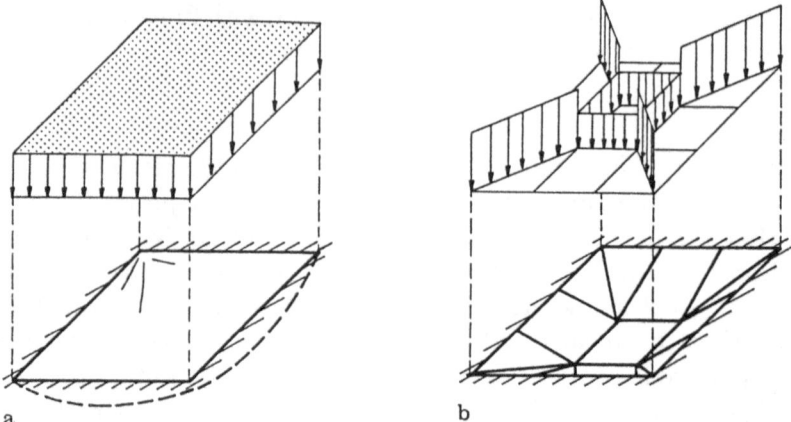

Figure 1.18 a-b. Lateral pressure on a membrane: **a** the original loadcase; **b** the loadcase solved with finite elements

$-N\Delta u_h^e = p_h^e$ would deform the elements. With linear elements the pressure is zero.

The traction t $(= N$-fold slope) of the FE-solution is the function

$$t_h(\boldsymbol{x}) = N\frac{\partial}{\partial n}u_h(\boldsymbol{x}) = Nu_i\frac{\partial}{\partial n}\varphi_i = \int_\Gamma (g_1 t_h - N\frac{\partial}{\partial\nu}g_1 u_h)\,ds\boldsymbol{y}$$

$$+ \sum_e \int_{\Omega_e} g_1 p_h^e \, d\Omega\boldsymbol{y} + \sum_m \int_{l_m} (g_1 t_\Delta - N\frac{\partial}{\partial\nu}g_1 u_\Delta)\,ds\boldsymbol{y}$$

This function differs from the influence function of the deflection only in that the fundamental solution g_0 is replaced by its normal derivative

$$g_1 = N\frac{\partial}{\partial n}g_0$$

However not only the FE-solution can be given in terms of an influence function but also the true solution

$$u(\boldsymbol{x}) = \int_\Gamma (g_0\,t - N\frac{\partial}{\partial\nu}g_0 u)\,ds\boldsymbol{y} + \int_\Omega g_0 p\,d\Omega\boldsymbol{y}$$

and its traction

$$t(\boldsymbol{x}) = N\frac{\partial u}{\partial n}(\boldsymbol{x}) = \int_\Gamma (g_1 t - N\frac{\partial}{\partial\nu}g_1 u)\,ds\boldsymbol{y} + \int_\Omega g_1 p\,d\Omega\boldsymbol{y}$$

Hence, the error in the FE-tractions is simply the difference between the two influence functions

$$t - t_h = \int\limits_{\Gamma} [\, g_1(t - t_h) - N\frac{\partial}{\partial \nu} g_1(u - u_h)]\, ds y + \sum_e \int\limits_{\Omega_e} g_1(p - p_h^e)\, d\Omega y$$

$$- \sum_m \int\limits_{l_m} (g_1 t_\Delta - N\frac{\partial}{\partial \nu} g_1 u_\Delta)\, ds y \, ,$$

and this shows that the error of an FE-solution has its origin in the discrepancy between the boundary layers, namely the difference between the tractions, $t - t_h$, and the deflections, $u - u_h$, on Γ, the jumps u_Δ and t_Δ between the elements and the fact that the FE-pressure p_h^e differs from the true pressure p.

To apply this formula we have to know both boundary values, u and t, of the true solution; so it seems of limited practical use. However, this formula teaches us on which quantities the error depends, which kernels determine the behavior of the error. Basically, the kernels are the interesting quantities. They are the "multipliers" which enlarge or diminish an error, depending on whether their shift is positive or negative, see section 1.9.

The integral representations of other FE-solutions are obtained in the same fashion. Let us consider for example elastic plates. FE-solutions of such structures normally use C^0- element, that is, the displacements are continuous across the element boundaries, but the tractions jump between the elements

$$t_{\Delta j} = t_j|_{\text{left}} + t_j|_{\text{right}} \, ,$$

and so we obtain for the error of a conforming FE-solution the expression

$$u_i(\boldsymbol{x}) - u_{hi}(\boldsymbol{x}) = \int\limits_{\Gamma} [U_{ij}(t_j - t_{hj}) - T_{ij}(u_j - u_{hj})]\, ds y$$

$$+ \sum_e \int\limits_{\Omega_e} U_{ij}(p_j - p_{hj}^e)\, d\Omega y - \sum_m \int\limits_{l_m} U_{ij} t_{\Delta j}\, ds y$$

Similarly the error of an FE-solution of a Kirchhoff plate can be expressed as

$$w(\boldsymbol{x}) - w_h(\boldsymbol{x}) = \int\limits_\Gamma [\, g_0(V - V_h) - \frac{\partial}{\partial \nu} g_0(M - M_h) - V_\nu(g_0)(w - w_h)$$

$$+ M_\nu(g_0)(\frac{\partial w}{\partial \nu} - \frac{\partial w_h}{\partial \nu})\,]\, ds_{\boldsymbol{y}}$$

$$+ \sum_m \int\limits_{l_m} [-M_\nu(g_0)\frac{\partial w_\Delta}{\partial \nu} - g_0 V_\Delta + \frac{\partial}{\partial \nu} g_0 M_\Delta]\, ds_{\boldsymbol{y}}$$

$$+ \sum_e \int\limits_{\Omega_e} g_0(\boldsymbol{y}, \boldsymbol{x})(p(\boldsymbol{y}) - p_h^e(\boldsymbol{y}))\, d\Omega_{\boldsymbol{y}} + \sum_k F_k\, g_0(\boldsymbol{y}^k, \boldsymbol{x})$$

$$+ \sum_c [\, g_0\{F(w) - F(w_h)\}(\boldsymbol{y}^c) - \{w - w_h\}(\boldsymbol{y}^c)F(g_0)]$$

Here the terms

$$\frac{\partial w_\Delta}{\partial \nu}, \quad M_\Delta, \quad V_\Delta$$

denote the interelement discontinuities and the F_k are nodal forces (which are not to be confused with the equivalent nodal forces). The force F_k is the sum of all the corner forces of the single elements that connect with node $\boldsymbol{x}_k$. If the derivative $w_{,xy}$ of the FE-solution is continuous at the nodes then the forces F_k are zero at the nodes. The domain functions

$$p_h^e = K \Delta\Delta w_h^e$$

are the element loads. They are simply the right-hand sides of the FE-solution on the single elements. The last term in the influence function is a sum over the corner points $\boldsymbol{y}^c$ of the plate. It represents the influence of the error in the corner forces and in the corner deflections.

Let us apply this error analysis to a rectangular hinged plate. Besides being supported on the boundary the plate rests, in addition, on four internal point supports, see Fig. 1.19 a. On account of the symmetry of the problem only a quarter of the plate had to be discretized. The load p is uniformly distributed over the plate.

The plots in Fig. 1.19 and 1.20 show which loadcase we really solved with finite elements. This loadcase involved the following forces:

the domain load $K \Delta\Delta w_h^e$
bending moments $M_{\Delta x}$
bending moments $M_{\Delta y}$
shear forces $V_{\Delta x}$
shear forces $V_{\Delta y}$, on the element boundaries l_m

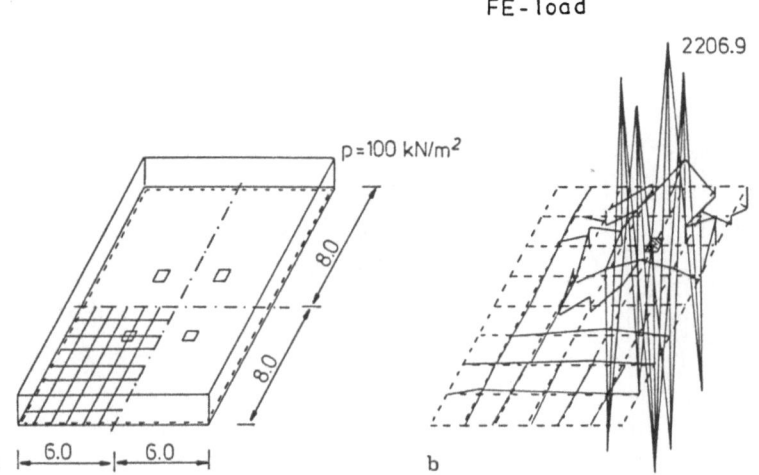

Figure 1.19 a-b. Plate under a uniform load: **a** the original uniform distribution of the load; **b** the FE-distribution (only the lower left quadrant is shown).

The lower index x or y respectively indicates that the moments or forces act along lines $x = $ constant or $y = $ constant.

This means that the FE-solution describes the equilibrium position of the plate under the action of strongly oscillating distributed forces $K\Delta\Delta w_h^e$, see Fig. 1.19b, and moments $M_{\Delta x}$, $M_{\Delta y}$ and forces $V_{\Delta x}$, $V_{\Delta y}$ which act along the element boundaries, see Fig. 1.20.

Looking at the strong oscillations of the domain forces it seems improbable that the internal actions which are caused by these exterior forces are close to the real internal actions. But this scepticism is unfounded as a comparison with a BE-solution reveals.

The BE-solution satisfies the equation $K\Delta\Delta w = p$ exactly and also the singularities which are caused by the point supports can be modelled very accurately by the BE-solution, so that we may use the BE-solution as a reference solution. Figure 1.21 shows a cross section of the plate and we notice that the bending moments of the FE-solution closely follow the bending moments of the BE-solution. There is no indication of the strong oscillations in the fourth order derivatives, $K\Delta\Delta w_h^e$, of the FE-solution anymore.

This smoothing of the oscillations is brought about by the kernels in the influence function for the bending moment,

$$M_n(w_h)(\boldsymbol{x}) = \int\limits_{\Gamma} [\, g_2 V_h - \frac{\partial}{\partial \nu} g_2 M_h - V_\nu(g_2) w_h + M_\nu(g_2) \frac{\partial}{\partial \nu} w_h]\, ds_y$$

$$+ \sum_m \int\limits_{l_m} [M_\nu(g_2) \frac{\partial w_\Delta}{\partial \nu} + g_2 V_\Delta - \frac{\partial}{\partial \nu} g_2 M_\Delta]\, ds_y$$

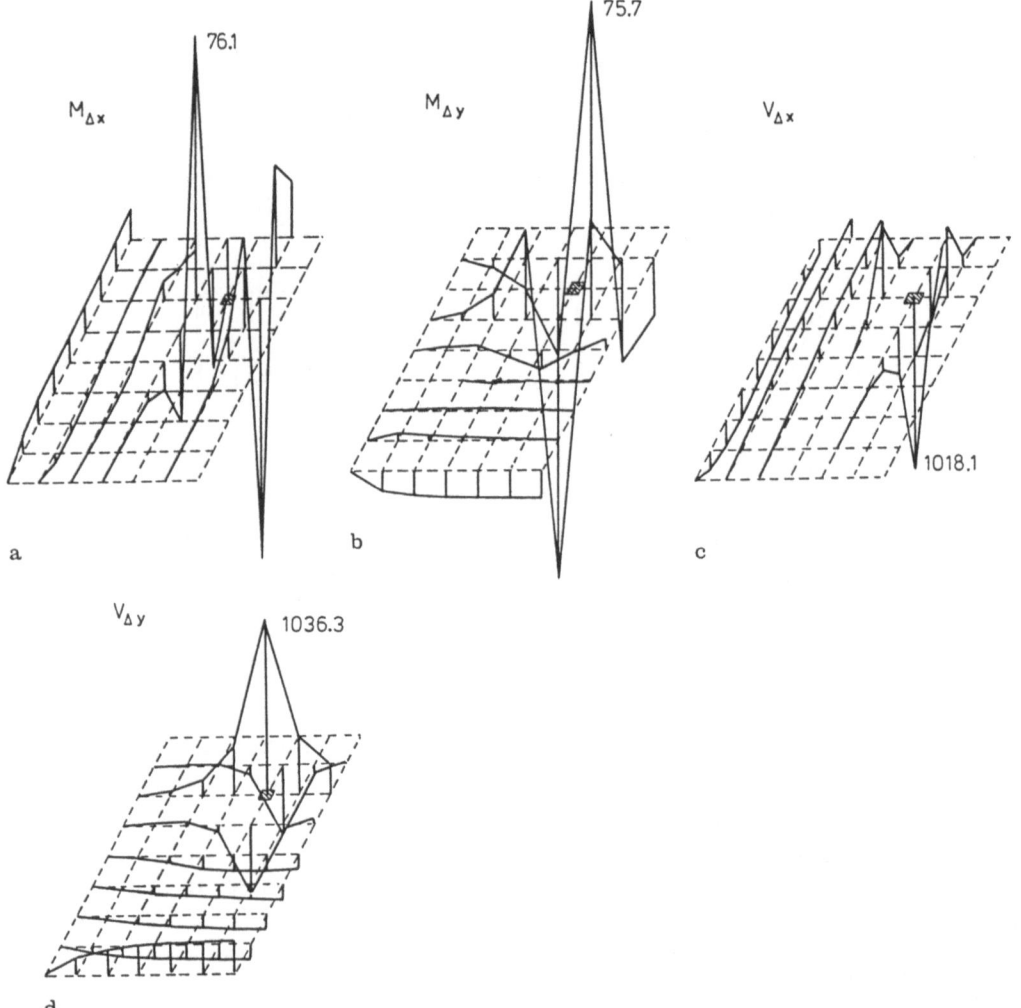

Figure 1.20 a-d. Along the element boundaries do act additional moments and forces: a moments M_{Δ_x}; b moments M_{Δ_y}; c forces V_{Δ_x}; d forces V_{Δ_y}; (only the lower left quadrant is shown)

$$+ \sum_e \int_{\Omega_e} g_2(\boldsymbol{y}, \boldsymbol{x}) p_h^e \, d\Omega_y + \sum_k F_k \, g_2(\boldsymbol{y}^k, \boldsymbol{x})$$

$$+ \sum_c [g_2(\boldsymbol{y}^c, \boldsymbol{x}) F(w_h)(\boldsymbol{y}^c) - w_h(\boldsymbol{y}^c) F(g_2)(\boldsymbol{y}^c, \boldsymbol{x})] \qquad (1.11)$$

These kernels provide depending on the domain of integration, Γ or Ω the following shifts

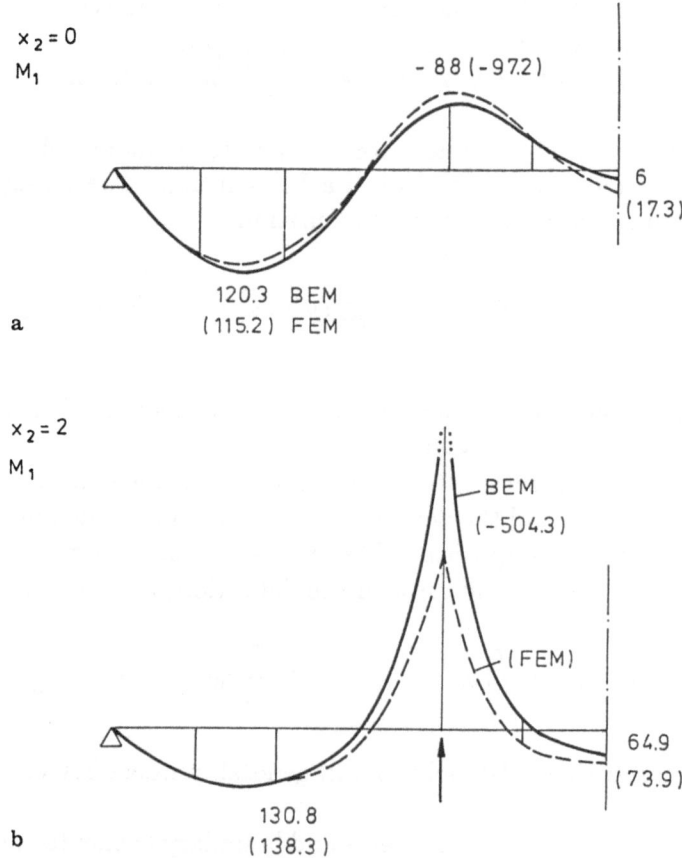

Figure 1.21 a-b. The distribution of the bending moments in two sections parallel to the x-axis: **a** in the middle of the plate; **b** in between the point supports; (from the edge to the centre of the plate)

kernel	shift (Γ)	shift (Ω)
g_2	-1	-2
$\partial g_2/\partial \nu$	0	
$M_\nu(g_2)$	1	
$V_\nu(g_2)$	2	

The worst terms, the oscillating domain forces p_h^e, are integrated twice by the kernel g_2 while the line forces V_Δ are shifted once. The negative effect of slope discontinuities $\partial w_\Delta/\partial \nu$ which occur between non-conforming elements also becomes apparent: the adjoint kernel $M_\nu(g_2)$ *differentiates* these discontinuities.

Normally we calculate the bending moments of a FE-solution by differentiating the solution elementwise. But this is the same as if we would substitute the exterior FE-forces (and, if present, also the slope discontinuities) into the influence function (1.11) of the bending moment. The result is the same at any time. *Hence, even if we do not think of influence functions, if we apply a "pure" finite element method, we are using influence functions and, thereby, their smoothing effects.*

Let us now apply the same error analysis to boundary elements to learn which terms contribute to the error of a BE-solution. The BE-approximation of the deflection of a membrane is the function

$$u_h(x) = \int_\Gamma (g_0\, t_{\mathrm{BE}} - N\frac{\partial}{\partial \nu} g_0\, u_{\mathrm{BE}})\, ds_y - \int_\Omega g_0 p\, d\Omega_y$$

where the boundary functions u_{BE} and t_{BE} are the piecewise polynomials which satisfy the discrete coupling conditions.

Because the BE-solution u_h satisfies the membrane equation $-N\Delta u_h = p$ and because the slope and the deflection are continuous in the interior no potentials are generated by jump-terms. The error stems only from the discrepancies on the boundary, namely the error in the deflection, $u - u_h$, and the traction, $t - t_h$,

$$u - u_h = \int_\Gamma [\, g_0(t - t_{\mathrm{BE}}) - N\frac{\partial}{\partial \nu} g_0(u - u_{\mathrm{BE}})]\, ds_y$$

This is the reason why a BE-solution, in general, renders better results than a FE-solution.

Let us summarize: Finite elements and boundary elements are similar. FE-solutions and BE-solutions are solutions in terms of the influence function of the governing equation. But unlike BE-solutions which consist only of boundary potentials (the volume potentials are usually transformed into equivalent boundary integrals) FE-solutions also consist of interior line potentials and volume potentials. The only significant difference between boundary elements and finite elements lies in the choice of the layers in the potentials. The FE-method determines these layers by applying the principle of virtual work and the BE-method by applying Betti's principle.

> finite elements = boundary potentials, line potentials and
> volume potentials, control equation:
> principle of virtual displacements

> boundary elements = boundary potentials
> control equation: Betti's principle

1.15 The scale

The scale is seemingly relative. Each quadratic membrane can be scaled down to the size of the unit square. By such a change of scale

$$r \longrightarrow \lambda r, \qquad 0 < \lambda,$$

the coupling condition between the Betti data

$$\frac{\Delta\varphi}{2\pi} u(\boldsymbol{x}) = -\frac{1}{2\pi N} \int_{\Gamma} [\ln r\, t(\boldsymbol{y}) - N\frac{r_\nu}{r} u(\boldsymbol{y})]\, ds_{\boldsymbol{y}},$$

transforms, because of

$$\ln(\lambda r) = \ln r + \ln \lambda$$

into the condition

$$\frac{\Delta\varphi}{2\pi} u(\boldsymbol{x}) = -\frac{1}{2\pi N} \left\{ \int_{\Gamma} [\ln r\, t(\boldsymbol{y}) - N\frac{r_\nu}{r} u(\boldsymbol{y})]\, ds_{\boldsymbol{y}} + \ln \lambda \int_{\Gamma} t(\boldsymbol{y})\, ds \right\}$$

(For ease of notation we do not distinguish between the transformed and the original quantities.) Hence, such a change of scale adds the equilibrium condition

$$\int_{\Gamma} t(\boldsymbol{y})\, ds = 0$$

to the coupling condition. If t is the correct traction then this integral is zero. Hence such a change of scale leaves the basic equation invariant.

But in practical applications it was observed that the scale can have some influence on the results. In particular when we treat plane problems and the kernel function is the logarithm, $\ln r$. In this case we should avoid the unit length 1. This has something to do with the fact that the logarithm has a zero at $r = 1$ and that the corresponding change of sign can lead to an extinction of opposing influences, [24], [25].

1.16 Trefftz's method

This method is the counterpart to Ritz's method. Ritz's method depends on basis functions which satisfy the geometric boundary conditions but not the differential equation. Trefftz's method, instead, uses basis functions which satisfy the differential equation but not the boundary conditions. The original idea of Trefftz was to determine their nodal values as in Ritz's method. Consider the boundary-value problem

$$\Delta u = 0 \quad \text{on } \Omega\,, \qquad u = \bar{u} \quad \text{on } \Gamma$$

Because the right-hand side of the differential equation is zero and because only displacements are prescribed on the boundary, the potential energy is simply the expression

$$\Pi_1(u) = \frac{1}{2}E(u,u) = \frac{1}{2}\int_\Omega \nabla u \cdot \nabla u \, d\Omega_y$$

According to Ritz we approximate u by

$$u_h = u_0 + u_i\Phi_i(\boldsymbol{x})\,, \qquad u_0 = \bar{u}\,, \quad \Phi_i = 0 \quad \text{on } \Gamma$$

and determine the coefficients by minimizing the energy. This is equivalent with the fact that

$$E(u - u_h, \Phi_i) = 0\,, \qquad \text{for } i = 1, 2, ..., n$$

Trefftz, [26], assumes now that the basis functions Φ_i are harmonic, that they satisfy the differential equation but not the boundary conditions. Integrating the energy by parts it follows

$$E(u - u_h, \Phi_i) = \int_\Gamma \frac{\partial \Phi_i}{\partial n}(\bar{u} - u_h)\, ds = 0\,, \qquad i = 1, 2...n$$

that is, the coefficients u_i are the solutions of the system

$$\boldsymbol{K}\boldsymbol{u} = \boldsymbol{b}$$

where

$$K_{ij} = \int_\Gamma \frac{\partial \Phi_i}{\partial n}\Phi_j\, ds\,, \qquad b_i = \int_\Gamma \frac{\partial \Phi_i}{\partial n}\bar{u}\, ds$$

Today we also speak of a Trefftz method if the selection of the coefficients is not based on an energy principle but, for example, on the least-square method as in [27]. In this sense the direct and indirect boundary element methods are variants of Trefftz's method. The basis functions (these are the potentials Φ_i generated by the single boundary layers φ_i) are homogeneous solutions of the governing equation and the selection of the coefficients is based on a collocation procedure on the boundary.

But more important than all these classifications is the fact that Trefftz's method provides an alternative to fundamental solutions, namely the concept of a complete system of functions which (theoretically at least) allows us to exhaust the solution space and thus, steered by error functionals on the boundary, to approximate the true solution more efficiently.

1.17 Construction of fundamental solutions

To apply the boundary element method we must know the fundamental solution of the governing operator.

To construct the fundamental solution of an *ordinary differential operator* we need a complete set of linearly independent homogeneous solutions. Such a set of functions are for example the functions that correspond to unit end displacements. These are not hard to find and so the construction of fundamental solutions in 1-D is rather simple. The situation becomes more complex in higher dimensions. For partial differential equations with constant coefficients Hörmander has developed an algorithm to construct fundamental solutions, [28]. Illustrative applications of this method can be found in [29, p.17]. If the coefficients are not constant then only in special circumstances can fundamental solutions be derived by employing integral transformations and similar techniques. A detailed list of fundamental solutions was published by Ortner, [30], [31]. A thorough discussion of the topic and further detailed information on the subject can be found in Ortner's contribution [32].

1.18 Mixed methods

In calculus we have learnt that we can split up any n-th order differential equation as

$$y^{(n)} = f$$

into a system of first-order equations

$$y_1' - y_2 = 0$$
$$y_2' - y_3 = 0$$
$$\cdots \quad \cdots$$
$$y_n' = f$$

This is the idea behind mixed methods. The second-order system of equations for the displacement field of an elastic plate is equivalent to the system of first-order equations

$$C\left[E(\boldsymbol{u})\right] - \boldsymbol{S} = \boldsymbol{0}$$

$$-\mathrm{div}\boldsymbol{S} = \boldsymbol{p} \qquad (1.12)$$

for the stress tensor $\boldsymbol{S} = [\sigma_{ij}]$ and the displacement field $\boldsymbol{u} = \{u_i\}$ where

$$E(u) = \frac{1}{2}(\nabla u + \nabla u^T) \qquad \nabla u = [u_{i,j}]$$

$$C[E] = 2\mu E + \lambda \, tr(E) I$$

$$\mathrm{div} S = \{\sigma_{ij,j}\}$$

Smooth displacement fields u and symmetric stress tensors S satisfy the equations

$$\int_\Omega E(u) \cdot S \, d\Omega = \int_\Omega -\mathrm{div}\, S \cdot u \, d\Omega + \int_\Gamma S n \cdot u \, ds$$

$$\int_\Omega \mathrm{div}\, u \, I \cdot S \, d\Omega = \int_\Omega -\mathrm{div}\, S \cdot u \, d\Omega + \int_\Gamma u \cdot n \, tr(S) \, ds$$

so that the second identity of the system (1.12) becomes

$$B(u, S; \hat{u}, \hat{S}) = \int_\Omega (C[E(u)] - S) \cdot \hat{S} \, d\Omega + \int_\Omega -\mathrm{div}\, S \cdot \hat{u} \, d\Omega$$

$$- \int_\Gamma S n \cdot \hat{u} \, ds - \int_\Gamma (2\mu \hat{S} n + \lambda \, tr(\hat{S}) I n) \cdot u \, ds$$

$$- \int_\Omega -(2\mu + \lambda) \, \mathrm{div}\, \hat{S} \cdot u \, d\Omega - \int_\Omega (E(\hat{u}) - \hat{S}) \cdot S \, d\Omega = 0$$

Note that the system (1.12) is not self-adjoint. By applying Hörmander's method Tosaka, [33], has constructed the fundamental solution of the adjoint system

$$E(u) - S = o$$

$$-(2\mu + \lambda) \mathrm{div}\, S = p$$

so that a mixed boundary element method can now be applied to the system (1.12).

1.19 Shells

Theoretically at least the extension of the boundary element method to shells is straightforward, [34], but in practice it is not because we have no analytical expressions for the fundamental solutions.

Up to now we can only apply the boundary element method to the theory of shallow shells where the governing equation is the biharmonic operator + lower terms, see Newton and Tottenham, [35], and Simmonds and Bradley, [36], Tepavitcharov, [37], Hansen [38], Hein [39] and Matsui and Matsuoka, [40]. In the framework of shallow-shell theory Hein has developed a BE-code that calculates the stresses in cylindrical shells with elliptical cutouts, see Fig. 1.22. The solution is a superposition of the solution for a cylinder without cutouts (obtained in the framework of classical shell theory) and a BE- solution. Figure 1.23 shows a cylinder with two elliptical holes under axial tension. Figure 1.24 shows the distribution of the membrane stresses and bending stresses at the edges of the holes.

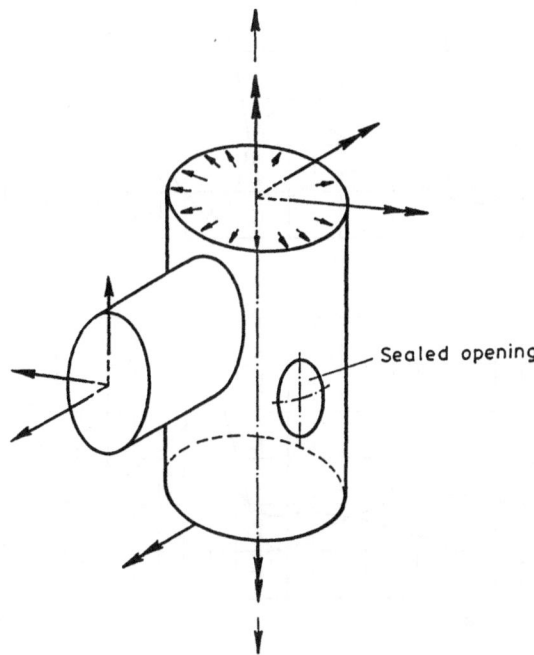

Figure 1.22 Intersection of two cylinders (Hein [39])

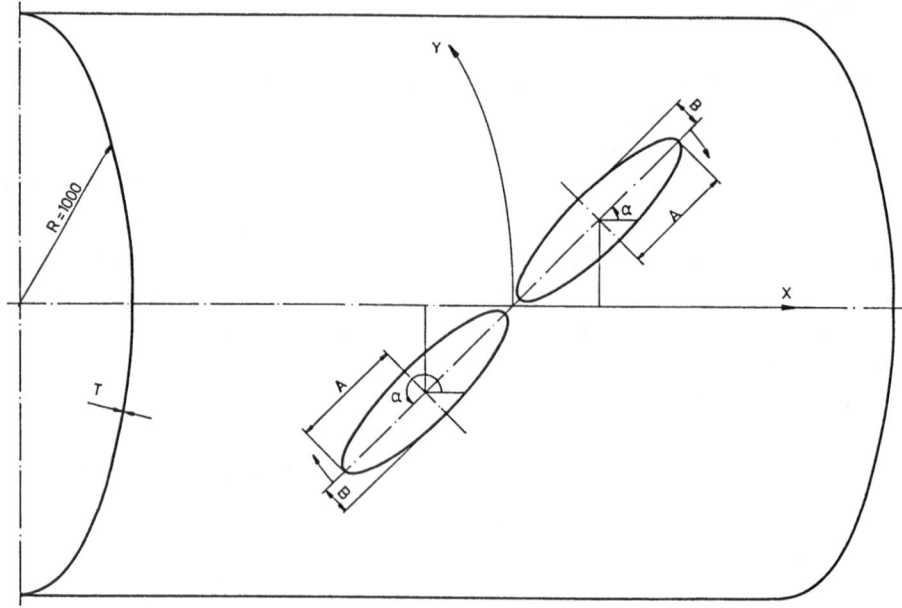

Figure 1.23 Cylinder with two elliptical cutouts

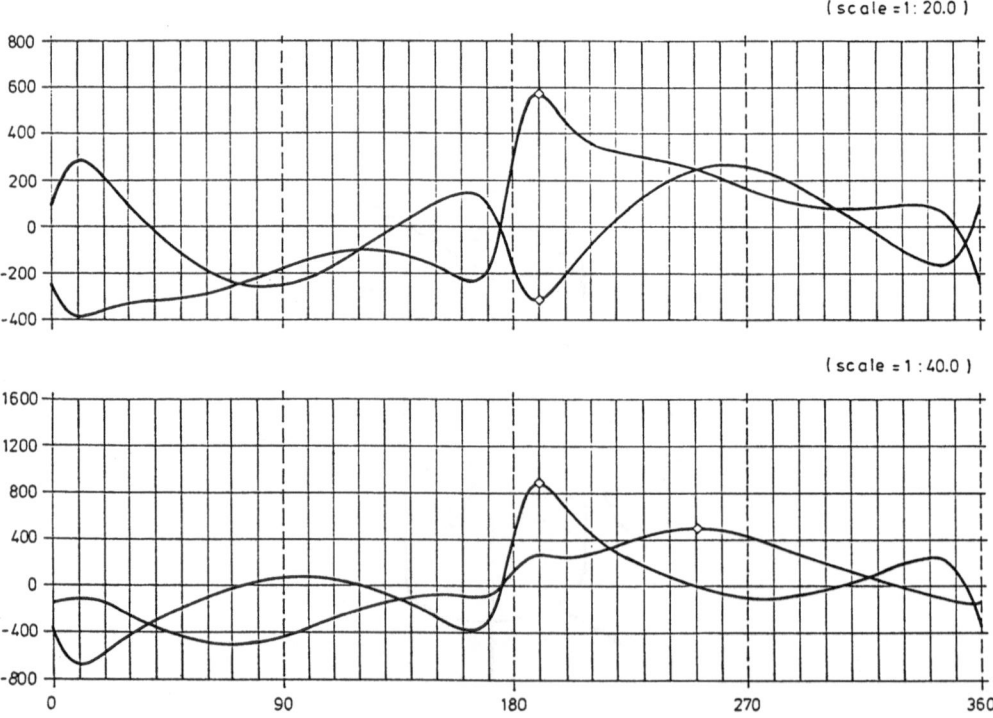

Figure 1.24 Distribution of membrane and bending stresses along the edge of the elliptical holes

Exercises

1. The fundamental solution g_0 of the equation $ax = b$ is $g_0 = 1/a$ since $a \times 1/a = 1$. Show that any solution x of the equation $ax = b$ is the 'scalar product' between the fundamental solution g_0 and the right-hand side b.

2. Calculate the inverse A^{-1} of the symmetric matrix

$$\begin{bmatrix} 3 & 2 & 7 & 1 \\ 2 & 1 & 12 & 5 \\ 7 & 12 & 8 & 6 \\ 1 & 5 & 6 & 9 \end{bmatrix}$$

and show that the rows r_i of the inverse A^{-1} are the fundamental solutions of the equation $Ax = b$, i.e. show that

$$Ar_i = e_i.$$

Show that the component x_i of the solution vector x is the scalar product between the fundamental solution r_i and the right-hand side b

$$x_i = r_i^T b.$$

3. The string in Fig. 1.25a is stressed by a normal force N. A concentrated lateral force $P = 1$ located at x will cause at a distant point y the deflection

$$G_0(y, x) = \frac{1}{Nl} \begin{cases} y(l - x), & y \le x, \\ x(l - y), & x \le y, \end{cases} \tag{1.13}$$

What will be the shape of the string under the action of a uniform lateral load p as in Fig. 1.25b and 5 equally spaced concentrated forces as in Fig. 1.25c?

4. The inverse of a differential operator is an integral operator

$$-Nu'' = p \qquad \Longleftrightarrow \qquad u = \int_0^l g_0(y, x) p(y) dy$$

whose kernel $g_0(y, x)$ is the fundamental solution of the differential equation

$$-N g_0''(y, x) = \delta_0(y - x)$$

that is the response of the infinite medium to a point load located at x. To obtain the inverse of a boundary value problem (= right-hand side p + bound-

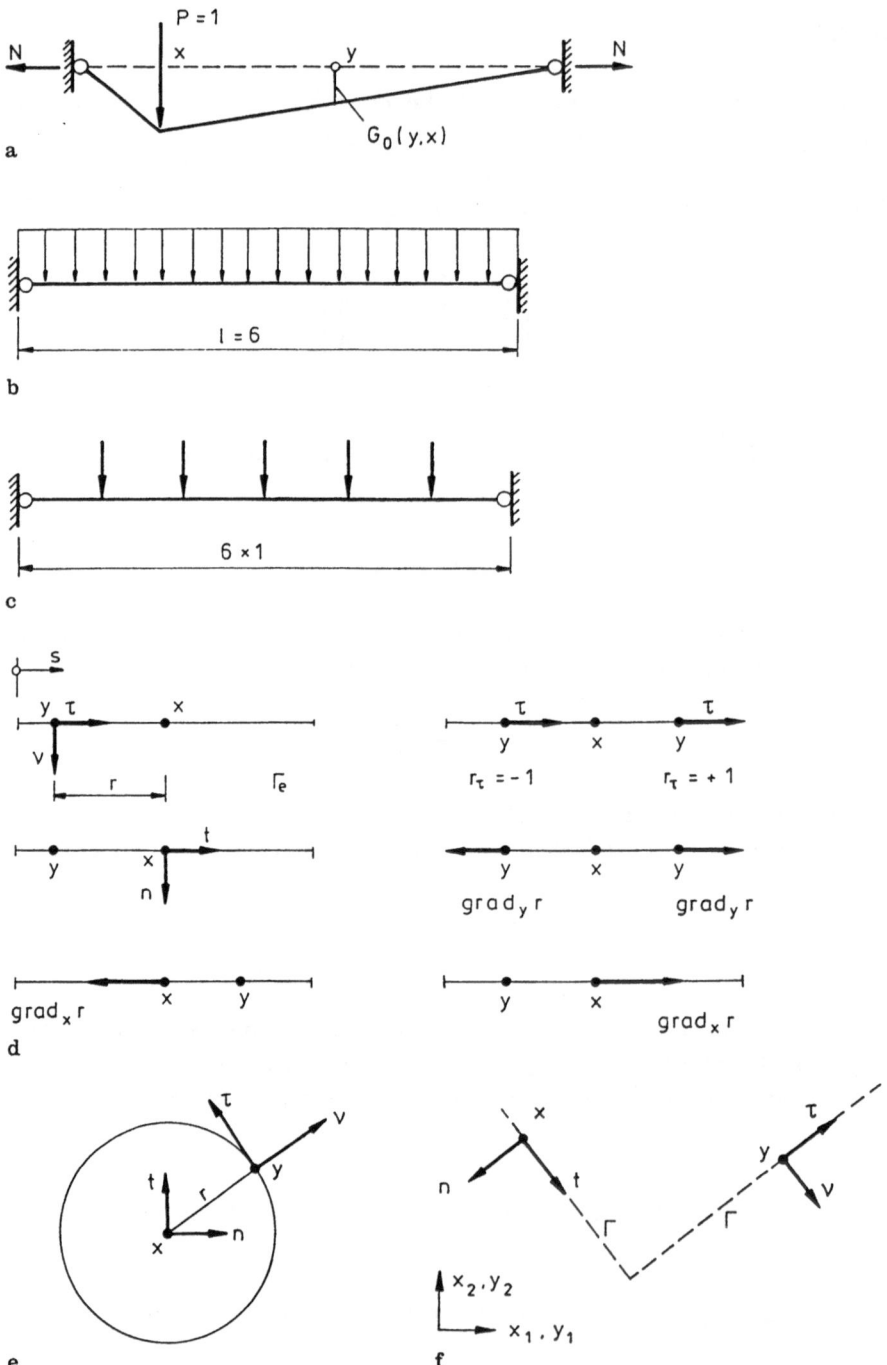

Figure 1.25 a-f Exercises: **a** Green's function of a tight string; **b** uniform load; **c** single forces; **d** x and y lie on the same element; **e** y lies on the circle and x is the centre point; **f** x and y lie both on the boundary.

ary values) we must replace the fundamental solution under the integral sign by a more refined tool, the Green's function $G_0(y, x)$ of the boundary value problem. (The Green's function differs from the fundamental solution in that it satisfies also the homogeneous boundary conditions).

5. A finite difference approximation of the boundary value problem

$$-Nu'' = p \qquad 0 < x < l = 6 \qquad u(0) = u(l) = 0$$

has lead to the system of equations

$$-N \begin{bmatrix} -2 & 1 & 0 & 0 & 0 \\ 1 & -2 & 1 & 0 & 0 \\ 0 & 1 & -2 & 1 & 0 \\ 0 & 0 & 1 & -2 & 1 \\ 0 & 0 & 0 & 1 & -2 \end{bmatrix} \begin{bmatrix} u_1 \\ u_2 \\ u_3 \\ u_4 \\ u_5 \end{bmatrix} = \begin{bmatrix} p_1 \\ p_2 \\ p_3 \\ p_4 \\ p_5 \end{bmatrix}$$

Calculate the inverse of this matrix and show that the inverse is just the Green's function (1.13)

$$a_{ij}^{(-1)} = G_0(x_i, x_j)$$

where $x_i = i\Delta x$, $i = 1, 2 \dots 5$, are the grid-points. In our case $\Delta x = 1$.

6. The distance $r = |y - x|$ between two points is a function of the coordinates of both points, x and y. Show that the normal derivative

$$r_\nu = \operatorname{grad}_y r \cdot \nu = r_{,1}\, \nu_1 + r_{,2}\, \nu_2 = 0$$

of r is zero if the two points x and y lie on the same element and the element is straight, see Fig. 1.25d. Show that under the same conditions the tangential derivative

$$r_\tau = \operatorname{grad}_y r \cdot \tau = r_{,1}\, \tau_1 + r_{,2}\, \tau_2$$

has the value -1 if the point y lies on the left-hand side of x (the point y approaches x so that r decreases) and that it has the value $+1$ if the point y lies on the right-hand side of x (the distance increases in the direction of the tangential vector τ).

Let the source point x be the centre of a circle Γ with radius r, see Fig. 1.25e. Show that at all points y on Γ the normal derivative r_ν of the distance r is $+1$ and that the value of the tangential derivative r_τ is zero.

Let the normal vector n and the tangent vector t at x be oriented as in Fig. 1.25e. Calculate the derivatives r_n and r_t, that is the rate of change of the distance r if the source point x moves into the direction of one of the two vectors while the point y on Γ is fixed.

Show that at all points on the circle $r_{,i} = \nu_i$.

Consider an arbitrary configuration of x and y as in Fig. 1.25f. Show that

$$r_\nu^2 + r_\tau^2 = 1.$$

(Hint: use $\tau_1 = -\nu_2, \tau_2 = \nu_1$, $r_{,1}^2 + r_{,2}^2 = 1$, $\nu_1^2 + \nu_2^2 = 1$). Verify that the same holds true for $r_n^2 + r_t^2 = 1$.
Show that

$$r_{\nu,n} = \frac{1}{r} r_\tau r_t \qquad\qquad r_{\tau,n} = -\frac{1}{r} r_\nu r_t.$$

7. A vertical force P on an elastic half-space will cause the stresses

$$\sigma_{xx} = -\frac{2P}{\pi} \frac{x^2 y}{(x^2+y^2)^2} \qquad \sigma_{yy} = -\frac{2P}{\pi} \frac{y^3}{(x^2+y^2)^2}$$

$$\sigma_{xy} = -\frac{2P}{\pi} \frac{xy^2}{(x^2+y^2)^2}$$

where it is understood that the force is located at the origin of the system of coordinates, see Fig. 1.26a. Show that these formulas are equivalent with

$$\sigma_{ij} = -\frac{2P}{\pi} \frac{r_{,i} r_{,j} r_{,2}}{r}.$$

Hint:

$$r_{,1} = \frac{y_1 - x_1}{r} = \frac{y_1 - 0}{r} = \frac{y_1}{r} = \frac{x}{r} \qquad \text{etc.}$$

8. Is flatland possible? The gravitational field which is generated by a solid homogeneous body Ω like the earth is given by the potential

$$u(\boldsymbol{x}) = \int_\Omega \frac{k}{r} d\Omega_y$$

where the constant k depends on the units employed. The gravitational force $\boldsymbol{F} = \nabla u(\boldsymbol{x})$ that pulls on a point mass m which moves in this field is the gradient of this potential at the location $\boldsymbol{x}$ of the point mass. The force $\boldsymbol{F}$ has the components

$$F_i = \int_\Omega \frac{k\, r_{,i}}{r^2} d\Omega_y$$

The force is finite as long as $\boldsymbol{x}$ lies in outer space because then the distance r never becomes zero. Does this integral also exist if the particle touches the surface? What happens if you replace Ω by a disk (flatland), see Fig. 1.26b. Could you walk on the rim of the disk or would you be smashed by infinite forces? Would the integral exist if in flatland the gravitational pull were of the order $1/r$?

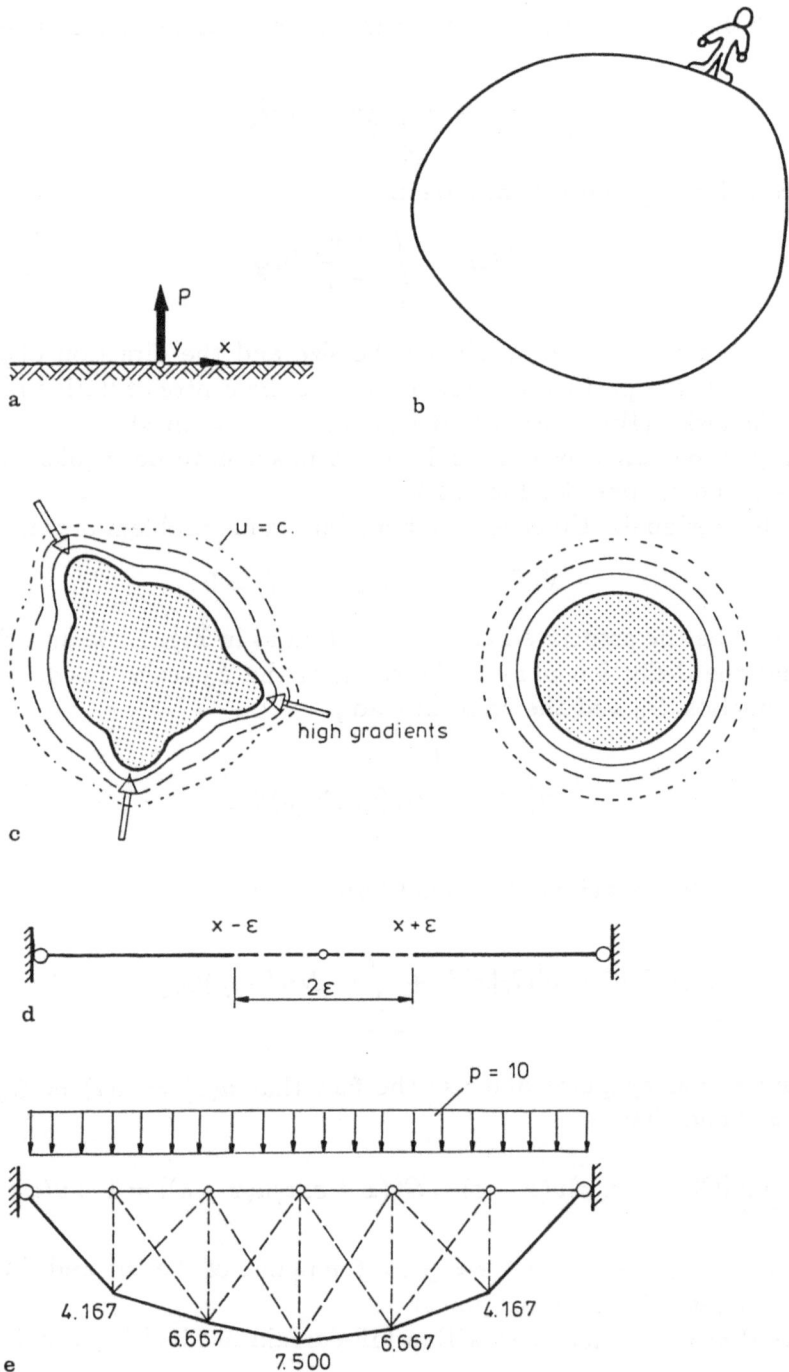

Figure 1.26 a-e Exercises: **a** Point load on the halfspace; **b** flatland; **c** the potential fields of solid bodies; **d** a cut around the source point x; **e** linear FE-approximation of the deflection

Let us then assume that the gravitational field in flatland is given by the potential

$$u(\boldsymbol{x}) = - \int\limits_{\Omega} \frac{k}{2\pi} \ln r \, d\Omega_{\boldsymbol{y}}$$

and therewith the gravitational force by

$$F_i(\boldsymbol{x}) = \int\limits_{\Omega} \frac{k}{2\pi} \frac{r_{,i}}{r} d\Omega_{\boldsymbol{y}}$$

Is that force finite? What would be the size and the direction of the force $F = \{F_1, F_2\}$ if the particle m were located at the centre of flatland, the midpoint of the disk? (Hint: recall that $r_{,1} = \cos\varphi$, $r_{,2} = \sin\varphi$).

Can you explain why the earth or the moon have no 'peaks', why they have this round shape? See Fig. 1.26c.

9. Weighted residuals. Consider the boundary value problem of a tight string

$$-Nu'' = p \qquad u(0) = u(l) = 0.$$

Let $G_0(y, x)$ be the associated Green's function as in Fig. 1.25a and Eq.(1.13). Show that the deflection $u(x)$ of the string is the L_2-scalar product between Green's function G_0 and the uniform load p

$$u(x) = \int\limits_0^l G_0(y, x) p(y) \, dy \, .$$

Hint: Start with the weighted residual formulation

$$\int\limits_0^{x-\varepsilon} (-Nu'' - p)G_0[x] dy + \int\limits_{x+\varepsilon}^l (-Nu'' - p)G_0[x] dy = 0$$

apply integration by parts and use the fact that $u(0) = u(l) = G_0(0, x) = G_0(l, x) = 0$ and that

$$\lim_{\varepsilon \to 0} [G_0'(x - \varepsilon, x)u(x - \varepsilon) - G_0'(x + \varepsilon, x)u(x + \varepsilon)] = 1 \times u(x) \, .$$

(The points $x - \varepsilon = y$ and $x + \varepsilon = y$ are the points of the cut and the single x is the coordinate of the source point).

Note that we cannot start with a full-domain residual formulation as

$$\int\limits_0^l (-Nu'' - p)G_0[x] dy = 0 \, .$$

While this integral exists its transformation with integration by parts would not be permissible because the Green's function G_0 has no continuous first derivative at x. This is why we excluded a neighborhood of the source point x from the interval, see Fig. 1.26d. Only later, after we perform the integration by parts, can we let the radius ε of the gap tend to zero.

Which result do you obtain if you use a fundamental solution $g_0(y, x)$ (i.e. a function which does not satisfy the homogeneous boundary conditions $u(0) = u(l) = 0$)?

10. The deflection of a tight string ($N = 1$) was approximated with 6 linear finite elements, see Fig. 1.26e. Show that the FE-solution u_h is the superposition of 6 Green's functions

$$u_h(x) = \sum_{i=1}^{6} G_0(x_i, x) P_i$$

(which confirms the view that FE-solutions, as BE-solutions, are solutions in terms of the influence function of the governing operator) and determine the forces P_i by calculating the jump in the first derivative at the nodes.

11. Let D denote a one-dimensional differential operator of order $2m$. Let $\varphi_i, i = 1 \ldots 2m$ be a complete set of linearly independent homogeneous solutions, $D\varphi_i = 0$. To calculate the Green's function of a boundary value problem we let

$$G_0(y, x) = \begin{cases} a_1\varphi_1 + a_2\varphi_2 + \cdots a_m\varphi_m, & y \leq x, \\ b_1\varphi_1 + b_2\varphi_2 + \cdots b_m\varphi_m, & x \leq y, \end{cases}$$

and we determine m of the $2m$ coefficients a_i and b_i by requiring that the derivatives $\partial_y^i G_0, i = 0 \ldots 2m - 2$, are continuous at the source point x

$$\partial^i G_0(x_-, x) - \partial^i G_0(x_+, x) = 0$$

and that the highest derivative

$$\partial^{2m-1} G_0(x_-, x) - \partial^{2m-1} G_0(x_+, x) = 1$$

exhibits a jump of magnitude 1. (The 'derivatives' ∂^i are the boundary operators in the Betti formula). The remaining m coefficients are then determined by satisfying the m homogeneous boundary conditions. Apply this technique to construct the Green's function of the prestressed string in (1.13) from the two solutions $\varphi_1 = (l - x)/l$ and $\varphi_2 = x/l$.

12. Green's function can also be obtained by integrating the right-hand side, see e.g. [100] p. 13. The trick is to introduce the Heaviside function

$$H(y, x) = \begin{cases} 1, y < x \\ 0, x < y \end{cases}$$

as the anti-derivative of the delta-function

$$EIw^{IV} = \delta(y,x) \qquad \rightarrow \qquad EIw''' = H(y,x) + c_1 .$$

Once this is done you can proceed as usual: integrate the right-hand side repeatedly and use the integration constants c_i to satisfy the boundary conditions.

2 One-dimensional problems

This chapter is intended as a boundary element primer. The method is explained by applying it to the one-dimensional problems of rods and beams.

2.1 Rods

We start with a short repetition of the basic equations for those readers who are not familiar with structural mechanics.

The longitudinal displacement $u(x)$ of a rod satisfies the differential equation $-EAu'' = p$. To this operator belong the two integral identities

$$p: \hat{u}, u \in C^2[0,l] \times C^1[0,l],$$

$$q: G(\hat{u}, u) = \int_0^l -EA\hat{u}''u\, dx + [\hat{N}u]_0^l - \int_0^l \frac{\hat{N}N}{EA} dx = 0 \qquad (2.1)$$

and

$$p: \hat{u}, u \in C^2[0,l],$$

$$q: B(\hat{u}, u) = \int_0^l -EA\hat{u}''u\, dx + [\hat{N}u - \hat{u}N]_0^l - \int_0^l \hat{u}(-EAu'')\, dx = 0 \quad (2.2)$$

The first identity formulates the principle of virtual displacements and the principle of virtual forces. The second identity formulates Betti's principle.

We say the function $g_0(y, x)$ is a solution of the differential equation

$$-EAg_0''(y, x) = \delta_0(y - x),$$

if the normal force

$$N(y, x) = EA\frac{d}{dy}g_0$$

suffers at x a jump discontinuity of magnitude 1

$$\lim_{\varepsilon \to 0}\{N(x + \varepsilon, x) - N(x - \varepsilon, x)\} = 1$$

If $u = g_0$ is such a fundamental solution and the source point x an interior point, $0 < x < l$, then the integral identities read

$$G(g_0, u) = u(x) + [N_0 u]_0^l - \int_0^l \frac{N_0 N}{EA}\, dy = 0, \tag{2.3}$$

$$B(g_0, u) = u(x) + [N_0 u - g_0 N]_0^l - \int_0^l g_0(-EAu'')\, dy = 0 \tag{2.4}$$

This ends the repetition of the basic equations and in the following we shall apply these results to practical engineering problems.

Assume that we want to find the displacement function $u(x)$ of the rod in Fig. 2.1b. In structural mechanics this problem is solved with the *principle of virtual forces*: we let a force $\hat{P} = 1$ act at x, see Fig. 2.1a, and we form the L_2-scalar product between the corresponding normal force $N_0(y, x)$ and the normal force $N(y)$ associated with the original distribution of forces

$$u(x) = \int_0^l \frac{N_0(y, x)N(y)}{EA}\, dy \tag{2.5}$$

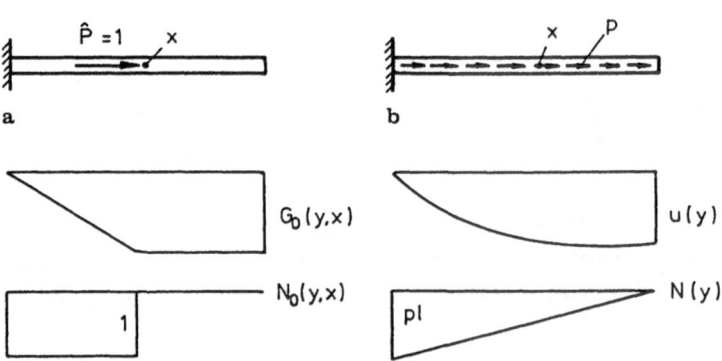

Figure 2.1 Calculation of the displacement $u(x)$ of the rod

Note that y is the integration variable and x the source point.

But the displacement function can also be obtained with *Betti's principle* by forming the L_2-scalar product between the distributed load p and *Green's function*

$$G_0(y, x) = \frac{1}{EA} \begin{cases} y, & y \leq x, \\ x, & x \leq y, \end{cases} \qquad (2.6)$$

Green's function is the displacement of the rod under the action of a concentrated force $\hat{P} = 1$. Betti's principle states that the work done by the distributed load $p(y)$ on acting through the displacement $G_0(y, x)$ is equal to the work of the concentrated force $\hat{P} = 1$ on acting through $u(x)$, that is we must have

$$W_{1,2} = 1 \times u(x) = \frac{p_0}{EA} \left\{ \int_0^x y \, dy + \int_x^l x \, dy \right\} = \frac{p_0}{EA} \left\{ lx - \frac{x^2}{2} \right\} = W_{2,1}$$

Now, let us change the setting: we do not let the concentrated force $\hat{P} = 1$ act on a rod with the same length and the same support conditions as the original rod but we let it act on an infinite rod. The corresponding displacement at a point y of the infinite rod, see Fig. 2.2., is

$$g_0(y, x) = \frac{1}{EA} \begin{cases} (1 - x)y + 1, & y \leq x, \\ (1 - y)x + 1, & x \leq y, \end{cases} \qquad (2.7)$$

The upper formula applies to points y which lie to the left of the source point x and the lower formula to the points to the right of it. The function $g_0(y, x)$ is called a *fundamental solution*.

The situation has now changed. Before we can apply Betti's principle we must first separate that subinterval of the infinite rod that coincides with the real rod from the infinite rod. The two displacements at the two cuts, the points $y = 0$ and $y = l$, we denote by

$$u_1[x] = 1, \qquad u_2[x] = (1 - l)x + 1,$$

and the end actions by

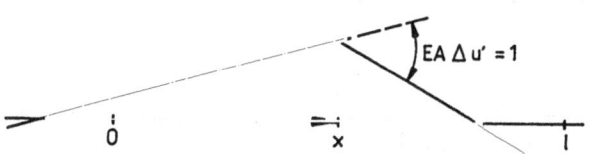

Figure 2.2 The discontinuity of the slope signals a jump of the normal force at the source point x

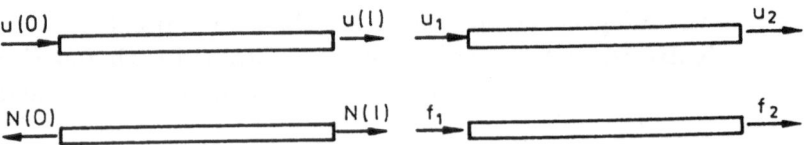

Figure 2.3 The old and the new notation for the end actions and end displacements

$$f_1[x] = x - 1, \qquad f_2[x] = -x,$$

The latter are simply the EA-fold first derivatives of the displacement function, the fundamental solution g_0,

$$N_0(y, x) = \begin{cases} 1 - x, & y \le x, \\ -x, & x \le y, \end{cases} \qquad \begin{array}{l} \text{(to the left of the source point } x\text{)}, \\ \text{(to the right of the source point } x\text{)} \end{array}$$

To obtain latter symmetric stiffness matrices we let all the boundary terms point in the same positive direction, see Fig. 2.3. Note that this leads to a change of sign: the end action f_1 has a different sign as $N(0)$. The trailing bracket $[x]$ is to denote that the boundary values at the two cuts depend on the position x of the source point; they change if we move the source point.

Analogously we separate the real rod from its support and denote the end displacements by

$$u_1 \qquad \text{and} \qquad u_2$$

and the end actions by

$$f_1 \qquad \text{and} \qquad f_2$$

Because both rods in Fig. 2.4 are in equilibrium the reciprocal work of their exterior forces must be the same

$$W_{1,2} = 1 \times u(x) + f[x]^T u = \int_0^l p(y)u(y, x)\, dy + f^T u[x] = W_{2,1} \qquad (2.8)$$

The left side is the work of the exterior forces of the subinterval of the infinite rod acting through the displacements of the real rod and the right side is the work of the distributed load of the real rod acting through the displacements of the subinterval of the infinite rod.

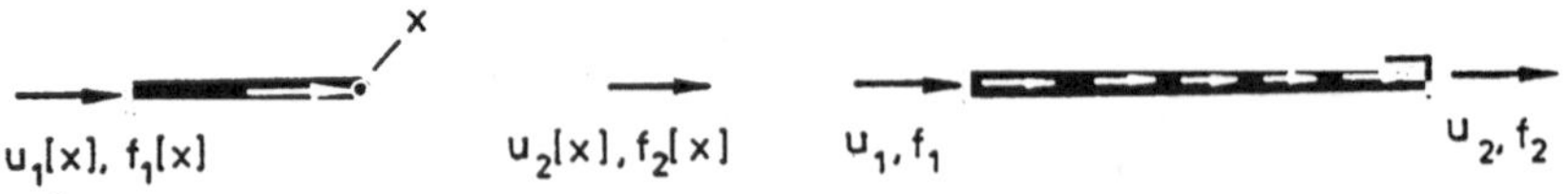

Figure 2.4 Each of the two rods is in equilibrium

If we solve this result for $u(x)$,

$$u(x) = -f[x]^T u + u[x]^T f + \int_0^l p(y)u(y,x)\,dy = (1-x)u_1 + x\,u_2$$

$$+ 1/EA\{1f_1 + [(1-l)x+1]f_2 + \int_0^x p(y)[(1-x)y+1]\,dy$$

$$+ \int_x^l p(y)[(1-y)x+1]\,dy\},\tag{2.9}$$

then we obtain an alternative influence function for the displacement $u(x)$ of the real rod. It is not as elegant as Eq.(2.5) it is much longer and, worse, it requires the explicit knowledge of *all* the boundary data

$$u_1,\ \ u_2,\ \ f_1,\ \ f_2$$

of the rod. This is always the case if we formulate Betti's principle with fundamental solutions instead of Green's functions or, in mechanical terms, if the support conditions of the auxiliary system are different from the support conditions of the primary system.

Green's function is the longitudinal displacement of the primary system under the action of a concentrated force. A *fundamental solution* or *free-space Green's function* instead is the displacement function of an *infinite* rod under the same load. The difference between the two solutions is that Green's function satisfies the support conditions while the fundamental solution does not. Common to both solutions is that their internal action, the normal force N, suffers at the source point a jump discontinuity of magnitude 1.

The formulation of Betti's principle with a Green's function G_0 renders

$$1 \times \text{displacement} = \int_0^l G_0 p\,dy$$

and with a fundamental solution g_0

$$1 \times \text{displacement} + \text{work on the boundary } (W_{1,2})$$

$$= \int_0^l g_0 p\,dy + \text{work on the boundary } (W_{2,1})$$

Typical for the second formulation is the appearance of *additional boundary terms*. If we use a Green's function then these terms are zero because of two conjugated boundary terms

$$force \times displacement$$

one is always zero, either the boundary term of Green's function or the conjugated term of the solution u. If we use a fundamental solution then this is no longer true and, therefore, the work done on the boundary must be considered as well. To calculate this work we must know all the boundary values of the primary system. (The boundary values of the auxiliary system are known.)

Therefore in principle, we would prefer to work with influence functions which are based on Green's functions because they require less information about the solution u; the only information required in fact are the prescribed boundary values. But the problem is that, in general, Green's function is unknown. This is the reason why we have to work with influence functions as

$$1 \times displacement = \int_\Omega g_0 p \, d\Omega + \text{work on the boundary } (W_{2,1})$$

$$- \text{work on the boundary } (W_{1,2})$$

This then leads to the next question: how do we obtain the boundary data which a formulation of Betti's principle with a free-space Green's function, a fundamental solution, requires? The answer is: by utilizing a coupling condition between the Betti data of the primary system.

To derive this coupling condition note that we can also calculate with the influence function (2.9) the boundary displacements $u(0) = u_1$ and $u(4) = u_2$. If we formulate Eq.(2.9) first at $x = 0$ and then at $x = 4$, we obtain

$$\begin{bmatrix} u_1 \\ u_2 \end{bmatrix} = \begin{bmatrix} 1 & 0 \\ -3 & 4 \end{bmatrix} \begin{bmatrix} u_1 \\ u_2 \end{bmatrix} + \frac{1}{EA} \begin{bmatrix} 1 & 1 \\ 1 & -11 \end{bmatrix} \begin{bmatrix} f_1 \\ f_2 \end{bmatrix} + \begin{bmatrix} d_1 \\ d_2 \end{bmatrix}, \qquad (2.10)$$

where the two terms d_i are the work of the distributed load p acting through the fundamental solutions

$$d_1 = \int_0^l g_0(y,0) p(y) \, dy = \frac{p_0}{EA} \int_0^4 1 \, dy = \frac{4}{EA} p_0 \,,$$

$$d_2 = \int_0^l g_0(y,4) p(y) \, dy = \frac{p_0}{EA} \int_0^4 [(1-4)y + 1] \, dy = -\frac{20}{EA} p_0 \,,$$

if the concentrated force $\hat{P} = 1$ is located at $x = 0$ or $x = 4$ respectively.

Now the end displacements u_i on the left side of Eq.(2.10) are the same as on the right side — they are simultaneously dependent variables as independent variables — hence, the two equations formulate two coupling conditions between the boundary data u_i and f_i of the rod.

Coupling on the boundary

The two vectors of end displacements $u = \{u_1, u_2\}^T$ and end actions $f = \{f_1, f_2\}^T$ are the end displacements and end actions of a rod if and only if they satisfy Eq.(2.10).

In principle this coupling condition is not new: if we put all displacement terms u_i on the left side

$$\begin{bmatrix} 0 & 0 \\ 3 & -3 \end{bmatrix} \begin{bmatrix} u_1 \\ u_2 \end{bmatrix} = \frac{1}{EA} \begin{bmatrix} 1 & 1 \\ 1 & -11 \end{bmatrix} \begin{bmatrix} f_1 \\ f_2 \end{bmatrix} + \frac{p_0}{EA} \begin{bmatrix} 4 \\ -20 \end{bmatrix}, \qquad (2.11)$$

and if we multiply the equation from the left with the inverse of the matrix on the right side, then follows

$$EA \begin{bmatrix} 0.25 & -0.25 \\ -0.25 & 0.25 \end{bmatrix} \begin{bmatrix} u_1 \\ u_2 \end{bmatrix} = \begin{bmatrix} f_1 \\ f_2 \end{bmatrix} + p_0 \begin{bmatrix} 2 \\ 2 \end{bmatrix}, \qquad (2.12)$$

or more simply

$$Ku = f + p$$

The components of the vector p are the negative end fixing forces and the matrix K is the stiffness matrix of the rod

$$K = \frac{EA}{l} \begin{bmatrix} 1 & -1 \\ -1 & 1 \end{bmatrix}, \qquad (l = 4 \text{ in Eq.(2.12)})$$

The end fixing forces are the support reactions when both ends of the rod are fixed, $u_i = 0$.

This equation appears in the literature usually only in the "homogeneous" form

$$Ku = f,$$

that is when the distributed forces are zero.

We thus recognize that a stiffness matrix formulates a coupling condition between the end displacements u_i and end actions f_i of a rod.

Let us illustrate this with a simple example. Assume we pull with a force $P = 50$ kN at a rod as in Fig. 2.5 and we measure the end displacement u_2 and find an elongation of 0.18 m. If this measurement is correct then the two end displacements

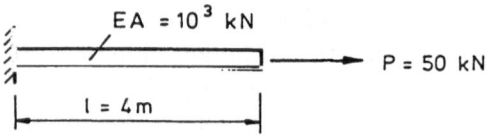

Figure 2.5 A stretched rod

$$u_1 = 0, \qquad u_2 = 0.18 \quad \mathrm{m}$$

and the two end actions of the rod

$$f_1 = -50 \quad \mathrm{kN}, \qquad f_2 = 50 \quad \mathrm{kN}$$

should satisfy the coupling condition (2.12),

$$\begin{bmatrix} 250 & -250 \\ -250 & 250 \end{bmatrix} \begin{bmatrix} 0 \\ 0.18 \end{bmatrix} = \begin{bmatrix} -50 \\ 50 \end{bmatrix},$$

But as the product $250 \cdot 0.18$ is 45 and not 50, the data u_i and f_i are not compatible: the measurement must be incorrect.

But the opposite is also true: A student might think that he or she can substitute any boundary values u_i and f_i into the influence function (2.9) and obtain a displacement function u which takes on these values. This cannot be. The boundary values u_i and f_i must be compatible, that is, they must satisfy the equation $\boldsymbol{Ku} = \boldsymbol{f} + \boldsymbol{p}$. Only then can they be sure that the "output equals the input".

What was demonstrated here with a rod holds true for other structural elements or, more generally speaking, for other differential equations as well. In many fields we can derive a formula as

$$1 \times \text{displacement} = \int_{\Omega} g_0 p \, d\Omega + \text{work on the boundary} - \text{work on the boundary.}$$

which connects the interior solution with the boundary data.

While the distributed load p is always given and m of the $2m$ end displacements and end actions are determined by the loading and support conditions it only remains to determine the missing, conjugated boundary data by solving the coupling condition. The solution is then simply obtained by substituting these data into the pertinent influence function.

Therefore, the two keywords of the boundary element method are

$$\mathrm{BEM} = coupling\ condition + influence\ function$$

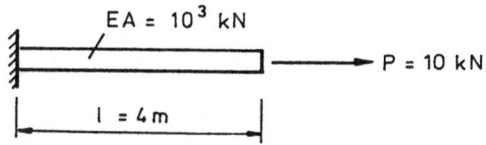

Figure 2.6 A simple boundary value problem

A simple example will illustrate this solution technique in more detail. Let us assume we want to find the displacement function $u(x)$ of the rod in Fig. 2.6. Of all the four Betti data u_i and f_i two, $u_1 = 0$ and $f_2 = 10$, are known,

$$u_1 = 0, \qquad u_2 = ?,$$

$$f_1 = ?, \qquad f_2 = 10$$

The unknown conjugated data u_2 and f_1 are obtained by solving the coupling condition (2.12)

$$\begin{bmatrix} 250 & -250 \\ -250 & 250 \end{bmatrix} \begin{bmatrix} 0 \\ u_2 \end{bmatrix} = \begin{bmatrix} f_1 \\ 10 \end{bmatrix}$$

This renders

$$u_2 = 0.04 \quad \text{m}, \qquad f_1 = -10 \quad \text{kN}$$

The set of Betti data is thus complete and it only remains to substitute the data into the influence function (2.9) for $u(x)$,

$$u(x) = (1 - x)0 + x\,0.04 + 1/EA\{1(-10) + [(1 - l)x + 1]10\}$$

$$= 10^{-2}x \quad \text{m}$$

The problem is solved.

If, in addition, forces $p(x)$ are distributed along the rod as in Fig. 2.7 then the coupling condition of the homogeneous case $(p = 0)$ is supplemented by a vector $\boldsymbol{p}$ whose components are the (negative) end fixing forces of the rod,

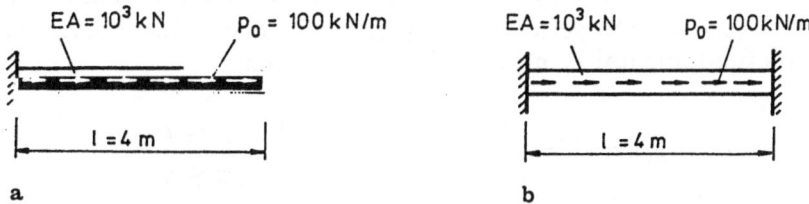

Figure 2.7 a-b. Possible loading conditions

$$\frac{EA}{l}\begin{bmatrix} 1 & -1 \\ -1 & 1 \end{bmatrix}\begin{bmatrix} u_1 \\ u_2 \end{bmatrix} = \begin{bmatrix} f_1 \\ f_2 \end{bmatrix} + \begin{bmatrix} p_1 \\ p_2 \end{bmatrix}$$

This is the only modification in the solution procedure. To find the displacement function $u(x)$ of the rod in Fig. 2.7a we solve the coupling condition

$$\begin{bmatrix} 250 & -250 \\ -250 & 250 \end{bmatrix}\begin{bmatrix} 0 \\ u_2 \end{bmatrix} = \begin{bmatrix} f_1 \\ 0 \end{bmatrix} + \begin{bmatrix} 200 \\ 200 \end{bmatrix}$$

for the unknown boundary values u_2 and f_1

$$u_2 = 0.8 \quad \text{m}, \qquad f_1 = -400 \quad \text{kN},$$

and substitute the then complete set of Betti data into the influence function (2.9)

$$u(x) = (1-x)0 + x\,0.8 + 1/EA\{-400 + [(1-4)x+1]0$$

$$+ 100\{ \int_0^x [(1-x)y+1]\,dy + \int_x^4 [(1-y)x+1]\,dy\}\}$$

$$= 0.4x - 0.05x^2 \quad \text{m}$$

Or consider the rod in Fig. 2.7b. The rod is fixed on both sides so that the end displacements are zero, $u_1 = u_2 = 0$, and therefore the left side of the coupling condition as well. Hence, in this case, the vector f is equal to the vector $-p$,

$$f_1 = -200, \qquad f_2 = -200$$

Substituting these data into the influence function (2.9) we obtain the displacement $u(x) = 0.2x - 0.05x^2$.

Before we end this section let us add a technical remark. If instead of the interval [0,4] you use the unit interval [0,1] then the matrix in Eq.(2.11) which multiplies with the u-terms is the zero matrix. The reason is that in this case you formulate Betti's principle twice with the same test function, namely $g_0(y,0) = 1$ $(y \geq 0)$ and $g_0(y,1) = 1$ $(y \leq 1)$. To avoid this linear dependency replace the fundamental in Eq.(2.7) by the solution

$$g_0(y,x) = \frac{1}{EA}\begin{cases} (1-ax)y + 1, & y \leq x, \\ (1-ay)x + 1, & x \leq y, \end{cases}$$

where a is any non-zero number $a \neq 1$. The choice $a = 1$ corresponds to

Eq.(2.7). If you choose, say, $a = 3$ then the second solution, $g_0(y, 1)$, becomes $g_0(y, 1) = -2y + 1$ which is different from $g_0(y, 0) = 1$. This point is also apt to illustrate that different fundamental solutions render different matrices $\boldsymbol{H}$ and $\boldsymbol{G}$

$$\boldsymbol{Hu} = \boldsymbol{Gf} + \boldsymbol{p}$$

but that their product $\boldsymbol{K} = \boldsymbol{G}^{-1}\boldsymbol{H}$ is always the same stiffness matrix.

2.2 Beams

This boundary element technique applies to beams as well, as we shall demonstrate in this section. But first we want to provide the necessary background material for readers who are not familiar with beam analysis.

The deflection $w(x)$ of a beam with a constant stiffness EI satisfies the differential equation

$$EIw^{IV}(x) = p$$

To this operator belong two integral identities

p: $\hat{w}, w \in C^4[0, l] \times C^2[0, l]$,

$$q: G(\hat{w}, w) = \int_0^l EI\hat{w}^{IV} w \, dx + [\hat{Q}w - \hat{M}w']_0^l - \int_0^l \frac{\hat{M}M}{EI} dx = 0 \quad (2.13)$$

and

p: $\hat{w}, w \in C^4[0, l]$,

$$q: B(\hat{w}, w) = \int_0^l EI\hat{w}^{IV} w \, dx + [\hat{Q}w - \hat{M}w' + \hat{w}'M - \hat{w}Q]_0^l$$

$$- \int_0^l \hat{w} EIw^{IV} \, dx = 0 \quad (2.14)$$

We say $g_0(y, x)$ or $g_1(y, x)$, respectively, is a solution of the differential equations

$$EIg_0^{IV}(y, x) = \delta_0(y - x) \qquad EIg_1^{IV}(y, x) = \delta_1(y - x),$$

if the shear force or the bending moment, respectively,

$$Q = -EI\frac{d^3}{dy^3}g_0(y, x), \qquad M = -EI\frac{d^2}{dy^2}g_1(y, x)$$

suffers at the source point x a jump discontinuity of magnitude 1.

$$\lim_{\varepsilon \to 0}\{Q(x+\varepsilon,x) - Q(x-\varepsilon,x)\} = 1,$$

$$\lim_{\varepsilon \to 0}\{M(x+\varepsilon,x) - M(x-\varepsilon,x)\} = 1$$

If g_0 is such a fundamental solution and x an interior point then the identities read

$$G(g_0,w) = w(x) + [Q_0 w - M_0 w']_0^l - \int_0^l \frac{M_0 M}{EI}\, dy = 0 \qquad (2.15)$$

$$B(g_0,w) = w(x) + [Q_0 w - M_0 w' + g_0' M - g_0 Q]_0^l$$

$$- \int_0^l g_0 EI w^{IV}\, dy = 0 \qquad (2.16)$$

In the case of the second solution g_1 we only replace the free term $w(x)$ by $w'(x)$. This concludes the introduction and we shall show in the following how these results are applied to solve beam problems.

The deflection of the beam in Fig. 2.8 b can be calculated by forming the L_2-scalar product between the bending moment $M_0(y,x)$ and $M(y)$

$$1 \times w(x) = \int_0^l \frac{M_0(y,x)M(y)}{EI}\, dy \qquad (2.17)$$

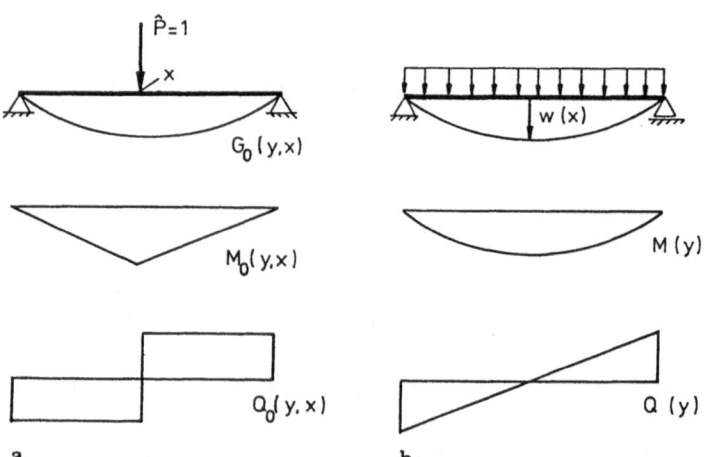

Figure 2.8 Calculation of the deflection $w(x)$ of the beam

or by forming the L_2-scalar product between the deflection $G_0(y, x)$ that is caused by a concentrated force $\hat{P} = 1$ acting at x and the constant load p

$$1 \times w(x) = \int_0^l G_0(y, x)p(y)\, dy \tag{2.18}$$

Equation (2.17) is based on the *principle of virtual forces*: the external work $1 \times w(x)$ is equal to the virtual internal strain energy, and Eq.(2.18) on *Betti's principle*: the reciprocal external work of two equilibrium systems is the same.

Betti's principle only requires that both systems are in equilibrium but it does not require that the support conditions are the same. Hence, it is admissible to choose as auxiliary system an infinite beam that is loaded with a concentrated force $\hat{P} = 1$. A function which describes the deflection of such an infinite beam is for example,

$$g_0(y, x) = \frac{1}{6EI} \times \begin{cases} \alpha(x)y - (1-x)y^3\,, & y \le x\,, \\ (y-x)^3 + \alpha(x)y - (1-x)y^3\,, & x \le y\,, \end{cases}$$

$$\alpha(x) = x(1-x)(2-x)$$

The upper part of the equation applies to the points y to the left of the source point and the lower part to the points y to the right of it.

The subinterval of the infinite beam which coincides with the real beam is then separated from the infinite beam and the internal actions at the cuts are declared exterior end actions. Analogously we separate the real beam from its supports and declare the support forces the end actions of the beam. The two beams then have the same length and both are they in equilibrium so that the reciprocal external work of their exterior forces must be the same:

$$W_{1,2} = 1 \times w(x) - Q_0(0, x)w(0) + M_0(0, x)w'(0) + Q_0(l, x)w(l)$$

$$- M_0(l, x)w'(l) = \int_0^l g_0(y, x)p(y)\, dy - Q(0)g_0(0, x)$$

$$+ M(0)g_0'(0, x) + Q(l)g_0(l, x) - M(l)g_0'(l, x) = W_{2,1}$$

On the left we find the work of the exterior forces of the auxiliary beam and on the right the work of the exterior forces of the real beam.

If we put the term $1 \times w(x)$ on the left side alone,

$$1 \times w(x) = \int_0^l g_0 p\, dy + \text{work on the boundary } (W_{2,1})$$

$$- \text{work on the boundary } (W_{1,2}),$$

then we obtain an influence function for the deflection $w(x)$. To calculate the rotation $w'(x)$ we let a concentrated couple $\hat{M} = 1$ act on the infinite beam the corresponding deflection is $g_1(y,x) = dg_0/dx$ and we repeat the formulation of Betti's principle

$$1 \times w'(x) = \int_0^l g_1 p\, dy + \text{work on the boundary } (W_{2,1})$$

$$- \text{work on the boundary } (W_{1,2}),$$

Summary: The term $w(x)$ or $w'(x)$, respectively, is equal to the work done by the distributed load p acting through the deflection g_0 or g_1, respectively, plus the work done on the boundary by the conjugated terms of the primary system and the auxiliary system. To calculate the work of these boundary terms it is necessary that the Betti data of the primary system as well as the auxiliary system are known. While the Betti data of the auxiliary system (the sub-interval of the infinite beam) are known our information with regard to the Betti data of the real beam as for example the beam in Fig. 2.9, is incomplete. Of two conjugated boundary data only one is prescribed, the other is unknown.

$$w(0) = 0, \qquad w'(0) = ?, \qquad w(l) = 0, \qquad w'(l) = ?,$$

$$Q(0) = ?, \qquad M(0) = 0, \qquad Q(l) = ?, \qquad M(l) = 0$$

To obtain the missing four terms we formulate a coupling condition between the boundary values, as follows:

The formulas for the deflection w and the rotation w' apply also at the left end, $x = 0$, and the right end, $x = l$, of the beam,

$$w(0) = \int_0^l g_0[0] p\, dy + \text{work on the boundary} - \text{work on the boundary},$$

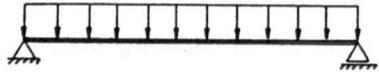

Figure 2.9 Hinged beam

$$w(l) = \int_0^l g_0[l]p\,dy + +\text{work on the boundary} - \text{work on the boundary},$$

$$w'(0) = \int_0^l g_1[0]p\,dy + \text{work on the boundary} - \text{work on the boundary},$$

$$w'(l) = \int_0^l g_1[l]p\,dy + +\text{work on the boundary} - \text{work on the boundary}$$

As the terms on the left side also appear on the right side, the displacement terms are simultaneously dependent variables and independent variables, that is, between the boundary terms a coupling condition exists.

To make this clear we, first, make all boundary terms the same positive direction, see Fig. 2.10, and we then put all the displacement terms

$$u_1 = w(0), \qquad u_2 = -w'(0), \qquad u_3 = w(l), \qquad u_4 = -w'(l),$$

on the left side and all the force terms

$$f_1 = -Q(0), \qquad f_2 = -M(0), \qquad f_3 = Q(l), \qquad f_4 = M(l),$$

on the right side. The resulting four equations can then be written as

$$H_{ij}\,u_j = G_{ij}\,f_j + d_i \tag{2.19}$$

where the coefficients H_{ij} and G_{ij} are work terms. The coefficient H_{ij} is the work done by that force term of the auxiliary beam, which is conjugated to u_j, on acting through $u_j = 1$ and the coefficient G_{ij} is the work done by the end action $f_j = 1$ of the real beam on acting through its conjugated auxiliary counterpart. The first index i indicates what kind of load acts on the infinite beam and where it acts, see Fig. 2.11 :

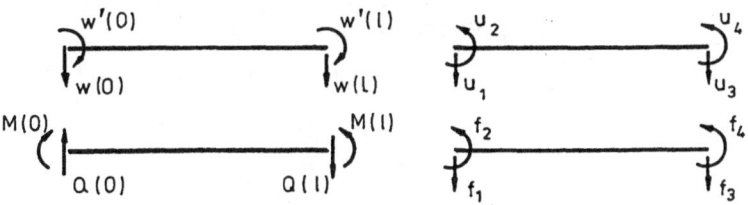

Figure 2.10 The old and the new notation for the end actions and end displacements

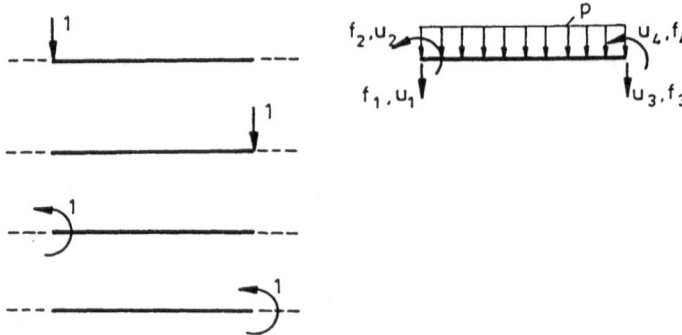

Figure 2.11 The four coupling conditions between the boundary data of the beam on the right are obtained by formulating Betti's principle $W_{1,2} = W_{2,1}$ consecutively with the four equilibrium positions of the beam on the left

$$i = 1 \qquad \text{the force } \hat{P} = 1 \text{ acts at } x = 0,$$

$$i = 2 \qquad \text{the force } \hat{P} = 1 \text{ acts at } x = l,$$

$$i = 3 \qquad \text{the couple } \hat{M} = 1 \text{ acts at } x = 0,$$

$$i = 4 \qquad \text{the couple } \hat{M} = 1 \text{ acts at } x = l$$

If we multiply Eq.(2.19) from the left with the inverse of $\boldsymbol{G}^{-1}$ then we obtain

$$\boldsymbol{G}^{-1}\boldsymbol{H}\boldsymbol{u} = \boldsymbol{f} + \boldsymbol{G}^{-1}\boldsymbol{d}$$

or

$$\boldsymbol{K}\boldsymbol{u} = \boldsymbol{f} + \boldsymbol{p}, \tag{2.20}$$

where $\boldsymbol{K}$ denotes the stiffness matrix of the beam. The components p_i of the vector $\boldsymbol{p} = \boldsymbol{G}^{-1}\boldsymbol{d}$ are the (negative) end fixing forces, that is, the support reactions when the ends are clamped.

For a beam of length $l = 4$ and which is loaded with uniformly distributed forces p these coupling conditions read

$$
\begin{bmatrix} 2 & 2 & -2 & -10 \\ 5 & 10 & -5 & -30 \\ 1 & 1 & -1 & -5 \\ 1 & 2 & -1 & -6 \end{bmatrix}
\begin{bmatrix} u_1 \\ u_2 \\ u_3 \\ u_4 \end{bmatrix}
= \frac{1}{6EI}
\begin{bmatrix} 8 & -12 & 280 & -156 \\ 280 & -180 & 1800 & -660 \\ 12 & -14 & 180 & -86 \\ 156 & -86 & 660 & -182 \end{bmatrix}
\begin{bmatrix} f_1 \\ f_2 \\ f_3 \\ f_4 \end{bmatrix}
$$

$$
+ \frac{p}{EI}
\begin{bmatrix} 384 \\ 3520 \\ 288 \\ 1504 \end{bmatrix}
$$

If we multiply this equation from the left with the inverse of the matrix on the right then we obtain the stiffness matrix of the beam

$$EI \begin{bmatrix} 0.1875 & -0.3750 & -0.1875 & -0.3750 \\ & 1.0000 & 0.3750 & 0.5000 \\ & & 0.1875 & 0.3750 \\ \text{sym.} & & & 1.0000 \end{bmatrix} \begin{bmatrix} u_1 \\ u_2 \\ u_3 \\ u_4 \end{bmatrix} = \begin{bmatrix} f_1 \\ f_2 \\ f_3 \\ f_4 \end{bmatrix} + p \begin{bmatrix} 2.00 \\ -1.33 \\ 2.00 \\ 1.33 \end{bmatrix},$$

and, as a simple calculation confirms, the vector p, indeed, is the vector of the (negative) end fixing forces.

Hence, the stiffness matrix of a beam formulates a coupling condition between the end displacements and end actions of a beam. In conjunction with the influence function for the deflection

$$w(x) = x\,w(l) + (1-x)w(0) + x(1-l)w'(l) + (1/6EI)\{[-3(l-x)^2$$

$$- \alpha(x) + 3l^2(1-x)]M(l) + \alpha(x)M(0) + [(l-x)^3 + \alpha(x)l$$

$$- (1-x)l^3]Q(l) + \int_0^x [\alpha(x)y - (1-x)y^3]p(y)\,dy$$

$$+ \int_x^l [(y-x)^3 + \alpha(x)y - (1-x)y^3]p(y)\,dy\},\tag{2.21}$$

$$\alpha(x) = x(1-x)(2-x),$$

all kind of beam problems can therefore be treated by the boundary element method. A simple example will demonstrate this.

Assume we are interested in the deflection of the beam in Fig. 2.12. The total number of displacements and force terms at the two end points of the beam is eight. Four of these are determined by support and loading conditions while the four conjugated terms are unknown:

$$w(0) = 0, \qquad w'(0) = 0, \qquad M(l) = 0, \qquad Q(l) = 0,$$

$$Q(0) = ?, \qquad M(0) = ?, \qquad w'(l) = ?, \qquad w(l) = ?$$

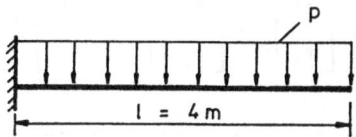

Figure 2.12 A cantilever beam

To determine these four unknown terms we first solve the 3rd and 4th equation of the coupling condition,

$$
\frac{EI}{l^3}
\begin{bmatrix}
12 & -6l & -12 & -6l \\
 & 4l^2 & 6l & 2l^2 \\
 & & 12 & 6l \\
\text{sym.} & & & 4l^2
\end{bmatrix}
\begin{bmatrix}
0 \\ 0 \\ w(l) \\ -w'(l)
\end{bmatrix}
=
\begin{bmatrix}
-Q(0) \\ -M(0) \\ 0 \\ 0
\end{bmatrix}
+ p
\begin{bmatrix}
2.00 \\ -1.33 \\ 2.00 \\ 1.33
\end{bmatrix}
$$

for the unknown displacement terms and then the 1st and 2nd equation for the unknown support forces,

$$
w(l) = \frac{32}{EI}p, \qquad w'(l) = \frac{10.67}{EI}p, \qquad M(0) = -8p, \qquad Q(0) = 4p
$$

and we finally substitute the complete set of Betti data into the influence function (2.21) and so obtain the solution

$$
w(x) = \frac{p}{24EI}(96x^2 - 16x^3 + x^4)
$$

2.3 Transfer matrices

If we arrange the coupling condition of a rod

$$
\frac{EA}{l}
\begin{bmatrix}
1 & -1 \\
-1 & 1
\end{bmatrix}
\begin{bmatrix}
u_1 \\ u_2
\end{bmatrix}
=
\begin{bmatrix}
f_1 \\ f_2
\end{bmatrix}
+
\begin{bmatrix}
p_1 \\ p_2
\end{bmatrix},
$$

in such a way that the displacement and force terms at the left end of the rod are placed on the left side of the equation and the terms at the right end of the rod on the right side then we obtain the expression

$$
\begin{bmatrix}
1 & -l/EA \\
0 & -1
\end{bmatrix}
\begin{bmatrix}
u_1 \\ f_1
\end{bmatrix}
+
\begin{bmatrix}
-l/EA & 0 \\
-1 & -1
\end{bmatrix}
\begin{bmatrix}
p_1 \\ p_2
\end{bmatrix}
=
\begin{bmatrix}
u_2 \\ f_2
\end{bmatrix}
$$

The first matrix is the *transfer matrix* of a rod. This equation allows us to calculate the displacement and force terms at the right end of a rod by evaluating the terms at the left end and by considering, in addition, the (negative) end fixing forces p_i. Because the length l is arbitrary this formula applies to any intermediate point $x = l$ alike (after an appropriate modification of the end fixing forces p_i).

Similarly the coupling condition (2.19) between the Betti data of a beam can be rearranged to render the transfer matrix of a beam

$$
\begin{bmatrix} 1 & -l & l^3/6EI & l^2/2EI \\ 0 & 1 & -l^2/2EI & -l/EI \\ 0 & 0 & -1 & 0 \\ 0 & 0 & -l & -1 \end{bmatrix} \begin{bmatrix} w(0) \\ -w'(0) \\ -Q(0) \\ -M(0) \end{bmatrix}
$$

$$
- \begin{bmatrix} l^3/6EI & l^2/2EI & 0 & 0 \\ -l^2/2EI & -l/EI & 0 & 0 \\ 1 & 0 & 1 & 0 \\ l & 1 & 0 & 1 \end{bmatrix} \begin{bmatrix} p_1 \\ p_2 \\ p_3 \\ p_4 \end{bmatrix} = \begin{bmatrix} w(l) \\ -w'(l) \\ Q(l) \\ M(l) \end{bmatrix}
$$

We can look at transfer matrices as the matrix formulation of influence functions. They therefore play an important part in computer codes for frame analysis.

2.4 Matrix-displacement method

The matrix-displacement method (MDM) is the boundary element method applied to frames, that is, structures which consist of many single rods and beams.

To the pair

BEM = *coupling conditions + influence functions*

corresponds the pair

MDM = *stiffness matrices + transfer matrices*

By considering the geometrical constraints at the nodes the stiffness matrices of the single rods and beams are assembled to form a global stiffness matrix

$$
\begin{bmatrix} K_{11} & K_{12} \\ K_{21} & K_{22} \end{bmatrix} \begin{bmatrix} \bar{u}_1 \\ u_2 \end{bmatrix} = \begin{bmatrix} f_1 \\ \bar{f}_2 \end{bmatrix} + \begin{bmatrix} p_1 \\ p_2 \end{bmatrix} \tag{2.22}
$$

We assume that this matrix is arranged in such a way that the vector $\bar{u}_1$ is the vector of the given nodal displacements and the vector u_2 the vector of the unknown nodal displacements. The vectors f_1 and $\bar{f}_2$, respectively, contain the corresponding nodal forces. The vector f_1 is unknown, the vector $\bar{f}_2$ is given. The vectors p_i are the vectors of the negative end fixing forces.

To solve this system of equations, we first solve the lower half for the vector u_2

$$K_{22}\, u_2 = \bar{f}_2 + p_2 - K_{21}\, \bar{u}_1 \,,$$

and then the upper half for the vector f_1 of nodal forces (at the supports),

$$f_1 = K_{11}\, \bar{u}_1 + K_{12}\, u_2 - p_1.$$

Often all supports of the frame are rigid supports hinges or fixed ends so that the vector $\bar{u}_1$ is the zero vector and thus the first equation simply becomes

$$K_{22}\, u_2 = \bar{f}_2 + p_2$$

The matrix K_{22} is called the *reduced stiffness matrix* because it is obtained on deleting all rows and columns in the global stiffness matrix which correspond to zero degrees of freedom.

Thus, if all the unknown nodal displacements u_2 and nodal forces f_1 are determined the deflection or the internal actions in the interior between the nodes can be obtained by evaluating the pertinent influence function or, more simple, by substituting the nodal values into the transfer matrices.

2.5 The general principle

The boundary element method, naturally, is not restricted to structural mechanics. In many areas of physics and mechanics the setting is the same: the problem is governed by a linear, self-adjoint differential operator D to which belongs an integral identity as

$$G(u,\hat{u}) = \int_0^l Du\,\hat{u}\,dx + \sum_{i=1}^m (-1)^i [\partial^{2m-i}u\,\partial^{i-1}\hat{u}]_0^l - E(u,\hat{u}) = 0$$

This identity is the basis of the *principle of virtual displacements*

$$G(u,\hat{u}) = 0, \qquad \text{for all } \hat{u} \,,$$

the *principle of virtual forces*

$$G(\hat{u},u) = 0, \qquad \text{for all } \hat{u} \,,$$

and *Betti's principle*

$$B(\hat{u}, u) = G(\hat{u}, u) - G(u, \hat{u}) = 0, \qquad \text{for all } \hat{u}, u$$

To such an operator belongs, furthermore, a set of $2m$ linearly independent homogeneous solutions φ_i, which we assume to be normalized in such a way that they correspond to unit end displacements. If $u(x)$ is a solution of the differential equation $Du = p$ and f_i and u_i are the force and displacement terms of the solution on the boundary then they are coupled by the equation

$$K u = f + p, \tag{2.23}$$

where

$$K_{ij} = E(\varphi_i, \varphi_j)$$

is the symmetric stiffness matrix and the components

$$p_i = \int\limits_0^l p\,\varphi_i\,dx$$

the (negative) end fixing forces or whatever meaning you want to attribute to these terms in the particular context. Simply by rearranging Eq.(2.23) a transfer matrix is obtained and, hence, all is set for the application of the matrix-displacement method.

Exercises

1. Boundary values u_i, f_i which do not satisfy the coupling condition

$$Ku = f + p$$

will spoil any influence function; the influence function will not reproduce these numbers. Verify this by substituting arbitrary boundary values (e.g. $u_1 = 1, u_2 = f_1 = f_2 = 0$) into the influence function of the rod (2.9) (for simplicity assume that the distributed load p is zero) and compare the boundary values of the function so constructed with the original boundary values. Repeat the procedure but this time make sure that the boundary values satisfy the coupling condition and check the result.

2. Calculate the displacement functions $u(x)$ of the two rods in Fig. 2.13a-b by the boundary element method as described in this chapter. Hint: recall that

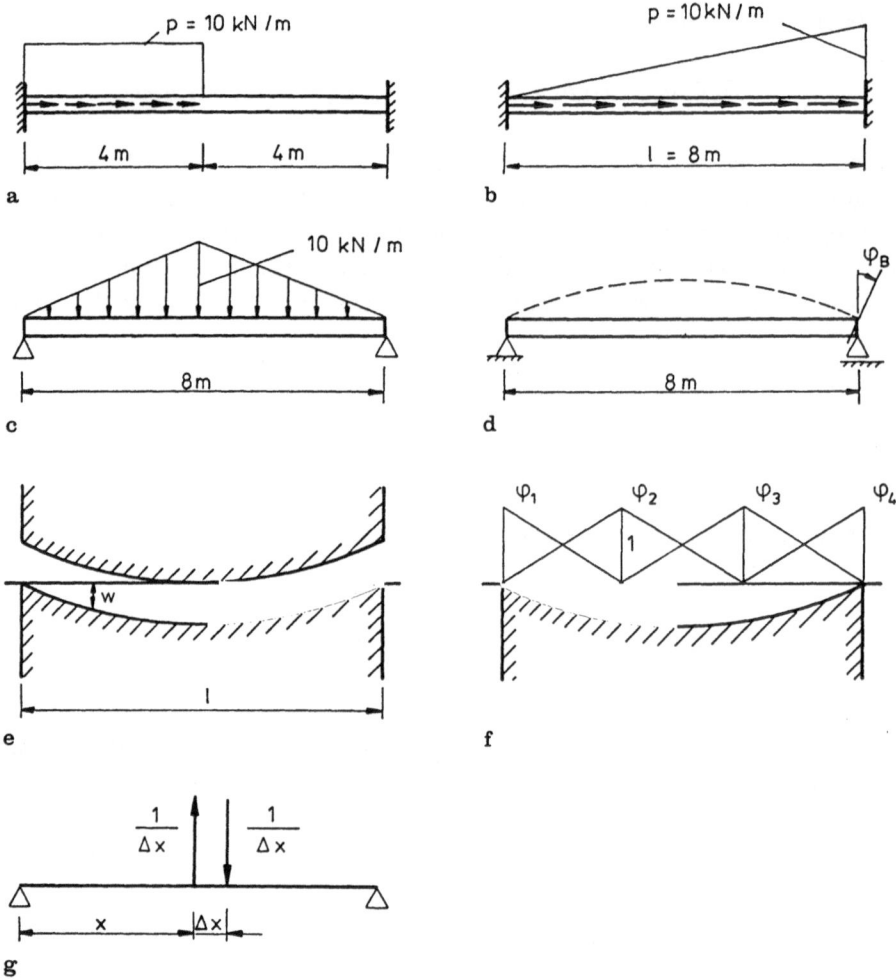

Figure 2.13 a-g Exercises: a-b longitudinal forces; c vertical forces; d end rotation; e a bar between two rigid dies; f piecewise linear approximation of the pressure p; g approximation of a couple by two point forces.

the fixed end actions p_i are the L_2-scalar product between the load and the normalized homogeneous solutions,

$$\varphi_1 = \frac{l-x}{l} \qquad \varphi_2 = \frac{x}{l}$$

so that

$$p_1 = \int_0^l \varphi_1 \, p \, dx = 30 \, \text{kN}, \qquad p_2 = \int_0^l \varphi_2 \, p \, dx = 10 \, \text{kN}$$

and $p_1 = 13.33 \, \text{kN}, p_2 = 26.66 \, \text{kN}$ in the second case.

3. In the same fashion calculate the deflection of the two beams in Fig. 2.13c-d. The normalized homogeneous solutions of the beam are

$$\varphi_1 = 1 - \frac{3x^2}{l^2} + \frac{2x^3}{l^3} \qquad \varphi_2 = -x + \frac{2x^2}{l} - \frac{x^3}{l^2}$$

$$\varphi_3 = \frac{3x^2}{l^2} - \frac{2x^3}{l^3} \qquad \varphi_4 = \frac{x^2}{l} - \frac{x^3}{l^2}$$

so that the fixed end actions of the first beam are $p_1 = 20\,\text{kN}$, $p_2 = -33.33\,\text{kNm}$, $p_3 = 20\,\text{kN}$, $p_4 = 33.33\,\text{kNm}$.

4. Forced deflection, [101] p. 28-6. A slender bar of length $l = 1$ is put between two matching rigid dies, see Fig. 2.13e. When the upper die is lowered a sinusoidal deflection $w(x) = 1/(\pi^4 EI)\sin\pi x$ of the bar is enforced. We ask for the distribution of the load $p(x)$ acting between the dies and the bar. The deflection at any point x is the sum of the influences of all the forces $p(y)dy$ acting on every line element dy of the bar

$$\int\limits_0^l G_0(y,x)p(y)dy = w(x) \tag{2.24}$$

where

$$G_0(y,x) = \frac{1}{6EIl}\begin{cases} y(x-l)(x^2-2lx+y^2), & y \le x, \\ x(y-l)(x^2-2ly+y^2), & x \le y, \end{cases} \tag{2.25}$$

is the Green's function. Equation (2.24) is an integral equation for the load $p(y)$. Approximate the load $p(y)$ by a series of hat-functions as in Fig. 2.13f, $p(y) = \sum p_i\varphi_i(y)$, and determine the nodal values p_i by a collocation procedure

$$\sum_j \int\limits_0^l G_0(y,x_i)p_j\varphi_j(y)dy = w(x_i) \qquad i = 1,2,3,4.$$

Demonstrate that this leads to the matrix equation $\mathbf{Ap} = \mathbf{w}$ where

$$a_{ij} = \int\limits_0^l G_0(y,x_i)\varphi_j(y)dy \qquad p_i = p(x_i) \qquad w_i = w(x_i).$$

Compare your solution with the exact solution $p = EIw^{IV}$.

5. Show that the Green's function due to a couple $M = 1$ is the derivative of the zero-order Green's function (= point load G. f.)

$$G_1(y,x) = \frac{d}{dx}G_0(y,x).$$

Hint: Approximate the couple by two opposite point loads as in Fig. 2.13g and consider the influence of these two point loads on a point y when the distance Δx tends to zero and the forces $P = 1/\Delta x$ tend to infinity

$$\lim_{\Delta x \to 0} \left[G_0(x + \Delta x, x)\frac{1}{\Delta x} - G_0(x, x)\frac{1}{\Delta x} \right].$$

6. Verify that the influence function (2.9) of the rod satisfies the differential equation $-EAu'' = p$. Hint: use the rule for the differentiation of an integral with respect to its parameter

$$\frac{d}{dx} \int\limits_{a(x)}^{b(x)} f(y, x)dy = \int\limits_{a(x)}^{b(x)} \frac{\partial f(y, x)}{\partial x}dy + f(b(x), x)b'(x) - f(a(x), x)a'(x)$$

In our case $f(y, x) = g_0(y, x)p(y)$.

7. A distributed load $p(y)$ on a beam will cause the deflection

$$w(x) = \int\limits_0^l G_0(y, x)p(y)\, dy \tag{2.26}$$

where $G_0(y, x)$ is the Green's function (2.25). Write a computer program which calculates $w(x)$. Use constant or linear elements to approximate the distributed load, see Fig. 2.14a, and evaluate the integral elementwise by numerical quadrature.

8. Show that numerical quadrature

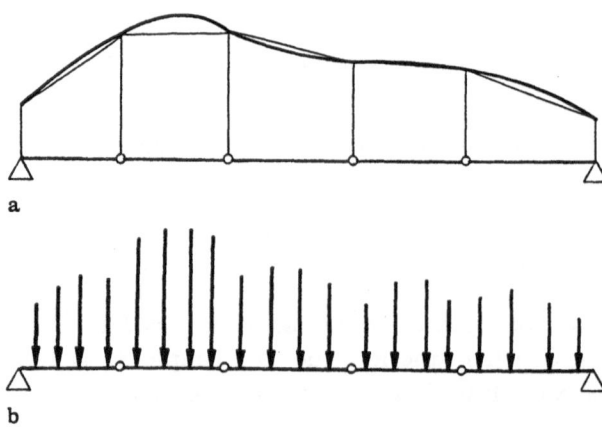

Figure 2.14 a-b Exercises: a distributed lateral load on a beam and the interpolation of the load with linear elements; b Numerical quadrature means effectively that we replace the distributed load by point forces.

$$\int\limits_0^l G_0(y,x)p(y)dy = \sum_i G_0(y_i,x)p(y_i)w_i + \text{quadrature error}$$

means that we replace the distributed load $p(y)$ on each element by point forces $p(y_i)w_i$ located at the Gauss points y_i, see Fig. 2.14b. Study the different integration formulas, *mid-point formula, trapezoidal rule, Gaussian integration*, and check the corresponding distribution of the point forces on each element and the size of these forces if the original distributed load is constant, $p = c$.

9. Formulate the first and second identity $G(u, \hat{u}) = 0$ and $B(u, \hat{u}) = 0$ respectively of the differential equation

$$u'' + \lambda^2 u = p$$

Find two homogeneous solutions $h_1(x), h_2(x)$ and use these to construct two 'normalized' homogeneous solutions

$$\varphi_1 = a_1 h_1(x) + a_2 h_2(x)$$
$$\varphi_2 = b_1 h_1(x) + b_2 h_2(x)$$

with the properties

$$\varphi_1(0) = 1, \qquad \varphi_1(l) = 0, \qquad \varphi_2(0) = 0 \qquad \varphi_2(l) = 1$$

and use these to calculate the stiffness matrix

$$K_{ij} = E(\varphi_i, \varphi_j)$$

where $E(\cdot, \cdot)$ is the energy of the differential operator as defined in the first identity.

A fundamental solution is $g_0(y, x) = (1/2\lambda)\sin\lambda r$, proof?

3 Membranes

A membrane is assumed to be a perfectly flexible, thin elastic fabric, which is uniformly stretched in all directions by a tension which has a constant value N per unit length along any section or boundary. The deflection $u \, (= u_3)$ satisfies the differential equation

$$-N(u_{,11} + u_{,22}) = -N\Delta u = p$$

where p is the lateral pressure. The traction across a cut is the product of the tension N and the derivative in the direction of the normal vector $\boldsymbol{n} = \{n_1, n_2\}^T$ of the cut,

$$t = N\frac{\partial u}{\partial n} = N(u_{,1}\, n_1 + u_{,2}\, n_2),$$

that is the N-fold normal derivative or N-fold slope. The close connection between the slope and the traction expresses Fig. 3.1. The greater the pressure the more the membrane will deflect and the greater the slope on the boundary and, therefore, also the traction t on the boundary.

Before we begin to discuss the boundary element approach let us, first, recall how we approximate the deflection of a membrane as in Fig. 3.1 with finite elements. We do so by subdividing the membrane into, say, linear finite elements. On each triangular finite element three local basis functions φ_i^e are defined which, together with the local basis functions of the neighboring elements, are patched together to form n global basis functions φ_i. Such a basis function has the value 1 at the node $\boldsymbol{x}^i$ and the value zero at all other nodes, see Fig. 3.2. It represents the shape of the membrane if the node $\boldsymbol{x}^i$ is pushed down by 1 unit of deflection and if the points in the vicinity of the node follow this movement to an extent that is proportional to their distance to the node $\boldsymbol{x}^i$. Only those elements will deflect which have the node $\boldsymbol{x}^i$ as one of their vertices. The main part of the membrane remains flat.

Because of the *slope = traction* analogy discontinuities of the slope are equivalent with discontinuities of the traction t and, therefore, such abrupt changes of the slope of the deflected membrane must be attributed to the action of line forces which are concentrated along the lines that connect the neighboring nodes with the active node and that connect the neighboring nodes among

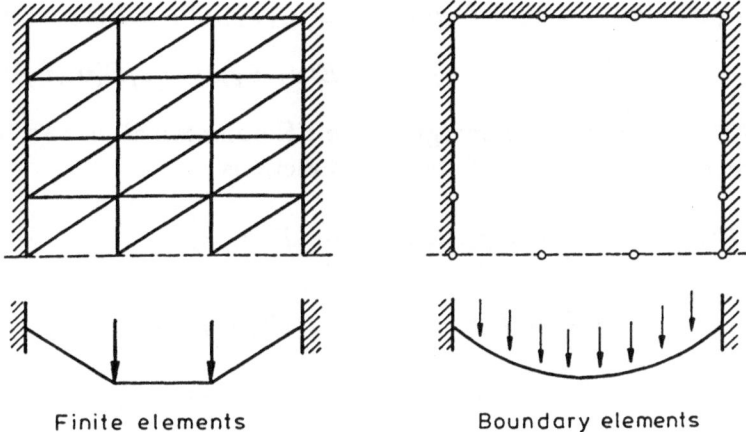

Finite elements   Boundary elements

Figure 3.1 The discretization of a membrane with finite elements and boundary elements

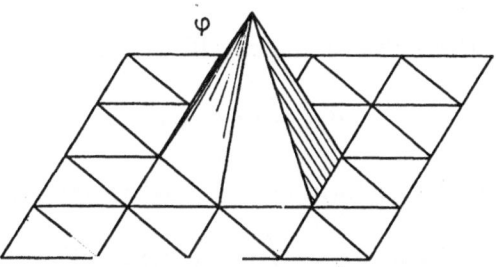

Figure 3.2 Hat-function

themselves. The magnitude of these forces is proportional to the discontinuity of the slope.

The potential energy of the membrane is the expression

$$\Pi_1(u) = \frac{1}{2}E(u,u) - \int_\Omega p\,u\,d\Omega$$

where

$$E(u,\hat{u}) = N \int_\Omega \nabla u \cdot \nabla \hat{u}\,d\Omega = N \int_\Omega (u_{,1}\,\hat{u}_{,1} + u_{,2}\,\hat{u}_{,2})\,d\Omega$$

denotes the *energy product* (= virtual strain energy) between two deflections u and $\hat{u}$. The notation $E(u,\hat{u})$ is to remind us that the energy is a *bilinear form*, i.e. that it a) depends on two functions and b) is linear in both,

$$E(a_1 u_1 + a_2 u_2, \hat{a}_1 \hat{u}_1 + \hat{a}_2 \hat{u}_2) = a_1 E(u_1, \hat{u}_1)\hat{a}_1 + a_1 E(u_1, \hat{u}_2)\hat{a}_2$$

$$+ a_2 E(u_2, \hat{u}_1)\hat{a}_1 + a_2 E(u_2, \hat{u}_2)\hat{a}_2 = a_i E(u_i, \hat{u}_j)\hat{a}_j\,,$$

Because of this property the energy of the FE-approximation can be given in terms of the energy products of the basis functions

$$\frac{1}{2}E(u_h, u_h) = \frac{1}{2}E(u_i\varphi_i, u_j\varphi_j) = \frac{1}{2}u_i E(\varphi_i, \varphi_j)u_j$$

$$= \frac{1}{2}\{u_1 E(\varphi_1, \varphi_1)u_1 + u_1 E(\varphi_1, \varphi_2)u_2$$

$$+ u_1 E(\varphi_1, \varphi_3)u_3 + \cdots + u_n E(\varphi_n, \varphi_n)u_n\}$$

so that the energy of an FE-approximation is a quadratic form of the nodal variables u_i,

$$\frac{1}{2}E(u_h, u_h) = \frac{1}{2}E(u_i\varphi_i, u_j\varphi_j) = \frac{1}{2}u^T K u\,,$$

and the elements of the stiffness matrix K the energy products of the basis functions

$$K_{ij} = E(\varphi_i, \varphi_j)$$

The external work is a linear form of the nodal variables,

$$\int_{\Omega} p u_h \, d\Omega = \int_{\Omega} p u_i\varphi_i \, d\Omega = f^T u\,,$$

with coefficients

$$f_i = \int_{\Omega} p \varphi_i \, d\Omega y\,, \tag{3.1}$$

so that on combining these results the potential energy of the FE-approximation can be expressed as

$$\Pi_1(u_h) = \frac{1}{2}u^T K u - f^T u$$

For the potential energy to become a minimum it is necessary that

$$\frac{\partial}{\partial u_i}\Pi_1(u_h) = 0\,, \qquad \text{for } i = 1, 2, \ldots, n$$

According to the chain rule we have

$$\frac{\partial}{\partial u_i}\Pi_1(u_h) = \frac{\partial}{\partial u_i}(\frac{1}{2}u^T K u - f^T u) = \frac{\partial}{\partial u_i}(\frac{1}{2}u_k K_{kj}u_j - f_j u_j)$$

$$= \frac{1}{2}(K_{ij}u_j + u_k K_{ki}) - f_i$$

$$= \frac{1}{2}(K_{ij}u_j + u_j K_{ji}) - f_i = 0$$

and because the stiffness matrix K is symmetric, $K_{ij} = K_{ji}$, this is equivalent to

$$K_{ij}u_j = f_i \qquad \text{for } i = 1, 2, \ldots, n$$

With regard to this equation it now holds that:

a) The *left* side of equation i is the energy product between the FE-solution u_h and the basis function φ_i,

$$K_{ij}u_j = E(\varphi_i, \varphi_1)u_1 + E(\varphi_i, \varphi_2)u_2 + \cdots + E(\varphi_i, \varphi_n)u_n$$

$$= E(\varphi_i, \varphi_1 u_1 + \varphi_2 u_2 + \cdots + \varphi_n u_n)$$

$$= E(\varphi_i, u_h)$$

b) Because the FE-solution, as any FE-solution, is an equilibrium solution, the virtual strain energy $E(\varphi_i, u_h) = \delta W_i$ is equal to the virtual external work δW_e of the FE-forces on acting through the virtual deflection φ_i,

$$K_{ij}u_j = E(\varphi_i, u_h) = \delta W_e \qquad \text{(first identity)}$$

c) The *right* side of equation i, the scalar f_i, is the work of the original exterior forces p on acting through φ_i, see Eq.(3.1), so that the equation

$$K_{ij}u_j = E(\varphi_i, u_h) = \delta W_e = f_i,$$

means that the virtual external work δW_e of the *FE-forces* on acting through φ_i is equal to the external work of the original exterior forces for any virtual deflection. Hence, to solve membrane problems with finite elements means we replace the original distribution of forces by forces which are equivalent in the sense that their virtual external work is equal to the work of the original forces,

$$\delta W_e(u_h, \varphi_i) = \delta W_e(u, \varphi_i),$$

and that we solve this loadcase instead of the original loadcase *exactly*.

Recall that the exterior forces which belong to the FE-solution are *not* the equivalent nodal forces f_i (these, as you will remember, are work terms, not forces). To satisfy the system $K_{ij}u_j = f_i$ does not mean to satisfy an equilibrium condition at the nodes but rather to tune the FE- forces in such a way that their virtual work is equal to the work of the lateral pressure p.

If the FE-solution really were the deflection of the membrane under the action of the n nodal forces f_i then the FE-solution would have a logarithmic singularity at each of the n nodes because concentrated forces cause such infinite deflections.

The FE-forces are the vertical forces t_Δ, which act along the element boundaries l_m and whose magnitude is proportional to the discontinuity of the slope between two neighboring elements. If we denote the traction on both faces by t_l and t_r, respectively, then the formula for these forces is

$$t_\Delta = t_l + t_r$$

The plus sign is correct because the minus sign, so to speak, is already cared for implicitly by the opposite directions of the two normals on the two faces of the cut.

Hence the statement that the virtual external work of the FE-solution and the true solution is the same means: the virtual work done by the line loads t_Δ is equal to the virtual work of the lateral pressure p for any virtual deflection φ_i,

$$\delta W_e(u_h, \varphi_i) = \sum_m \int_{l_m} t_\Delta \varphi_i \, ds = \int_\Omega p \, \varphi_i \, d\Omega = \delta W_e(u, \varphi) \qquad \text{for all } \varphi_i$$

How do we solve the same problem with boundary elements? The BE-solution is based on the influence function for the deflection

$$u(\boldsymbol{x}) = \int_\Omega [g_0(\boldsymbol{y}, \boldsymbol{x})t(\boldsymbol{y}) - N \frac{\partial}{\partial \nu} g_0(\boldsymbol{y}, \boldsymbol{x})u(\boldsymbol{y})] \, ds_{\boldsymbol{y}} + \int_\Omega g_0(\boldsymbol{y}, \boldsymbol{x})p(\boldsymbol{y}) \, d\Omega_{\boldsymbol{y}} \quad (3.2)$$

An immediate application of this function is hampered by the fact that on any part of the boundary of two conjugated boundary values one is prescribed and one is unknown: along the free edge, where the traction is zero, $t = 0$, the deflection is unknown and along the fixed edge, where the deflection is zero, $u = 0$, the traction is unknown.

To determine these unknown functions we first replace them by, say, piecewise quadratic polynomials whose nodal values we determine in a second step by requiring that Betti's principle applies. The basic auxiliary equilibrium state with which we control the validity of Betti's principle is the equilibrium state of an infinite membrane when one of the n nodes is loaded with a concentrated force $\hat{P} = 1$.

By placing the force consecutively at the n nodes and formulating Betti's principle each time we obtain n equations which can be solved for the n unknown nodal values of the real membrane. With the influence function (3.2) we then obtain the deflection at any internal point.

Summary: To solve the membrane problem with linear finite elements means to approximate the curved surface of the deflected membrane with a patchwork of flat triangles. From a mechanical point of view this means that the action of the lateral pressure p is approximated by forces which are concentrated along the element boundaries. The geometrical boundary condition $u = 0$ on the fixed edge is satisfied exactly but the statical boundary condition $t = 0$ on the free edge only approximately. The linear elements at the edge do not lie flat; they will have a small inclination towards the free edge and this means that the slope, and therefore the traction on the boundary, is not zero.

In contrast to the piecewise linear FE-solution the BE-solution is smooth. The BE-solution satisfies the differential equation exactly and therefore the surface has a continuous curvature. Discrepancies only occur on the boundary. But these now involve both boundary terms, the deflection and the traction. On the fixed edge the boundary condition $u = 0$ is satisfied only at the collo-

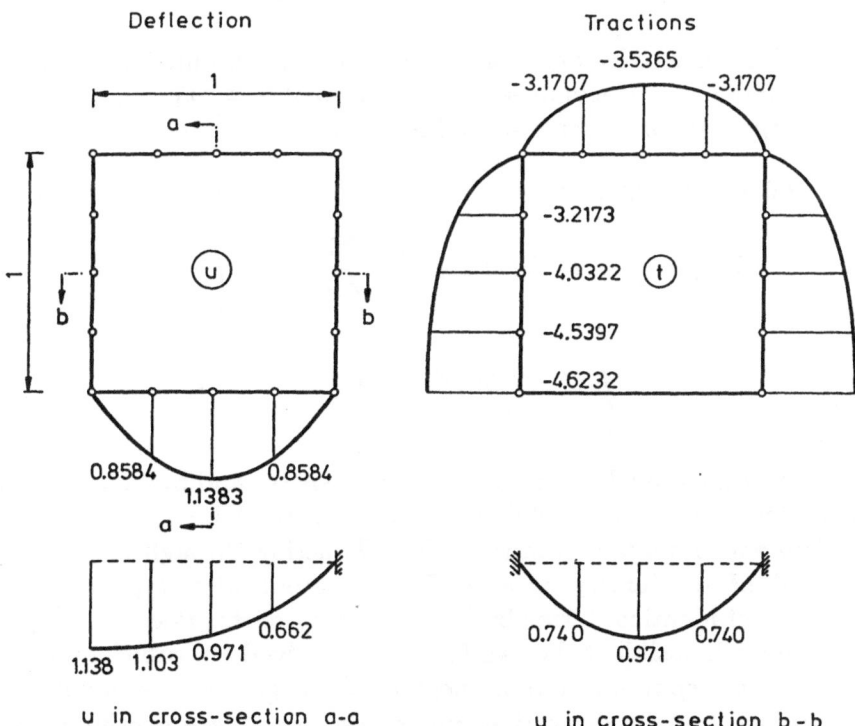

Figure 3.3 The BE-solution of the membrane problem for a lateral pressure $p = 10$ and a prestressing force $N = 1$

cation points and along the free edge the traction t will not be identical zero. But according to St.Venant's principle we may hope that these errors on the boundary will have only a minor influence on the quality of the solution in the interior.

Figure 3.3. shows the distribution of the boundary functions u and t for a square membrane whose lower horizontal edge is free. These boundary values were obtained with the program BE-LAPLACE (two quadratic elements on each side, $p = 10$, $N = 1$) and then substituted into the influence function (3.2) to calculate the deflection along the cross sections a–a and b–b.

3.1 The influence function for the deflection $u(x)$

The derivation of an influence function is done with the help of a fundamental solution of the governing operator and Betti's principle.

The fundamental solution of a membrane

$$g_0(y, x) = -\frac{1}{2\pi N} \ln r, \qquad r = |y - x|,$$

gives the deflection at the point $y = (y_1, y_2)$ of an infinite membrane if a concentrated force $\hat{P} = 1$ acts at some distant point $x = (x_1, x_2)$.

Betti's principle, or Green's second identity,

p: $\hat{u}, u \in C^2(\bar{\Omega})$,

q: $B(\hat{u}, u) = \int\limits_{\Omega} -N \Delta \hat{u}\, u\, d\Omega + \int\limits_{\Gamma} N \frac{\partial \hat{u}}{\partial n} u\, ds - \int\limits_{\Gamma} \hat{u} N \frac{\partial u}{\partial n}\, ds$

$$- \int\limits_{\Omega} \hat{u}(-N \Delta u)\, d\Omega = 0, \tag{3.3}$$

states that the work done by the domain forces $-N \Delta \hat{u}$ and boundary tractions $N \partial \hat{u}/\partial n$ acting through the deflection u is equal to the work of the domain forces $-N \Delta u$ and boundary tractions $N \partial u/\partial n$ acting through $\hat{u}$.

We speak of an identity because Green's second identity, basically, is a transformation of identical terms by integration by parts. Because integration by parts is only admissible if the two functions involved are sufficiently smooth we mention a corresponding requirement at the beginning. A function is in $C^2(\bar{\Omega})$, if u itself and its derivatives up to the order 2 are continuous in the closed domain $\bar{\Omega}$. Closed domain means the same as closed interval $[a, b]$ (the boundary is a part of the domain) this is indicated by the bar. A hat-function is not in $C^2(\bar{\Omega})$. In such a case we would formulate Betti's principle for each element separately and then add the single results.

At the source point the deflection and the traction of the fundamental solution are infinite. This we understand if we consider that the equilibrium condition requires that the integral of the traction t ($=$ slope $\times$ the tension N) over circles of shrinking size centred at x must tend to 1,

$$\lim_{\varepsilon \to 0} \int_{\Gamma_{N_\varepsilon}(x)} N\frac{\partial u}{\partial \nu} \, ds_y = 1$$

When the perimeter $p = 2\pi\varepsilon$ of the circles $\Gamma_{N_\varepsilon}(x)$ tends to zero then the traction t must tend to $+\infty$ to balance the opposite tendency of the perimeter that is the traction must behave as $t = 1/2\pi\varepsilon$. This behavior excludes the fundamental solution $g_0(y, x)$ from the class $C^2(\bar{\Omega})$, the class of two-times continuously differentiable domain functions and, hence, Betti's principle is not applicable at least not in the naive sense.

To circumvent this restriction we proceed as follows: We isolate the singularity from the domain Ω by removing a small circular neighborhood

$$N_\varepsilon(x) = \{ y \in \Omega \mid |y - x| \leq \varepsilon \}$$

of the source point, see Fig. 3.4, and, to keep everything in balance, we let the prior internal tractions become exterior forces that act along the edge of the hole. The elastic state of the punctured membrane

$$\Omega_\varepsilon(x) = \Omega - N_\varepsilon(x),$$

is now a regular elastic state because the tractions and the deflection $g_0(y, x)$ are bounded. Hence we may — if also the other equilibrium position u is regular — formulate Betti's principle on the punctured domain

$$B(g_0[x], u)_{\Omega_\varepsilon} = \int_{\Omega_\varepsilon} -N\Delta g_0 u \, d\Omega_y + \int_{\Gamma_\varepsilon} N\frac{\partial}{\partial \nu} g_0 u \, ds_y$$

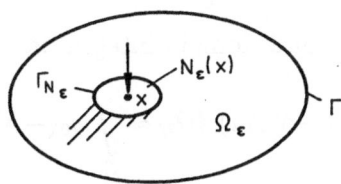

Figure 3.4 The punctured domain

$$-\int_{\Gamma_\epsilon} g_0 N \frac{\partial u}{\partial \nu}\, ds_{\boldsymbol{y}} - \int_{\Omega_\epsilon} g_0(-N\Delta u)\, d\Omega_{\boldsymbol{y}} = 0 \qquad (3.4)$$

We reason now as follows: because the left-hand side of this equation is zero for all $\epsilon > 0$ the limit must be zero as well

$$\lim_{\epsilon \to 0} B(g_0[\boldsymbol{x}], u)_{\Omega_\epsilon} = \lim_{\epsilon \to 0} \{ \int_{\Omega_\epsilon} -N\Delta g_0 u\, d\Omega_{\boldsymbol{y}} + \int_{\Gamma_\epsilon} N\frac{\partial}{\partial \nu} g_0 u\, ds_{\boldsymbol{y}}$$

$$-\int_{\Gamma_\epsilon} g_0 N \frac{\partial u}{\partial \nu}\, ds_{\boldsymbol{y}} - \int_{\Omega_\epsilon} g_0(-N\Delta u)\, d\Omega_{\boldsymbol{y}} \} = 0 \quad (3.5)$$

This limit is the extension of Betti's principle to singular solutions.

It remains to determine the limits of the *single* integrals in Equ.(3.5). Right now we only know that their sum must be zero.

a) The domain integrals

Before we start we place the origin of the system of coordinates at the source point $\boldsymbol{x} = \boldsymbol{o}$ and we switch to polar coordinates

$$y_1 = r \cos\varphi, \qquad y_2 = r \sin\varphi$$

We begin the investigation with the two domain integrals. Because the fundamental solution is a homogeneous solution in the punctured domain

$$-N\Delta g_0 = 0 \qquad \text{in } \Omega_\epsilon$$

the first domain integral is zero. With regard to the second integral we note that

$$g_0(\boldsymbol{y}, \boldsymbol{x}) = -\frac{1}{2\pi N} \ln r, \quad |N\Delta u| < \infty \ \text{ in } \Omega, \quad d\Omega = r\, dr\, d\varphi,$$

and

$$\lim_{r \to 0} r \ln r = 0, \qquad (3.6)$$

so that the limit of the second domain integral is

$$\lim_{\epsilon \to 0} \int_{\Omega_\epsilon} g_0(-N\Delta u)\, d\Omega_{\boldsymbol{y}} = \int_{\Omega} g_0(-N\Delta u)\, d\Omega_{\boldsymbol{y}},$$

i.e. the limit is simply the integral over the full domain.

b) Boundary integrals

The boundary Γ_ε of the punctured domain consists of the boundary Γ of the membrane and the edge $\Gamma_{N\varepsilon}(x)$ of the hole so that each boundary integral splits into two integrals

$$\int_{\Gamma_\varepsilon} \cdots ds_y = \int_\Gamma \cdots ds_y + \int_{\Gamma_{N\varepsilon}(x)} \cdots ds_y$$

The integrals over the exterior boundary Γ do not depend on ε so that we can concentrate on the study of the two boundary integrals

$$\int_{\Gamma_{N\varepsilon}} g_0 N \frac{\partial u}{\partial \nu} ds_y , \qquad \int_{\Gamma_{N\varepsilon}} N \frac{\partial}{\partial \nu} g_0 \, u \, ds_y \tag{3.7}$$

over the internal boundary $\Gamma_{N\varepsilon}(x)$. If the radius of the hole shrinks to zero then the integrand in the first integral tends, because of

$$g_0(y, x) = -\frac{1}{2\pi N} \ln r , \qquad |\frac{\partial u}{\partial \nu}| < \infty, \qquad ds = \varepsilon \, d\varphi ,$$

and Eq.(3.6), to zero and, therefore, the integral as well

$$\lim_{\varepsilon \to 0} \int_{\Gamma_{N\varepsilon}} g_0 N \frac{\partial u}{\partial \nu} ds_y = \lim_{\varepsilon \to 0} \frac{-1}{2\pi N} \varepsilon \ln \varepsilon \int_0^{2\pi} \frac{\partial u}{\partial \nu} d\varphi = 0$$

The second integral depends on the normal derivative of the fundamental solution

$$N \frac{\partial}{\partial \nu} g_0(y, x) = -\frac{1}{2\pi r} (r_{,y_1} \nu_1(y) + r_{,y_2} \nu_2(y)) ,$$

Because of

$$r_{,y_1} = \frac{y_1 - x_1}{r} = \frac{y_1}{\varepsilon} = \cos \varphi , \qquad r_{,y_2} = \frac{y_2 - x_2}{r} = \frac{y_2}{\varepsilon} = \sin \varphi ,$$

and

$$\nu_1 = -\cos \varphi , \quad \nu_2 = -\sin \varphi , \quad y_1 = \varepsilon \cos \varphi , \quad y_2 = \varepsilon \sin \varphi ,$$

the normal derivative is constant on $\Gamma_{N\varepsilon}(x)$

$$N \frac{\partial}{\partial \nu} g_0 = \frac{1}{2\pi\varepsilon} (\cos \varphi \cos \varphi + \sin \varphi \sin \varphi) = \frac{1}{2\pi\varepsilon}$$

The position vectors of the points y on the circle $\Gamma_{N\varepsilon}(x)$ can be split into two vectors,

$$y = x + \varepsilon \nabla_y r, \quad \nabla_y r = \{r_{,y_1}, r_{,y_2}\}^T = \{\cos\varphi, \sin\varphi\}^T,$$

(the vector $\nabla_y r$ is the unit vector which points from x to y) and, hence, it follows easily that

$$\lim_{\varepsilon \to 0} \int_{\Gamma_{N\varepsilon}} N\frac{\partial}{\partial\nu} g_0\, u(y)\, ds_y = \lim_{\varepsilon \to 0} \left\{ \frac{1}{2\pi\varepsilon} \int_0^{2\pi} u(x + \varepsilon\nabla_y r)\varepsilon\, d\varphi \right\} = u(x)$$

Summing it up, we have

$$\lim_{\varepsilon \to 0} B(g_0[x], u)_{\Omega_\varepsilon} = u(x) + \int_\Gamma (N\frac{\partial}{\partial\nu} g_0[x]u - g_0[x]N\frac{\partial u}{\partial\nu})\, ds_y$$

$$- \int_\Omega g_0[x](-N\Delta u)\, d\Omega_y = 0,$$

or if we solve for $u(x)$

$$u(x) = \int_\Gamma (g_0[x]N\frac{\partial u}{\partial\nu} - N\frac{\partial}{\partial\nu} g_0[x]u)\, ds_y + \int_\Omega g_0[x](-N\Delta u)\, d\Omega_y \qquad (3.8)$$

This is the influence function for the deflection at interior points. It remains to consider the case that the point x lies on the boundary.

Then the boundary, see Fig. 3.5,

$$\Gamma_\varepsilon = \Gamma'_\varepsilon \cup \Gamma_{N_\varepsilon}(x)$$

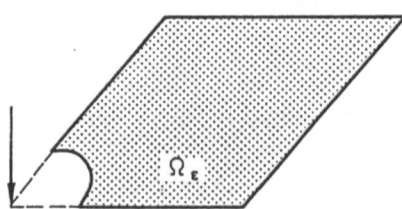

Figure 3.5 The source point is located at a corner point

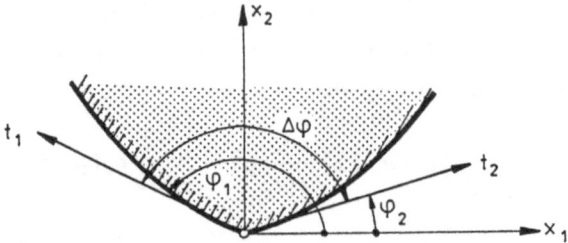

Figure 3.6 Notations at a corner point

consists of the part Γ'_ε, this is the original boundary Γ minus that part which is cut off when we remove a segment $N_\varepsilon(\boldsymbol{x})$ from the domain, and the circular arc $\Gamma_{N_\varepsilon}(\boldsymbol{x})$. With regard to the two integrals (3.7) over the circular arc the previous considerations hold true. The first integral tends to zero and the second integral has, because we integrate now only over a section of the unit circle, the limit

$$\lim_{\varepsilon \to 0} \int\limits_{\Gamma_{N_\varepsilon}} N \frac{\partial}{\partial \nu} g_0 \, u(\boldsymbol{y}) \, ds\boldsymbol{y} = \lim_{\varepsilon \to 0} \left\{ \frac{1}{2\pi\varepsilon} \int\limits_{\varphi_1}^{\varphi_2} u(\boldsymbol{x} + \varepsilon \nabla \boldsymbol{y} r) \varepsilon \, d\varphi \right\} = \frac{\Delta\varphi}{2\pi} u(\boldsymbol{x})$$

This limit depends on the angle $\Delta\varphi = \varphi_1 - \varphi_2$ between the two tangents at the boundary point, see Fig. 3.6.

The limits of the two integrals over the outer boundary

$$\lim_{\varepsilon \to 0} \int\limits_{\Gamma'_\varepsilon} g_0 N \frac{\partial u}{\partial \nu}(\boldsymbol{y}) \, ds\boldsymbol{y} = \int\limits_{\Gamma} g_0 N \frac{\partial u}{\partial \nu}(\boldsymbol{y}) \, ds\boldsymbol{y} \,,$$

$$\lim_{\varepsilon \to 0} \int\limits_{\Gamma'_\varepsilon} N \frac{\partial}{\partial \nu} g_0 \, u(\boldsymbol{y}) \, ds\boldsymbol{y} = \int\limits_{\Gamma} N \frac{\partial}{\partial \nu} g_0 \, u(\boldsymbol{y}) \, ds\boldsymbol{y} \,,$$

(3.9)

are bounded: the integrand in the first integral is only weakly singular ($\ln r$) and the integrand in the second integral,

$$N \frac{\partial}{\partial \nu} g_0(\boldsymbol{y}, \boldsymbol{x}) = -\frac{r_\nu}{2\pi r} \,, \quad r_\nu = \nabla \boldsymbol{y} r \cdot \boldsymbol{\nu} = r_{,1} \nu_1 + r_{,2} \nu_2 \,,$$

is (if the boundary is straight) zero on both sides of the source point, because the normal vector and the gradient are orthogonal so that their scalar product, the normal derivative r_ν, is zero. It follows that the limit also exists if the boundary is curved because the curvature adds a "positive" factor $O(r^\alpha)$, $\alpha > 0$, to the integrand.

If the concentrated force is located outside of the domain of the membrane then it contributes no work so that in such a case the work $1 \times u(\boldsymbol{x})$ is zero.

Collecting our results we have

$$c(\boldsymbol{x})u(\boldsymbol{x}) = \int_{\Gamma} [g_0(\boldsymbol{y},\boldsymbol{x})N\frac{\partial u}{\partial \nu}(\boldsymbol{y}) - N\frac{\partial}{\partial \nu}g_0(\boldsymbol{y},\boldsymbol{x})u(\boldsymbol{y})]\,ds\boldsymbol{y}$$

$$+ \int_{\Omega} g_0(\boldsymbol{y},\boldsymbol{x})(-N\Delta u)\,d\Omega \boldsymbol{y} \qquad (3.10)$$

The characteristic function

$$c(\boldsymbol{x}) = \begin{cases} 1, & \boldsymbol{x} \in \Omega, \\ \Delta\varphi/2\pi, & \boldsymbol{x} \in \Gamma, \\ 0, & \boldsymbol{x} \in \Omega^c, \quad \text{(complement)}, \end{cases} \qquad (3.11)$$

is the work done by the concentrated force $\hat{P} = 1$ on acting through $u = 1$. If the source point $\boldsymbol{x}$ is an interior point then $c(\boldsymbol{x}) = 1$, if the force acts on the boundary then only that part of $\hat{P}$ which lies inside of Ω contributes to the work, $c(\boldsymbol{x}) = \Delta\varphi/2\pi$, and if the source point lies outside of the membrane, the complement, then $\hat{P}$ contributes no work, that is $c(\boldsymbol{x}) = 0$.

Equation (3.10) is the complete influence function of the membrane. According to this equation the deflection $u(\boldsymbol{x})$ in the interior depends on the deflection $u(\boldsymbol{y})$ and the traction $t = N\partial u(\boldsymbol{y})/\partial \nu$ on the boundary and the lateral pressure $-N\Delta u = p$ in the domain. As on any part of the boundary only one of these two boundary functions is prescribed, either the deflection u or the traction $N\partial u/\partial \nu$, we must determine the pertinent unknown function by a boundary element procedure. To this end we place in Eq.(3.10) the point $\boldsymbol{x}$ on the boundary

$$\frac{\Delta\varphi}{2\pi}u(\boldsymbol{x}) = \int_{\Gamma} [g_0(\boldsymbol{y},\boldsymbol{x})N\frac{\partial u}{\partial \nu}(\boldsymbol{y}) - N\frac{\partial}{\partial \nu}g_0(\boldsymbol{y},\boldsymbol{x})u(\boldsymbol{y})]\,ds\boldsymbol{y} + \int_{\Omega} g_0(\boldsymbol{y},\boldsymbol{x})p\,d\Omega \boldsymbol{y},$$

so that the function u on the left side is the same function u as on the right side, that is, it is simultaneously the dependent as well as the independent variable. Rearranging this equation according to displacement and force terms we obtain the integral equation

$$\frac{\Delta\varphi}{2\pi}u(\boldsymbol{x}) + \int_{\Gamma} N\frac{\partial}{\partial \nu}g_0(\boldsymbol{y},\boldsymbol{x})u(\boldsymbol{y})\,ds\boldsymbol{y} = \int_{\Gamma} g_0(\boldsymbol{y},\boldsymbol{x})N\frac{\partial u}{\partial \nu}(\boldsymbol{y})\,ds\boldsymbol{y}$$

$$+ \int_{\Omega} g_0(\boldsymbol{y},\boldsymbol{x})p\,d\Omega \boldsymbol{y} \qquad (3.12)$$

which formulates a coupling condition between the boundary displacement u and the traction $N\partial u/\partial \nu$ of a membrane. Two functions u and $N\partial u/\partial \nu$ are the boundary values of a membrane if and only if they satisfy this integral equation. This compatibility condition in turn offers the possibility to determine the unknown boundary value by solving the integral equation.

3.2 Discretization

The coupling condition (3.12) represents an infinite system of equations because the boundary consists of infinitely many points x. To solve this system approximately we interpolate the deflection u and the traction t by piecewise constant, linear or quadratic functions whose n nodal values we then determine by satisfying the coupling condition at K collocation points, see Fig. 3.7.

The basis functions φ_i and ψ_i which approximate the boundary functions u and t,

$$u(x) = u_i\varphi_i(x), \qquad t(x) = t_i\psi_i(x),$$

have a finite support: only in a small part of the boundary are they different from zero. The function $\varphi_i(x)$ has the value 1 at the node x^i and at all other nodes the value zero,

$$\varphi_i(x^j) = \delta_{ij}$$

It represents, so to speak, a unit deflection of the node x^i: the node x^i is pushed down by one unit of deflection, $u_i = 1$, while all other nodes remain fixed.

Similarly the functions ψ_i represent unit tractions. At corner points the traction has two degrees of freedom, t_j and t_{j+1}, and so it must be interpolated at such points by two functions, ψ_j and ψ_{j+1}. These two functions are the two

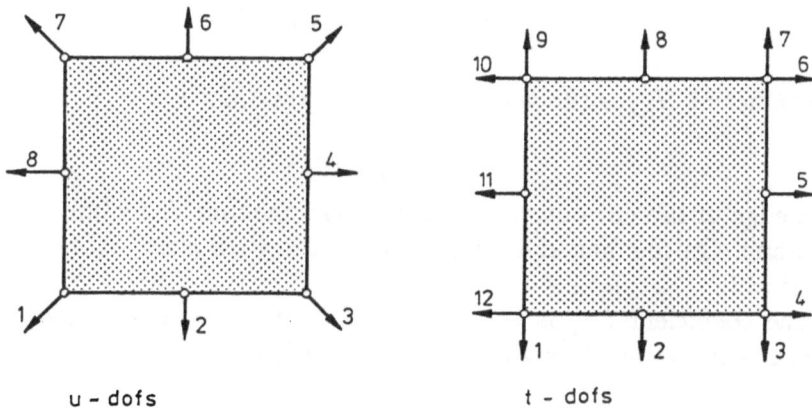

Figure 3.7 Degrees-of-freedom

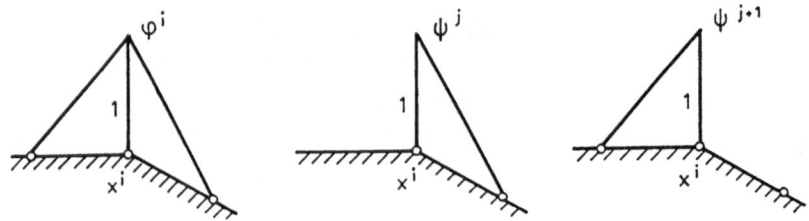

Figure 3.8 Continuous and discontinuous basis functions at a corner point

halves (in the case of a linear approximation) of a hat-function, see Fig. 3.8. This is also the reason why the number of functions ψ_i is always greater than the number of functions φ_i.

The basis functions φ_i and ψ_i are pieced together from *local* basis functions φ_i^e which are, in general, for φ_i and ψ_i the same.

The unknown nodal values u_i and t_i are determined by satisfying integral equation (3.12) at K collocation points; the number of points must be equal to the number of unknowns. The coupling condition (3.12) thus becomes a linear system of equations for the nodal values

$$Hu = Gt + p$$

where

$$H_{kj} = c(x^k)\delta_{kj} + \int_\Gamma N \frac{\partial}{\partial \nu} g_0(y, x^k)\varphi_j(y)\, ds_y\,,$$

$$G_{kj} = \int_\Gamma g_0(y, x^k)\psi_j(y)\, ds_y\,,$$

and the vector

$$p_k = \int_\Omega g_0(y, x^k)p(y)\, d\Omega_y$$

The coefficient H_{kj} is the work done by the traction

$$T = N \frac{\partial}{\partial \nu} g_0(y, x^k)$$

at the edge of the cut-out on acting through the unit deflection φ_j of the real membrane. On the diagonal we add to this the work $c(x^k)$ of the concentrated force $\hat{P} = 1$.

The coefficient G_{kj} is the work done by the unit traction ψ_j of the real membrane on acting through the edge deflection

$$U = g_0(y, x^k)$$

of the cut-out.

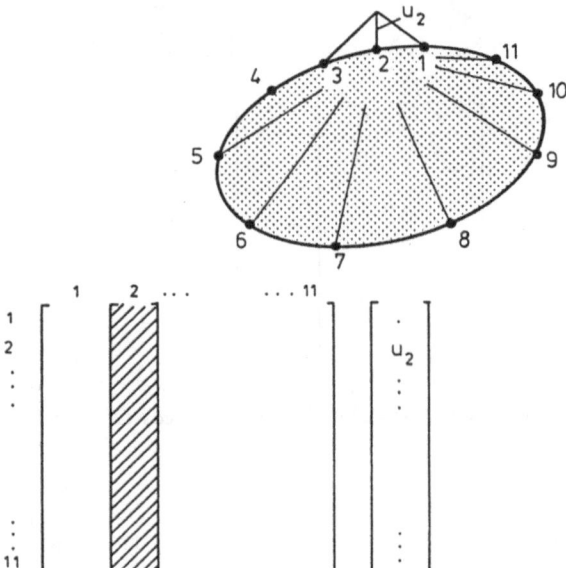

Figure 3.9 The influence of the nodal degree-of-freedom u_2 on the single collocation points is represented by a vector

Considering their origin, the terms H_{kj} and G_{kj} are work terms, but with regard to their meaning they are influence coefficients. The column j of matrix H and G, respectively, represents the influence of the boundary layer φ_j and ψ_j, respectively, on the collocation points x^k, see Fig. 3.9. The influence depends on the kernels $N\partial g_0(y, x^k)/\partial\nu$ and $g_0(y, x^k)$.

3.3 Element matrices

In nearly the same way as stiffness matrices are assembled from element matrices K^e the two matrices H and G are assembled from element matrices H^e and G^e, see Fig. 3.10.

$$H_{kj}^e = \bar{c}(x^k)\varphi_j^e(x^k) + \int_{\Gamma_e} N\frac{\partial}{\partial\nu}g_0(y, x^k)\varphi_j^e(y)\, ds_y,$$

$$G_{kj}^e = \int_{\Gamma_e} g_0(y, x^k)\varphi_j^e(y)\, ds_y, \quad j = 1,\ldots,A, \quad k = 1,\ldots,K$$

Such an element matrix has A columns and K rows where A is equal to the number of local basis functions on the single element and K equal to the number

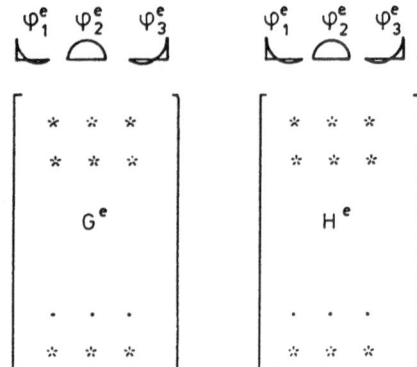

Figure 3.10 The element matrices G^e and H^e

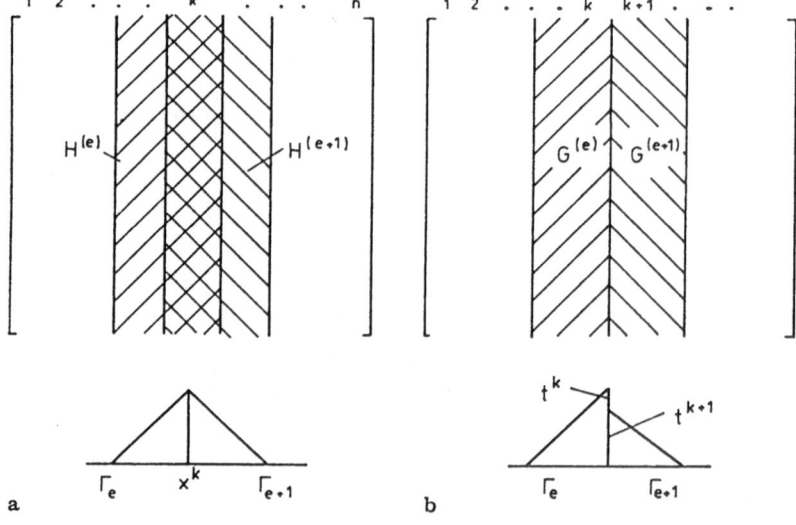

Figure 3.11 The assemblage of the element matrices G^e and H^e reflects the interelement continuity of the functions u and t at the element boundaries

of nodes on the boundary, so that an element matrix is a slender rectangle that has the same height as the matrix G or H. The entries in the first column of the element matrix G^e represent the influence, which the first local basis function exerts on the K collocation points x^k according to the weight of the kernel $g_0[x^k]$. The second column represents the influence of the second local basis function, etc.

The assemblage of the element matrices G^e and H^e depends on the interelement continuity of the functions u and t at the element boundaries, see Fig. 3.11a and b. The element matrices H^e overlap because the deflection u

is continuous. This is also the reason why we multiply the c-term of the first node and the last node of each element by $1/2$

$$\bar{c}(\boldsymbol{x}^k) = \begin{cases} \dot{c}(\boldsymbol{x}^k) & \boldsymbol{x}^k = \text{mid-node}, \\ \frac{1}{2}\dot{c}(\boldsymbol{x}^k) & \boldsymbol{x}^k = \text{first or last node} \end{cases}$$

After the assemblage the correct value will then appear on the diagonal of the matrix $\boldsymbol{H}$,

$$\dot{c}(\boldsymbol{x}^k) = \frac{1}{2}\dot{c}(\boldsymbol{x}^k) + \frac{1}{2}\dot{c}(\boldsymbol{x}^k)$$

If the traction is discontinuous at the interelement node then the two element matrices $\boldsymbol{G}^e$ lie side by side, see Fig. 3.11b.

3.4 The master element

The integration is done on a master element $[0,1]$, see Fig. 3.12. With respect to its coordinate ξ the basis functions are defined as, see Fig. 3.13:

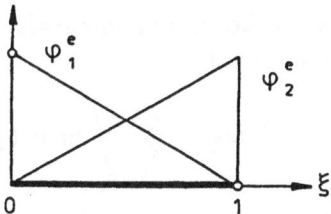

Figure 3.12 Master element with two basis functions

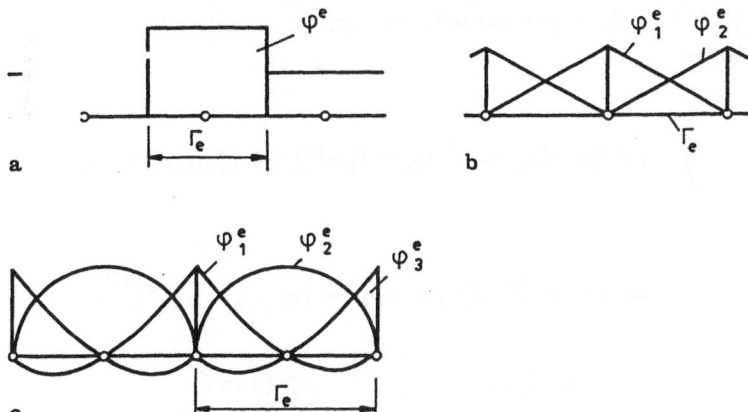

Figure 3.13 a-c. Basis functions: a constant; b linear; c quadratic basis functions

Constant function

$$\varphi^e = 1$$

Linear functions

$$\varphi_1^e = 1 - \xi, \qquad \varphi_2^e = \xi$$

Quadratic functions

$$\varphi_1^e = (1 - \xi)(1 - 2\xi), \quad \varphi_2^e = 4\xi(1 - \xi), \quad \varphi_3^e = 2\xi(\xi - 0.5)$$

The shape function of a straight boundary element with the two nodes (y_1^a, y_2^a) and (y_1^b, y_2^b) is

$$y_1^e = y_1^a \varphi_1^e(\xi) + y_1^b \varphi_2^e(\xi) = y_1^a(1 - \xi) + y_1^b \xi,$$

$$y_2^e = y_2^a \varphi_1^e(\xi) + y_2^b \varphi_2^e(\xi) = y_2^a(1 - \xi) + y_2^b \xi,$$

and, therefore, the differential of the arc-length is

$$ds = ((y_{1,\xi}^e)^2 + (y_{2,\xi}^e)^2)^{1/2} \, d\xi = l_e \, d\xi, \quad l_e = \text{ length of the element}$$

To explain the reduction of the boundary integrals to integrals over the master element let us consider the integral

$$\int_\Gamma \ln r \, t(\mathbf{y}) \, ds_\mathbf{y} = \sum_e \int_{\Gamma_e} \ln r \, t(\mathbf{y}) \, ds_\mathbf{y}$$

The function

$$r = \left[(y_1 - x_1)^2 + (y_2 - x_2)^2 \right]^{1/2}$$

is the distance between the integration point $\mathbf{y} = (y_1, y_2)$ and the source point $\mathbf{x} = (x_1, x_2)$. If t is elementwise a linear function with the nodal values t_1^e and t_2^e then we have

$$\int_{\Gamma_e} \ln r \, t(\mathbf{y}) \, ds_\mathbf{y} = \int_0^1 \ln r \, (t_1^e \varphi_1^e(\xi) + t_2^e \varphi_2^e(\xi)) l_e \, d\xi,$$

where

$$r = r(\xi) = \left[(y_1^e(\xi) - x_1)^2 + (y_2^e(\xi) - x_2)^2 \right]^{1/2}$$

and

$$\varphi_1^e(\xi) = 1 - \xi, \qquad \varphi_2^e(\xi) = \xi$$

The functions $y_i^e(\xi)$ are the forementioned shape functions of the element.

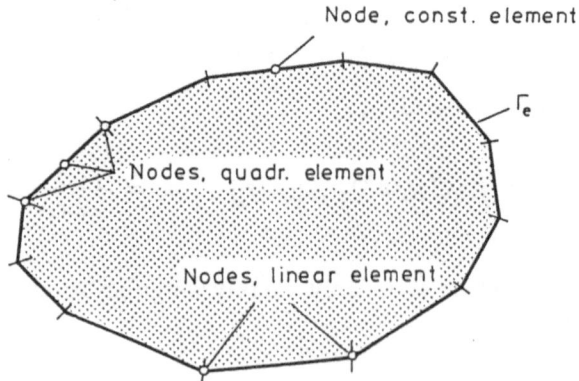

Figure 3.14 Position of the nodes

If the source point x is located on a different element, see Fig. 3.14, then the integrand

$$f(\xi) = \ln r(\xi)\,(t_1^e \varphi_1^e(\xi) + t_2^e \varphi_2^e(\xi))\,l_e$$

is regular and the integral can be evaluated by Gaussian quadrature.

$$\int_0^1 f(\xi)\,d\xi = \sum_i f(\xi_i) w_i$$

The actual number of weights w_i and points ξ_i will depend on the distance between the source point x and the element Γ_e. The author uses in his programs 4, 6, 8 or 10 points respectively.

3.5 Singular integrals

If the source point x lies on the element Γ_e itself then we evaluate the influence analytically.

With regard to the integral

$$H_{kj}^e = \bar{c}(x^k)\varphi_j^e(x^k) + \int_{\Gamma_e} N\frac{\partial}{\partial \nu} g_0(y, x^k)\varphi_j^e(y)\,ds_y, \quad N\frac{\partial}{\partial \nu} g_0 = -\frac{1}{2\pi}\frac{r_\nu}{r},$$

things are simple. Because the gradient $\nabla_y r$ and the normal vector $\nu = \nu(y)$ are orthogonal if y and x lie on the same (straight!) element the function

$$r_\nu = \nabla_y r \cdot \nu$$

is zero so that

$$H_{kj}^e = \bar{c}(\boldsymbol{x}^k)\varphi_j^e(\boldsymbol{x}^k)$$

The integrand in the integral

$$G_{kj}^e = \int_\Gamma g_0(\boldsymbol{y}, \boldsymbol{x}^k)\varphi_j^e(\boldsymbol{y})\,ds\boldsymbol{y}, \quad g_0 = -\frac{1}{2\pi N}\ln r,$$

does not vanish; however, it is only weakly singular so that we easily obtain the following results:

a) Constant basis functions

$\boldsymbol{x} = \boldsymbol{x}^m$ (the collocation point is the centre point)

$$\int_{\Gamma_e} \ln r\,\varphi^e\,ds\boldsymbol{y} = l_e(\lambda_e - \ln 2 - 1),$$

$$l_e = \text{length of the element}, \quad \lambda_e = \ln l_e$$

b) Linear basis functions

$\boldsymbol{x} = \boldsymbol{x}^a$ (the collocation point is the first point of the element)

$$\int_{\Gamma_e} \ln r\,\varphi_1^e\,ds\boldsymbol{y} = \frac{l_e}{4}(2\lambda_e - 3),$$

$$\int_{\Gamma_e} \ln r\,\varphi_2^e\,ds\boldsymbol{y} = \frac{l_e}{4}(2\lambda_e - 1)$$

c) Quadratic basis functions

$\boldsymbol{x} = \boldsymbol{x}^a$

$$\int_{\Gamma_e} \ln r\,\varphi_1^e\,ds\boldsymbol{y} = \frac{l_e}{36}(6\lambda_e - 17),$$

$$\int_{\Gamma_e} \ln r\,\varphi_2^e\,ds\boldsymbol{y} = \frac{l_e}{9}(6\lambda_e - 5),$$

$$\int_{\Gamma_e} \ln r\,\varphi_3^e\,ds\boldsymbol{y} = \frac{l_e}{36}(6\lambda_e + 1)$$

$$x = x^m$$

$$\int_{\Gamma_e} \ln r \, \varphi_1^e \, ds_y = l_e(\frac{\lambda_e}{6} - 0.17108),$$

$$\int_{\Gamma_e} \ln r \, \varphi_2^e \, ds_y = 8l_e(\frac{\lambda_e}{12} - 0.1688733)$$

Due to the inherent symmetry these results, naturally, also apply if the collocation point is the end point of the element.

3.6 The treatment of the system of equations

The assembled system of equations has the form

$$\begin{bmatrix} H \end{bmatrix} \begin{bmatrix} u \end{bmatrix} = \begin{bmatrix} G \end{bmatrix} \begin{bmatrix} t \end{bmatrix} + \begin{bmatrix} p \end{bmatrix}$$

The quadratic H-matrix has K (= number of collocation points) columns and the rectangular G-matrix T columns, where $T > K$ is the number of t-dofs (degrees-of-freedom).

Next this system is arranged in the following form, see Fig. 3.15,

$$\begin{bmatrix} A \end{bmatrix} \begin{bmatrix} x \end{bmatrix} = \begin{bmatrix} b \end{bmatrix}$$

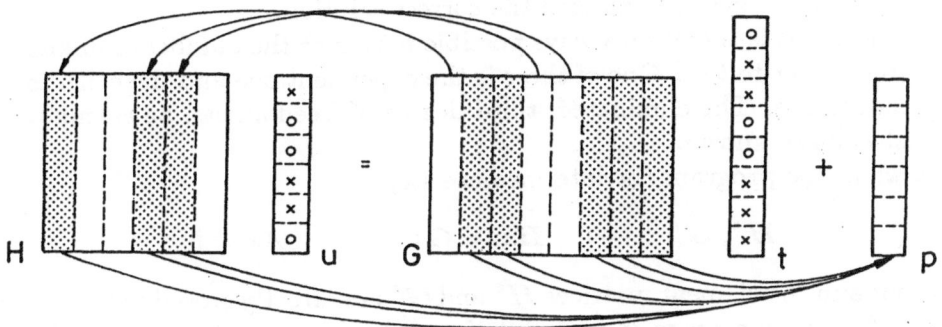

Figure 3.15 The rearrangement of the equations

where the vector $\boldsymbol{x}$ contains the unknown degrees-of-freedom, the matrix $\boldsymbol{A}$ the associated influence vectors and the vector $\boldsymbol{b}$ is the sum of all those influence vectors (multiplied with the pertinent nodal values) which correspond to given degrees-of-freedom.

On first glance it might seem that such a even-handed reduction n equations for n unknowns is not possible because at corner points we have three degrees-of-freedom

$$u_i, \quad t_j, \quad t_{j+1},$$

instead of the usual two. But the opposite is true. Corner points are points where, in principle, the solution is overdetermined. If, say, the deflection u is prescribed on the left-hand side and the right-hand side of the corner point then the tangential derivative on both sides, u_t^l and u_t^r, can be determined approximately with finite differences and, therefore, also the gradient

$$u_{,1} = \frac{1}{D}(t_2^r u_t^l - t_2^l u_t^r), \qquad D = t_1^l t_2^r - t_1^r t_2^l,$$

$$u_{,2} = \frac{1}{D}(-t_1^r u_t^l + t_1^l u_t^r),$$

and the normal derivatives as well

$$t^l = N(u_{,1}\, n_1^l + u_{,2}\, n_2^l),$$

$$t^r = N(u_{,1}\, n_1^r + u_{,2}\, n_2^r)$$

Such boundary conditions would, therefore, not require a collocation point at the corner. For programming reasons we will nevertheless do so and treat during the assemblage one of the degrees-of-freedom as an 'unknown' degree-of-freedom. Prior to the solution of the system of equations the off-diagonal terms of that row which corresponds to the integral equation of the corner point will be set to zero, the entry on the diagonal will be set to 1 and the term on the right-hand side will be set equal to the prescribed value. Not surprisingly the thus modified system will give the correct solution.

Basically, it is therefore always possible to reduce the number of unknowns at a corner point to two. One of these is the dependent one and the other is the independent one. The eliminated, third degree-of-freedom can be calculated if the first two are known.

Remark: In the program the intermediate step

$$\boldsymbol{H}^e, \boldsymbol{G}^e \quad \longrightarrow \quad \boldsymbol{H}\boldsymbol{u} = \boldsymbol{G}\boldsymbol{t} \quad \longrightarrow \quad \boldsymbol{A}\boldsymbol{x} = \boldsymbol{b}$$

is left out and the element matrices $\boldsymbol{H}^e$ and $\boldsymbol{G}^e$ are directly stored in the system matrix $\boldsymbol{A}$ to save storage space. But if you do this step, for example in the testing phase of a program, then do not store, first, all columns of the matrix H

in the system matrix A and then all columns of G, but rather store the single columns in their "natural" order, that is in the order in which the unknowns appear on the boundary. Otherwise the condition number deteriorates.

3.7 The domain integral

The domain integral in Eq.(3.10), which represents the influence of the lateral pressure p can, if the pressure p is harmonic, $\Delta p = 0$, be transformed into a boundary integral as follows:

The function

$$U = -\frac{1}{8\pi N^2} r^2 (\ln r - 1)$$

is the integral of the fundamental solution with respect to the Laplacian

$$-N\Delta U = -\frac{1}{2\pi N} \ln r$$

If we substitute in Green's second identity (3.3) for $\hat{u}$ and u the functions $\hat{u} = U$ and $u = p$, then it follows

$$-\frac{1}{2\pi N} \int_{\Omega} \ln r \, p \, d\Omega_y = -\int_{\Gamma} (N\frac{\partial U}{\partial \nu} p - UN\frac{\partial p}{\partial \nu}) \, ds_y ,$$

which, in case the pressure p is constant, simplifies further to

$$-\frac{1}{2\pi N} \int_{\Omega} \ln r \, p \, d\Omega_y = -\int_{\Gamma} N\frac{\partial U}{\partial \nu} p \, ds_y = -\frac{p}{8\pi N} \int_{\Gamma} r(2\ln r - 1) r_\nu \, ds_y \quad (3.13)$$

If a concentrated force P acts at an internal point ξ then the domain integral (3.10) is replaced by

$$g_0(\xi, x) P ,$$

and if line loads p act along a curve γ then the domain integral is replaced by

$$\int_{\gamma} g_0(y, x) p(y) \, ds_y \equiv \sum_e \int_{\Gamma_e} g_0(y, x)(p_1^e \varphi_1^e(y) + p_2^e \varphi_2^e(y)) \, ds_y$$

To approximate this integral we shall, as indicated, model the curve γ by a series of linear elements Γ_e and interpolate the distribution p, locally, by linear basis functions φ_i^e.

3.8 Internal actions

The influence functions for the two derivatives are, see section 3.11,

$$u_{,x_i}(x) = \int_{\Gamma} [\frac{\partial}{\partial x_i} g_0(y, x) t(y) - \frac{\partial}{\partial x_i} N \frac{\partial}{\partial \nu} g_0(y, x) u(y)] \, ds_y$$

$$+ \int_{\Omega} \frac{\partial}{\partial x_i} g_0(y, x)(-N \Delta u) \, d\Omega_y$$

where

$$\frac{\partial}{\partial x_i} g_0(y, x) = -\frac{1}{2\pi N} \frac{1}{r} r_{,x_i} \,,$$

$$\frac{\partial}{\partial x_1} N \frac{\partial}{\partial \nu} g_0(y, x) = \frac{1}{2\pi} \frac{1}{r^2} (r_{,x_1} r_\nu - r_\tau r_{,x_2}) \,,$$

$$\frac{\partial}{\partial x_2} N \frac{\partial}{\partial \nu} g_0(y, x) = \frac{1}{2\pi} \frac{1}{r^2} (r_{,x_2} r_\nu + r_\tau r_{,x_1})$$

If the domain integral is transformed into a boundary integral then the derivatives of the kernel function in Eq.(3.13) are also required

$$\frac{\partial}{\partial x_1} \{\frac{1}{8\pi N} [r(-2\ln r + 1) r_\nu]\} = \frac{1}{8\pi N} \{-2\ln r(r_{,x_1} r_\nu + r_{,x_2} r_\tau)$$

$$- r_{,x_1} r_\nu + r_{,x_2} r_\tau\} \,,$$

$$\frac{\partial}{\partial x_2} \{\frac{1}{8\pi N} [r(-2\ln r + 1) r_\nu]\} = \frac{1}{8\pi N} \{-2\ln r(r_{,x_2} r_\nu - r_{,x_1} r_\tau)$$

$$- r_{,x_1} r_\tau - r_{,x_2} r_\nu\}$$

3.9 Examples

3.9.1 Shear and torsion

According to the membrane analogy the shear stresses within a rod are proportional to the slope of *Prandtl's stress function* Φ. This function is the deflection of a membrane if you blow up the membrane by a uniform pressure of magnitude $p = 2$

$$-\Delta\Phi = 2 \quad \text{in } \Omega, \quad \Phi = 0 \quad \text{on } \Gamma$$

The shear stresses are

$$\tau_{xy} = -G\frac{\vartheta}{l}\Phi_{,z}, \qquad \tau_{zz} = G\frac{\vartheta}{l}\Phi_{,y},$$

$$G = \text{shear modulus}, \quad l = \text{length}, \quad \vartheta = \text{twist},$$

and the torsional stiffness J_T of the cross section is given by twice the volume under the membrane

$$J_T = 2\int_\Omega \Phi\, d\Omega$$

If the cross section contains openings then additional constraints must be considered at the internal boundaries to make the solution unique. This is why we prefer to calculate the shear stresses by solving the boundary-value problem

$$\Delta\varphi = 0 \quad \text{in } \Omega \quad \frac{\partial\varphi}{\partial n} = -x_1 n_2 + x_2 n_1 \quad \text{on } \Gamma$$

for the warping function φ. The boundary condition means that the slope must be equal to the scalar product between the radius vector $\boldsymbol{x}$ of the pertinent boundary point and the tangent vector $\boldsymbol{t} = \{-n_2, n_1\}$ at this point. This boundary condition applies at external and internal boundaries alike.

In terms of the warping function the shear stresses are

$$\tau_1 = -G\vartheta'(\varphi_{,1} - x_2) \quad \tau_2 = -G\vartheta'(\varphi_{,1} + x_1)$$

and the torsional stiffness

$$I_t = I_r - \int_\Gamma (x_2 n_1 - x_1 n_2)\,ds \quad I_r = \int_\Omega (x_1^2 + x_2^2)\,d\Omega$$

The coordinates of the shear centre are

$$x_1^s = \frac{I_{12}R_2 - R_1 I_2}{I_1 I_2 - I_{12}^2} \quad x_2^s = \frac{R_2 I_1 - I_{12}R_1}{I_1 I_2 - I_{12}^2}$$

where

$$R_1 = \int_\Omega \varphi x_2\, d\Omega, \quad R_2 = \int_\Omega \varphi x_1\, d\Omega,$$

The integrals I_1, I_2, I_{12} are the moments of inertia of the cross section with respect to the centroid.

The warping rigidity with respect to the shear centre is

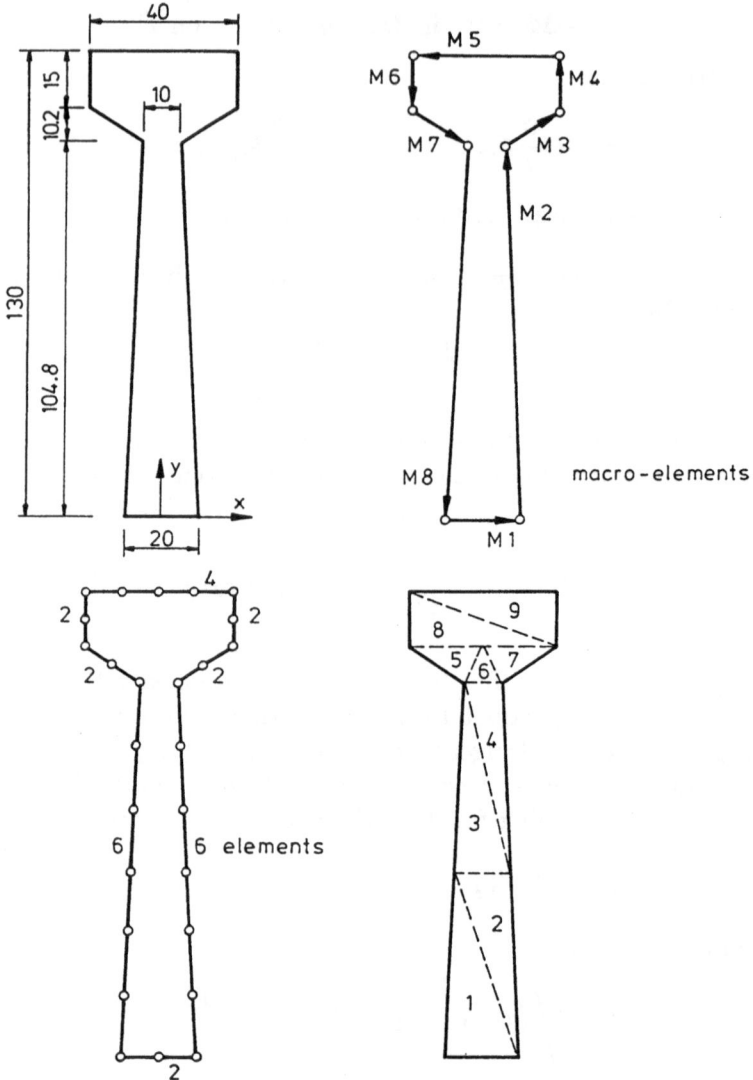

Figure 3.16 Cross section with interior elements

$$C_s = C + 2x_1^s R_1 - 2x_2^s - 2x_1^s x_2^s I_{12} + (x_1^s)^2 I_1 + (x_2)^2 I_2$$

where

$$C = \int_\Omega \varphi^2 d\Omega$$

While the domain integrals R_i can easily be transformed into boundary integrals (use the first identity of the Laplace operator) the integral of φ^2 over the cross section allows no such reduction and, therefore, a triangulation of the cross section is necessary to calculate the warping rigidity.

Figure 3.16 indicates what such a triangulation looks like and which effect the grading of the mesh has on the warping rigidity. Figure 3.17 shows a cross section with an opening.

These two problems were solved with the program BE-TORSION. This program is an offspring of the Laplace program and provides some additional features, as semi-automatic mesh generation, etc.

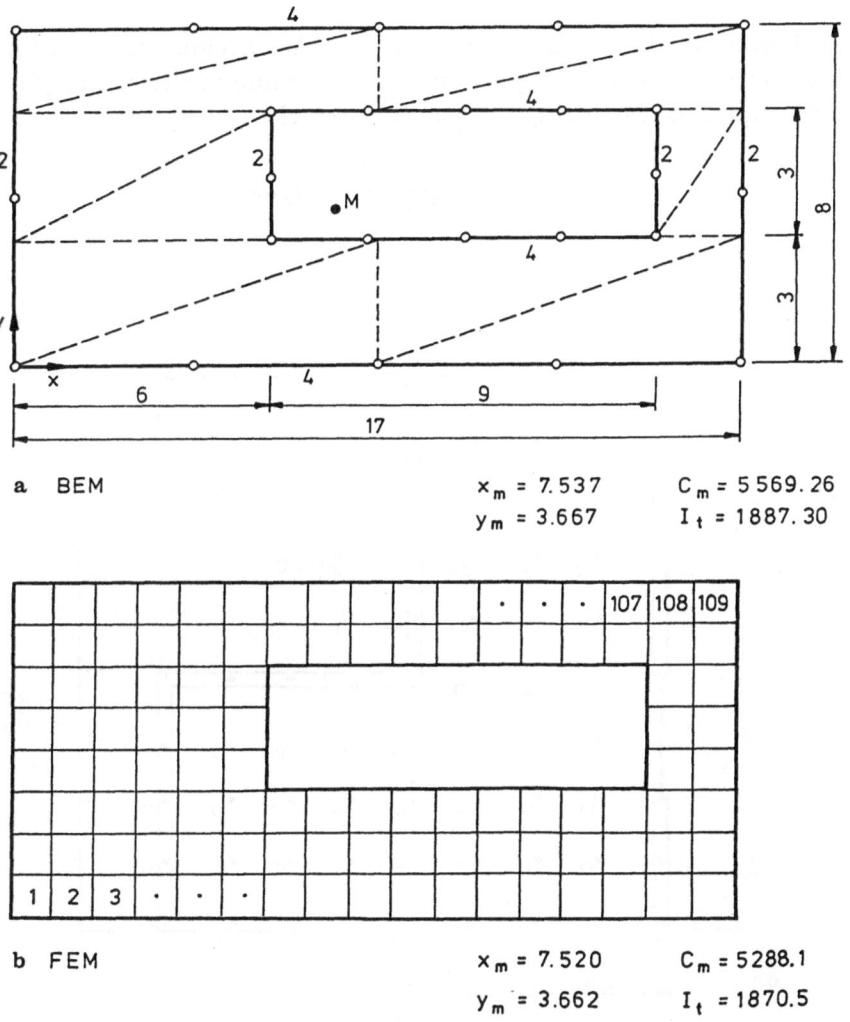

a BEM

$x_m = 7.537 \qquad C_m = 5\,569.26$
$y_m = 3.667 \qquad I_t = 1887.30$

b FEM

$x_m = 7.520 \qquad C_m = 5288.1$
$y_m = 3.662 \qquad I_t = 1870.5$

Figure 3.17 a-b. Cross section with openings: **a** BE-model; **b** FE-model

3.9.2 Temperature distribution

The next example is a study of the temperature distribution in an L-shaped
living room, see Fig. 3.18. The walls of the living-room are considered to
be perfectly isolated so that the flux at the walls is zero, $\partial T/\partial n = 0$. The
temperature of the window panes is assumed to be 10° centigrade and the
temperature of the chimney at the upper end of the living room is assumed to
be 50° centigrade.

The temperature distribution T in the living room is the function that
satisfies these boundary conditions and the differential equation

$$\Delta T = 0$$

Any such solution is called a *harmonic function* because the solution in our
case the temperature has the property that its value at the centre of any circle
$\Gamma_{N\epsilon}(\boldsymbol{x})$ is equal to the average value on the circle

$$T(\boldsymbol{x}) = \frac{1}{2\pi\epsilon} \int\limits_{\Gamma_{N\epsilon}} T(\boldsymbol{y})\,ds$$

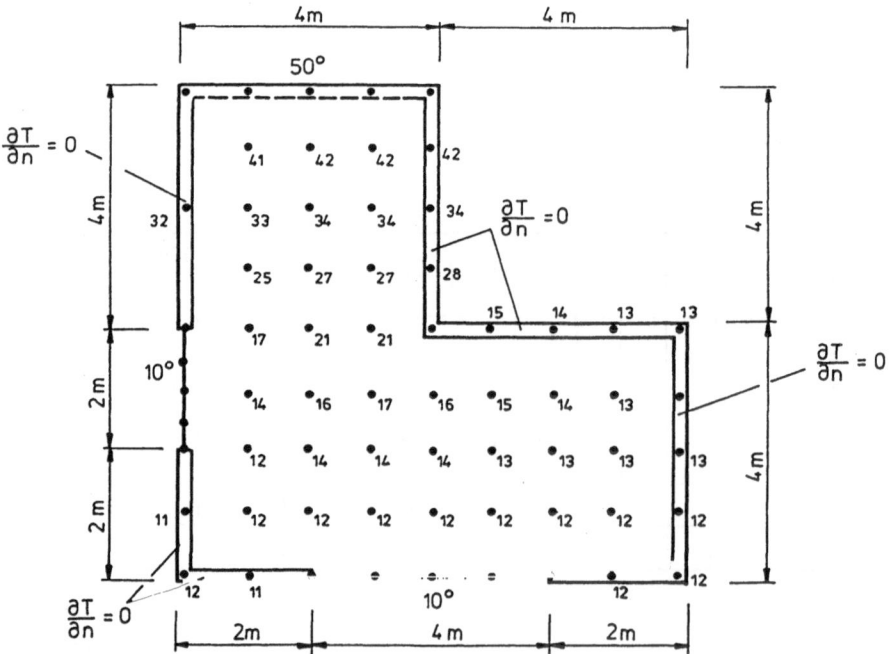

Figure 3.18 Temperature distribution in a living room

Figure 3.18 depicts the temperature distribution as calculated with the program BE-LAPLACE. The boundary was divided into 16 quadratic elements of varying length. The points on the boundary indicate the positions of the nodes.

3.9.3 Brownian motion

The connection between probability theory and potential theory is an example that seemingly unrelated branches of physics are in some sense mathematically equivalent, [41].

To demonstrate this connection we consider the Brownian motion of molecules in a plane domain Ω. The domain is bounded by an internal, Γ_i, and an external curve Γ_e, see Fig. 3.19a, and we ask: what is the probability $p(x)$ that a molecule which starts at a point x will hit the internal wall Γ_i before it hits the outer wall Γ_e?

For molecules which reside on Γ_i the probability is 1. For molecules which reside on the outer wall Γ_e the probability is 0. For all other molecules x the probability $p(x)$ is identical with the solution of the boundary-value problem

$$\Delta p = 0, \quad p = 1 \quad \text{on } \Gamma_i, \quad p = 0 \quad \text{on } \Gamma_e$$

If the domain is bounded by two concentric circles, see Fig. 3.19b, $R =$ outer radius, $\varepsilon =$ inner radius, then the solution is

$$p(r) = \frac{\ln R}{\ln R - \ln \varepsilon} - \frac{\ln r}{\ln R - \ln \varepsilon}$$

If the inner circle has the radius $\varepsilon = 1$, then this simplifies to

$$p(r) = 1 - \frac{\ln r}{\ln R}$$

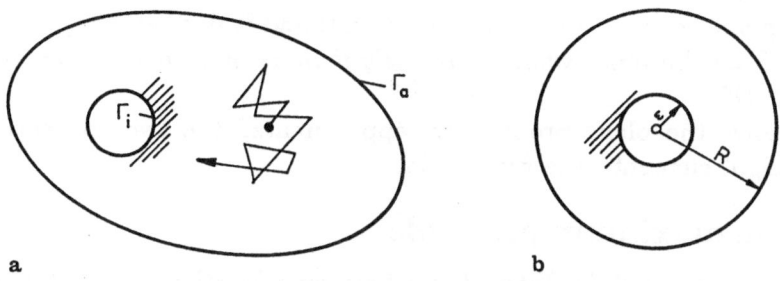

a   b

Figure 3.19 a-b. Brownian motion: a path of a molecule; b the model domain

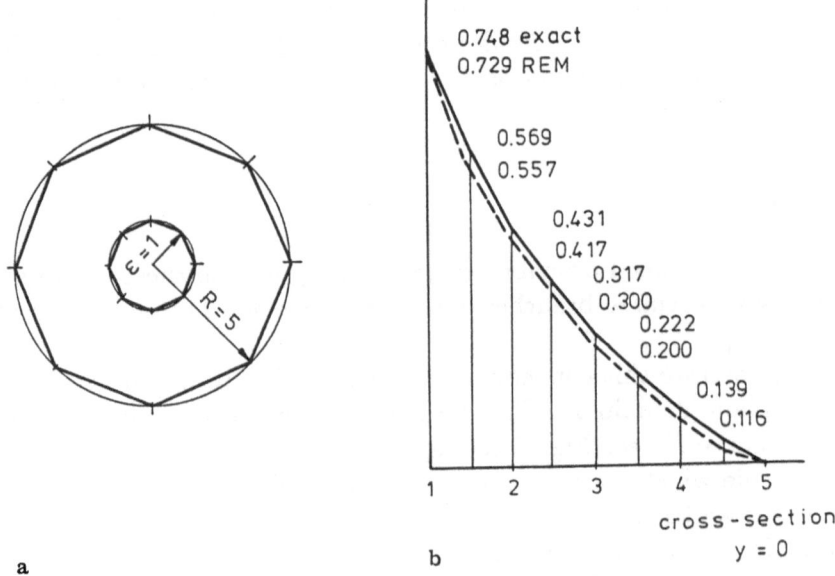

Figure 3.20 a-b. Solution of the model problem: a BE-approximation of the domain; b the exact and the approximate solution

Hence, if the radius R of the outer circle tends to infinity then the probability tends to 1. Wherever a molecule starts its random walk it will always hit the inner wall and *never* drift off.

In three-dimensional space the same problem has the solution

$$p(r) = (\frac{1}{R} - \frac{1}{r})/(\frac{1}{R} - \frac{1}{\varepsilon})$$

In the special case $\varepsilon = 1$ and $R \to \infty$ it becomes

$$p(r) = \frac{1}{r}$$

This is a different result as in 2-D. Now the probability is less than one (if the starting point does not lie on the inner wall) and it decreases rapidly with the distance from the inner wall. *"Obviously there is more room to escape in 3-D than in 2-D"*.

To solve the plane problem we approximated the two concentric circles with straight elements, see Fig. 3.20a.

3.10 The maximum principle

The maximum principle states that a harmonic function, $\Delta u = 0$, assumes its maximum value on the boundary. This implies that the error $u - u_h$ of a BE-

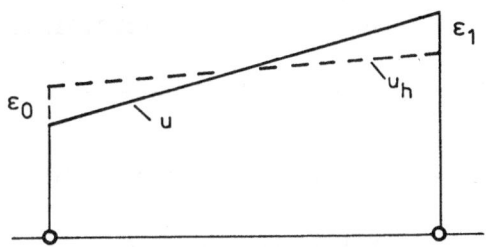

Figure 3.21 The error $u - u_h$ of the BE-solution has its maximum on the boundary.

solution has its maximum on the boundary: The BE-solution u_h, Eq.(3.10), of the boundary-value problem

$$-N\Delta u = p \quad \text{in } \Omega, \quad u = \bar{u} \quad \text{on } \Gamma_1, \quad t = \bar{t} \quad \text{on } \Gamma_2$$

satisfies the differential equation

$$-N\Delta u_h = p,$$

(if we neglect for once quadrature errors) and, hence, the error is a harmonic function

$$-N\Delta(u - u_h) = p - p = 0,$$

which takes on its maximum value on the boundary.

Figure 3.21 demonstrates this property with a small one-dimensional example. The differential equation $u'' = 0$ is the one-dimensional analog of the differential equation $\Delta u = 0$ and to solve the boundary value problem

$$u'' = 0, \quad u(0) = u_0, \quad u(1) = u_1$$

by the boundary element method means connecting the two points $u_0 + \varepsilon_0$ and $u_1 + \varepsilon_1$ with a straight line. The terms ε_0 and ε_1 represent the errors on the boundary which invariably creep in when the problem is to be solved on a two- or three-dimensional domain. Evidently the maximum error occurs on the boundary.

Corresponding maximum norm error estimates for the plate equation were given by Schulze and Wildenhain, [42], and for plates with pieceswise smooth boundaries by Maz'ja and Plamenevskij, [43].

Concerning elastic plates and bodies we refer to the contributions of Fichera, [44], and Adler, [45].

3.11 The influence function for the normal derivative

If we formulate Betti's principle with the fundamental solution

$$g_1(y, x) = N \frac{\partial}{\partial n_x} g_0(y, x) = -\frac{1}{2\pi r}(r_{,x_1} n_1 + r_{,x_2} n_2)$$

and the function $u(y) - u(x)$ then we obtain, after the usual limiting process

$$\lim_{\varepsilon \to 0} B(g_1[x], u(y) - u(x))_{\Omega_\varepsilon} = 0,$$

the expression

$$c_1(x)u_{,1}(x) + c_2(x)u_{,2}(x) = \int_\Gamma [g_1(y, x) N \frac{\partial u}{\partial \nu}(y)$$

$$- N \frac{\partial}{\partial \nu} g_1(y, x)(u(y) - u(x))] \, ds_y + \int_\Omega g_1(y, x) p(y) \, d\Omega_y, \quad (3.14)$$

where the functions c_i are the two characteristic functions

$$c_i(x) = \begin{cases} n_i, & x \in \Omega, \\ \dot{c}_i(x), & x \in \Gamma, \\ 0, & x \in \Omega^c, \end{cases}$$

whose boundary values are

$$\dot{c}_1(x) = \frac{1}{2\pi}[(\varphi + \frac{1}{2} \sin 2\varphi)n_1 + \sin^2 \varphi \, n_2]_{\varphi_2}^{\varphi_1},$$

$$\dot{c}_2(x) = \frac{1}{2\pi}[\sin^2 \varphi \, n_1 + (\varphi - \frac{1}{2} \sin 2\varphi)n_2]_{\varphi_2}^{\varphi_1},$$

Employing the unit matrix I and the matrix

$$J(\varphi) = \begin{bmatrix} 1/2 \sin 2\varphi & \sin^2 \varphi \\ \sin^2 \varphi & -1/2 \sin 2\varphi \end{bmatrix}$$

the boundary value of the left side of Eq.(3.14) can also be written as

$$\dot{c}_1(x)u_{,1}(x) + \dot{c}_2(x)u_{,2}(x) = \nabla u(x)^T \left\{ \frac{\Delta\varphi}{2\pi} I + \frac{1}{2\pi}(J(\varphi_1) - J(\varphi_2)) \right\} n$$

This very same matrix $J(\varphi)$ will also appear at the same spot in the formulation of the influence functions of plate flexure and plate stretching.

Note that, because the normal derivative of the function $u(\boldsymbol{y}) - u(\boldsymbol{x})$ is identical with that of the function $u(\boldsymbol{y})$, we subtracted from $u(\boldsymbol{y})$ the constant term $u(\boldsymbol{x})$. This manoeuvre facilitates the numerical handling of Eq.(3.14) because the strong singularity $O(r^{-2})$ of the kernel $\partial g_1 / \partial \nu$ is tamed by the artificially introduced zero of the layer $f(\boldsymbol{y}) = u(\boldsymbol{y}) - u(\boldsymbol{x})$.

3.12 Substructures

If the factor N in the differential equation $-N \Delta u = p$ is only piecewise constant as in Fig. 3.22 then we proceed as follows:

We first formulate the coupling condition for both subdomains separately

$$
\begin{bmatrix}
H_{11} & H_{12} & 0 & 0 \\
H_{21} & H_{22} & 0 & 0 \\
0 & 0 & H_{33} & H_{34} \\
0 & 0 & H_{43} & H_{44}
\end{bmatrix}
\begin{bmatrix}
u_1 \\ u_2 \\ u_3 \\ u_4
\end{bmatrix}
=
\begin{bmatrix}
G_{11} & G_{12} & 0 & 0 \\
G_{21} & G_{22} & 0 & 0 \\
0 & 0 & G_{33} & G_{34} \\
0 & 0 & G_{43} & G_{44}
\end{bmatrix}
\begin{bmatrix}
t_1 \\ t_2 \\ t_3 \\ t_4
\end{bmatrix}
$$

and, after considering the interface conditions

$$
u_2 - u_3 = 0, \qquad t_2 + t_3 = 0
$$

we add them together

$$
\begin{bmatrix}
H_{11} & H_{12} & 0 \\
H_{21} & H_{22} & 0 \\
0 & H_{33} & H_{34} \\
0 & H_{43} & H_{44}
\end{bmatrix}
\begin{bmatrix}
u_1 \\ u_2 \\ u_4
\end{bmatrix}
=
\begin{bmatrix}
G_{11} & G_{12} & 0 \\
G_{21} & G_{22} & 0 \\
0 & -G_{33} & G_{34} \\
0 & -G_{43} & G_{44}
\end{bmatrix}
\begin{bmatrix}
t_1 \\ t_2 \\ t_4
\end{bmatrix}
$$

Note that unlike in finite element methods the single blocks do not overlap but only slide beneath each other. This means that the number of equations does not diminish. Of all six vectors u_i and t_i only two are determined by boundary conditions (e.g. u_1 and t_4), so that exactly four unknown vectors must be determined by solving four matrix equations. Given a compound domain as in Fig. 3.23, the system of equations reads

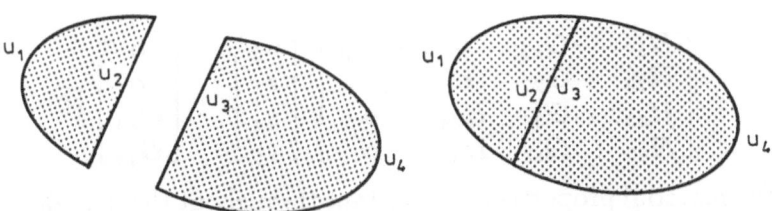

Figure 3.22 Two domains with different material properties

136

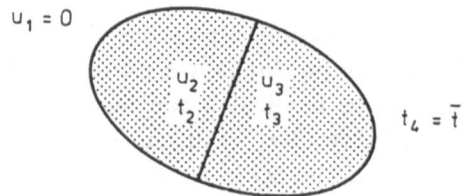

Figure 3.23 Boundary-value problem for a compound domain

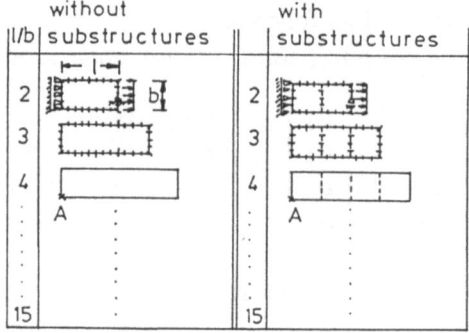

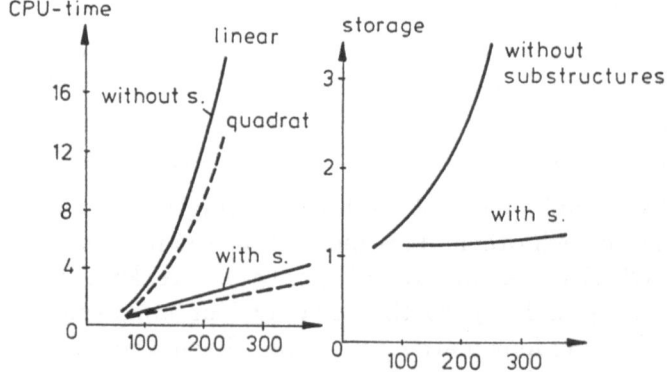

Figure 3.24 Substructures reduce the computer time

$$\begin{bmatrix} H_{12} & -G_{12} & o & -G_{11} \\ H_{22} & -G_{22} & o & -G_{21} \\ H_{33} & G_{33} & H_{34} & o \\ H_{43} & G_{43} & H_{44} & o \end{bmatrix} \begin{bmatrix} u_2 \\ t_2 \\ u_4 \\ t_1 \end{bmatrix} = \begin{bmatrix} o \\ o \\ G_{34} \\ G_{44} \end{bmatrix} [\bar{t}]$$

Even if the material properties are constant throughout the domain an artificial subdivision can be useful to give the system matrix a banded structure and to, thus, reduce the solution time, see Fig. 3.24, [46].

3.13 Alternatives to substructures

Substructures require rather complicated data structures in a boundary element code. We therefore want to discuss in this section two simpler methods to handle equations with discontinuous coefficients.

For our model problem we choose a membrane, see Fig. 3.25, which is uniformly stretched over the subdomain Ω_a by a force N_a and over the subdomain Ω_b by a different force N_b. The deflection u of the membrane satisfies the equations

$$-N_a \Delta u = p \quad \text{in } \Omega_a$$

$$-N_b \Delta u = p \quad \text{in } \Omega_b$$

$$u = 0 \quad \text{on } \Gamma$$

$$u_a - u_b = 0 \quad t_a + t_b = 0 \qquad \text{on } \Gamma_i \text{ (the interface)}$$

(If you have difficulties imagining how the prestressing force N in a membrane can be piecewise constant then you are free to switch to other models, as the temperature distribution in a compound domain with a piecewise constant thermal diffusity or to the model of a cross section with different elastic properties.)

To set the fundamental solutions of the two subdomains apart we label them as follows

$$g_0^a(y, x) = -\frac{1}{2\pi N_a} \ln r, \qquad g_0^b(y, x) = -\frac{1}{2\pi N_b} \ln r$$

and similarly the different parts of Γ,

$$\Gamma_a = \text{exterior boundary of } \Omega_a$$

$$\Gamma_b = \text{exterior boundary of } \Omega_b$$

$$\Gamma_i = \text{interface between } \Omega_a \text{ and } \Omega_b$$

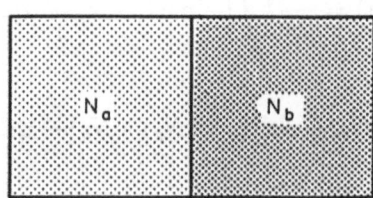

Figure 3.25 The coefficient N of the differential equation is different for the two subdomains

The influence of the Betti data of the domain Ω_a on a point x is

$$c_a(x)u(x) = \int_{\Gamma_a} [g_0^a t - N_a \frac{\partial}{\partial \nu} g_0^a u] ds y$$

$$+ \int_{\Gamma_i} [g_0^a t^a - N_a \frac{\partial}{\partial \nu_b} g_0^a u^a] ds y + \int_{\Omega_a} g_0^a p \, d\Omega y$$

(3.15)

and the influence of the Betti data of the domain Ω_b is

$$c_b(x)u(x) = \int_{\Gamma_b} [g_0^b t - N_a \frac{\partial}{\partial \nu} g_0^b u] ds y$$

$$+ \int_{\Gamma_i} [g_0^b t^b - N_b \frac{\partial}{\partial \nu_a} g_0^b u^b] ds y + \int_{\Omega_b} g_0^b p \, d\Omega y$$

(3.16)

where the functions $c_a(x)$ and $c_b(x)$ are the characteristic functions of Ω_a and Ω_b respectively. The indices on the normal derivative

$$\frac{\partial}{\partial \nu_a} \cdots \qquad \frac{\partial}{\partial \nu_b}$$

are to indicate the direction of the normal vector which points to Ω_a or Ω_b respectively. Due to these opposite directions and

$$N_a g_0^a = N_b g_0^b$$

we have

$$N_a \frac{\partial}{\partial \nu_b} g_0^a = -N_b \frac{\partial}{\partial \nu_a} g_0^b$$

We now multiply each equation with the pertinent stiffness, that is Eq.(3.15) with N_a and Eq.(3.16) with N_b, and we add the two equations. We thus obtain

$$(N_a c_a(x) + N_b c_b(x))u(x) = \frac{1}{2\pi} \{ \int_{\Gamma_a} [-\ln r \, t + N_a \frac{r_\nu}{r} u] ds y$$

$$+ \int_{\Gamma_b} [-\ln r \, t + N_b \frac{r_\nu}{r} u] ds y \int_{\Gamma_i} (N_a - N_b) \frac{r_\nu}{r} u \, ds y$$

(3.17)

$$+ \int_\Omega \ln r \, p \, d\Omega y \}$$

where it is understood that the normal vector on the interface points into Ω_b. At (smooth) points on the interface the left-hand side is

$$(N_a c_a(\boldsymbol{x}) + N_b c_b(\boldsymbol{x}))u(\boldsymbol{x}) = \frac{1}{2}(N_a + N_b)u(\boldsymbol{x})$$

Equation (3.17) is the influence function for the compound domain. The solution now also depends on the deflection u of the interface. To determine this function we only have to place additional collocation points at the interface. This simple extension makes, in our opinion, the alternative attractive and particularly well suited for small computers.

Originally we had the idea to simply add the two equations (3.15) and (3.16). It was only later that we learnt from colleagues the trick with the factor N. If we simply add the two equations then this leads to the expression

$$(c_a(\boldsymbol{x}) + c_b(\boldsymbol{x}))u(\boldsymbol{x}) = \int\limits_{\Gamma_a} [g_0^a t - N_a \frac{\partial}{\partial \nu} g_0^a u] ds y + \int\limits_{\Gamma_b} [g_0^b t - N_b \frac{\partial}{\partial \nu} g_0^b u] ds y$$

$$+ \int\limits_{\Gamma_i} g_0^i t \, ds y + \int\limits_{\Omega_a} g_0^a p \, d\Omega y + \int\limits_{\Omega_b} g_0^b p \, d\Omega y$$

$$(3.18)$$

where

$$g_0^i = -\frac{1}{2\pi N_i} \ln r, \qquad N_i = \frac{N_a N_b}{N_b - N_a}$$

and t on Γ_i stands for

$$t = t^a = -t^b$$

The unknown interface function is now the traction t and to find this function t we need the influence function for the traction.

If we denote the boundary integrals over the outer boundary Γ_a by

$$B^a(\boldsymbol{x}) = \int\limits_{\Gamma_a} [g_0^a t - N_a \frac{\partial}{\partial \nu} g_0^a u] ds y$$

and the domain integral by

$$P^a(\boldsymbol{x}) = \int\limits_{\Omega_a} g_0^a p \, d\Omega y$$

(the functions $B^b(\boldsymbol{x})$ and $P^b(\boldsymbol{x})$ have the analogous meaning) then the influence function for the traction t in the domain Ω_a is

$$t = N_a \frac{\partial u}{\partial n}(\boldsymbol{x}) = N_a \{B_n^a + B_n^b + P_n^a + P_n^b\} + \int\limits_{\Gamma_i} N_a \frac{\partial}{\partial n} g_0^i t \, ds y \qquad (3.19)$$

where the index n denotes the normal derivative. The last integral in Eq.(3.19), the interface integral,

$$I(x) = \int\limits_{\Gamma_i} N_a \frac{\partial}{\partial n} g_0^i t \, ds_y = -\frac{1}{2} \frac{N_b - N_a}{N_b} \int\limits_{\Gamma_i} \frac{1}{r} r_n t \, ds_y$$

is a potential of the second kind which is discontinuous at the interface. Its limit on approaching the interface from the left, from within Ω_a, is

$$\lim_{x \to \Gamma} I(x) = \frac{1}{2} \frac{N_b - N_a}{N_b} t(x) - \frac{1}{2} \frac{N_b - N_a}{N_b} \int\limits_{\Gamma_i} \frac{1}{r} r_n \, ds_y$$

All the other potentials in Eq.(3.19) are continuous at the interface so that, if we take the limit of both sides, we obtain

$$t(x) = \frac{1}{2} \frac{N_b - N_a}{N_b} t(x) + \dots$$

or

$$\frac{1}{2} \frac{N_b + N_a}{N_b} t(x) = \int\limits_{\Gamma_i} N_a \frac{\partial}{\partial n} g_0^i t \, ds_y + N_a \{B_n^a + B_n^b + P_n^a + P_n^b\} \qquad (3.20)$$

Equations (3.18) and (3.20) constitute a pair of coupled integral equations for the traction t on the outer boundary and the traction t at the interface.

The two approaches we introduced here differ by the unknown function at the interface. In the first case (multiply with N) the unknown is the deflection and in the second case (naive approach) the unknown is the traction. The advantage of the N-approach is that the integral equation on the interface is the same integral equation as on the outer boundary. Its disadvantage is that the traction at the interface must be calculated separately, i.e. you have to evaluate the influence function for t at points on the interface. So you run, only later, into the same difficulty as in the naive approach: you have to integrate the strongly singular integrals in the influence function for t analytically. In the naive approach the sequence is reversed: here the traction is a system variable, and to set up its integral equation you have to integrate the strongly singular kernels analytically, while the displacement at the interface must be evaluated afterwards separately by an influence function. Only that the kernels of this influence function are better behaved than those in the influence function for t. But the main and probably decisive disadvantage of the naive approach is that the interface integral in (3.20) does not exist at non-smooth points of the interface (if the layer has not a zero at the point), because it formulates a directional derivative. Though this can be overcome by introducing a double-node and by continuing the layers beyond this point by zero.

We applied the first technique, the N-approach, to a membrane which was stretched over a rectangular domain, see Fig. 3.26, and where the tension varied from subdomain to subdomain according to the ratios $N_1 : N_2 : N_3 = 3 : 1 : 3$, the lateral pressure being $p = 10$ units. The boundary was divided into 32 quadratic elements and each interface into 4 such elements. The two dashed lines in Fig. 3.26 indicate the deflection of the membrane if the tension N were the same in all three subdomains: the upper line corresponds to a uniform tension $N = 3$ and the lower line to a uniform tension $N = 1$.

To verify the second technique we applied it to the membrane in Fig. 3.27. At the corner point we had to introduce a double node to circumvent the singularity of the normal derivative of the interface potential. The values in parentheses were obtained with finite elements.

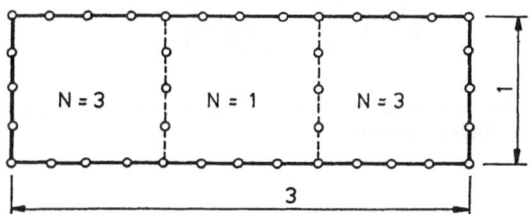

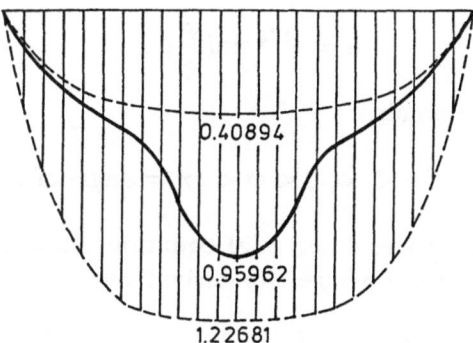

Figure 3.26 The deflection of the membrane

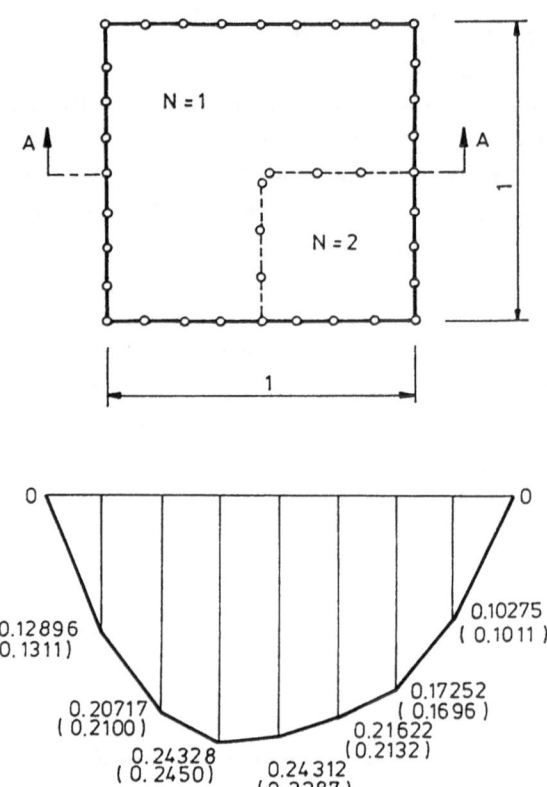

Figure 3.27 At the interior corner point a double node was introduced to sidestep the singularity of the double-layer potential.

3.14 Singularities

The function

$$u = r^{1/q} \sin(\frac{\varphi}{q})$$

is in the wedge-shaped domain

$$\Omega = \{(r, \varphi) \mid 0 < r < \infty,\ 0 < \varphi < q\pi = \text{angle of the wedge}\}$$

a homogeneous solution of the differential equation $\Delta u = 0$ and has the value $u = 0$ on the boundary of the wedge. Its derivatives

$$u_{,i} = \frac{1}{q} r^{1/q - 1} \left[\sin(\frac{\varphi}{q}) \mp \cos(\frac{\varphi}{q}) \frac{x_i}{r} \right]$$

are of the order $O(r^{1/q-1})$. Hence, the regularity of this function depends on the parameter q and, therefore, on the angle $q\pi$ of the wedge. If the angle is less than $180°$ $(q < 1)$ then the tangent at u at the point $r = 0$ has a horizontal direction (like the edge of a bath tub), if the angle is greater than $180°$ $(q > 1)$, then the tangent has a vertical direction (like the edge of a canyon), see Fig. 3.28. An angle 2π corresponds to a slit; in that case the singularity is of the order $O(r^{-0.5})$.

This dependancy on the size of the angle is typical for the solutions of elliptic problems. To cope with these singularities different strategies are possible:

a) *refinement of the boundary element mesh,*
b) *introduction of singular elements, see* [47],[48]
c) *subtraction of the singularity,*
d) *application of special fundamental solutions.*

As a model problem to test the strategies a,b and c we consider a membrane with a slit, see Fig. 3.29. The deflection of this membrane is the function that satisfies the equations

$$\Delta u = 0 \qquad \text{in } \Omega,$$

$$u = 0 \qquad \text{along HA (upper part of left edge) ,}$$

$$u = 1000 \qquad \text{along AE (lower part of left edge) ,}$$

$$t = 0 \qquad \text{on the rest of the boundary.}$$

Due to the symmetry of the problem the function $u - 500$ is antimetric with respect to the horizontal axis AB and therefore u is in the lower half of the square the solution of the problem

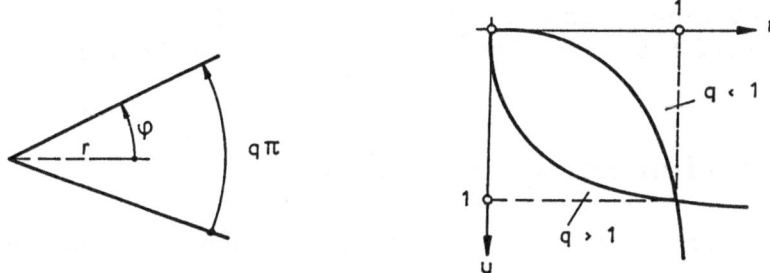

Figure 3.28 The regularity of the solution depends on how the function approaches zero

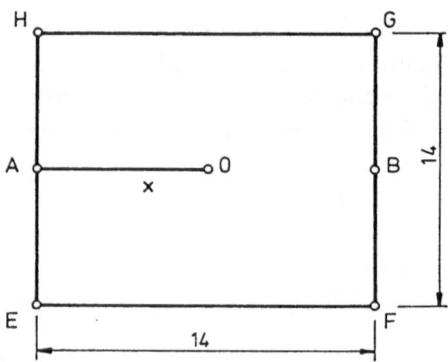

Figure 3.29 A membrane with a slit

$$\Delta u = 0 \qquad \text{on EFBOA}$$

$$u = 1000 \qquad \text{along AE (lower part of the left edge),}$$

$$u = 500 \qquad \text{along BO (to the right of the slit),}$$

$$t = 0 \qquad \text{on the remaining boundary.}$$

At the point x on the slit which has a distance $r = 0.75$ units from the tip of the slit the "exact" solution has the value $u = 634.45$. If the single boundary segments are subdivided in 5 elements each the bottom edge is subdivided into 10 elements the different methods render the following results

a_0	a_1	b	c	"exact"
618.62	632.02	634.85	634.45	634.45

The values a_0 and a_1 were obtained with the program BE-LAPLACE. The first value a_0 (5 elements model along the slit) is interpolated because no node coincides with the point x. The second value a_1 is based on a subdivision of the slit into 9 elements and the result is the value of u at the point $r = 0.77$ units (instead of 0.75 units). The numbers b and c were obtained by Atkinson, [49].

3.15 Three-dimensional problems

The integral representation of the solution of the differential equation

$$-\Delta u = -(u_{,11} + u_{,22} + u_{,33}) = p$$

is

$$c(\boldsymbol{x})u(\boldsymbol{x}) = \int\limits_{\Gamma} [\, g_0(\boldsymbol{y}, \boldsymbol{x}) \frac{\partial u}{\partial \nu}(\boldsymbol{y}) - \frac{\partial}{\partial \nu} g_0(\boldsymbol{y}, \boldsymbol{x}) u(\boldsymbol{y})] \, ds_{\boldsymbol{y}} + \int\limits_{\Omega} g_0(\boldsymbol{y}, \boldsymbol{x}) p(\boldsymbol{y}) \, d\Omega_{\boldsymbol{y}} \,,$$

where

$$g_0(\boldsymbol{y}, \boldsymbol{x}) = \frac{1}{4\pi r} \,, \qquad \frac{\partial}{\partial \nu} g_0(\boldsymbol{y}, \boldsymbol{x}) = -\frac{1}{4\pi r^2} r_\nu \,,$$

and

$$c(\boldsymbol{x}) = \begin{cases} 1, & \boldsymbol{x} \in \Omega \,, \\ \Delta\varphi(\boldsymbol{x})/4\pi \,, & \boldsymbol{x} \in \Gamma \,, \\ 0, & \boldsymbol{x} \in \Omega^c \end{cases}$$

The term $\Delta\varphi(\boldsymbol{x})$ is the angle of the boundary point.

The influence function of the gradient reads

$$c_{ij}(\boldsymbol{x})u_{,j}(\boldsymbol{x}) = \int\limits_{\Gamma} [\frac{1}{4\pi r^2} r_{,i} \frac{\partial u}{\partial \nu}(\boldsymbol{y}) + \frac{1}{4\pi r^3}(3\, r_{,i}\, r_{,j}$$

$$- \delta_{ij})\nu_j(u(\boldsymbol{y}) - u(\boldsymbol{x}))] \, ds_{\boldsymbol{y}} + \int\limits_{\Omega} \frac{1}{4\pi r^2} r_{,i}\, p(\boldsymbol{y}) \, d\Omega_{\boldsymbol{y}} \,,$$

with

$$c_{ij}(\boldsymbol{x}) = \begin{cases} \delta_{ij} \,, & \boldsymbol{x} \in \Omega \,, \\ \dot{c}_{ij}(\boldsymbol{x}) \,, & \boldsymbol{x} \in \Gamma \,, \\ 0, & \boldsymbol{x} \in \Omega^c \,, \end{cases}$$

The boundary values $\dot{c}_{ij}$ of the characteristic functions c_{ij} are the integrals

$$\dot{c}_{ij}(\boldsymbol{x}) = \frac{1}{4\pi} \int\limits_{S_1(\boldsymbol{x}, \Omega)} 3\, r_{,i}\, r_{,j} \, dS_1$$

where $S_1(\boldsymbol{x}, \Omega)$ is that portion of the unit sphere, which lies inside the tangent cone.

Analytical expressions for the singular integrals can be found in [50].

For an application we consider an example from electrostatics, see Fig. 3.30, which was solved with the program BEASY. The problem was to find the potential Φ that takes on the values + 1 and - 1, respectively, on the surfaces of the two tori and which satisfies the equation

$$\Delta\Phi = 0$$

at all other points.

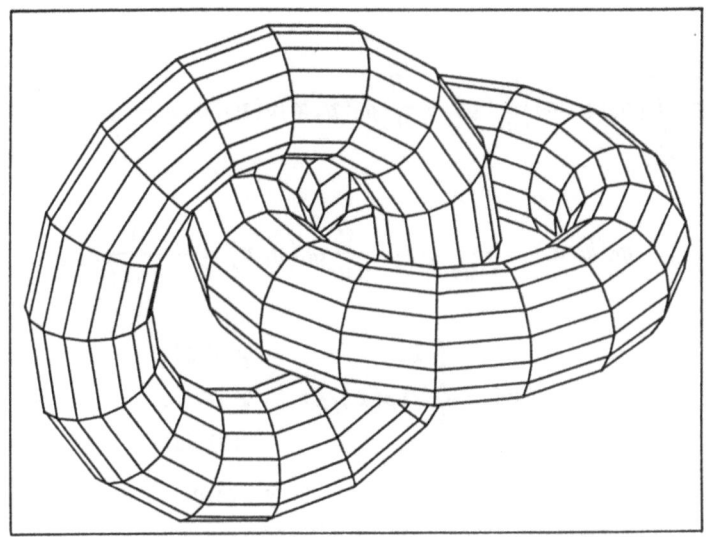

Figure 3.30 Two tori with surface potentials Φ of + 1 and - 1 respectively

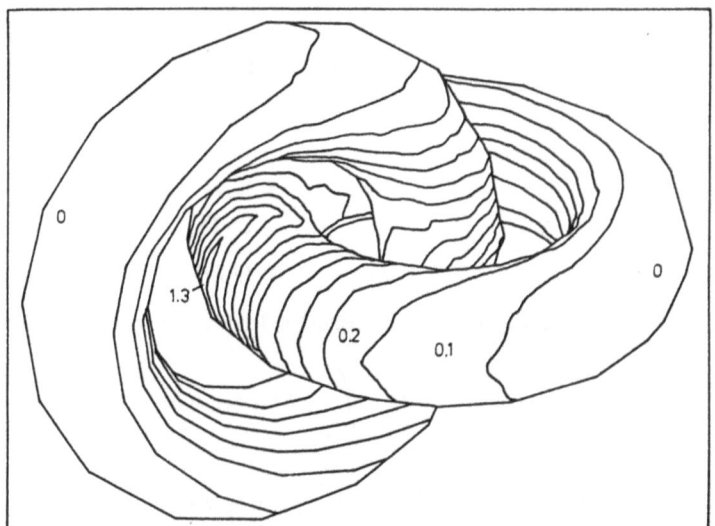

Figure 3.31 The iso-lines of the electric field $\boldsymbol{E}$

The interesting quantity is the gradient of the solution. It represents the electric field of the potential

$$E = -\nabla\Phi$$

The intensity of the field has its peak where the distance between the two tori is at its minimum, see Fig. 3.31.

Exercises

1. Show that the fundamental solution of the Laplacian, $g_0(y, x) = -1/(2\pi) \ln r$, satisfies the equation $\Delta g_0 = 0$. Hint: use

$$\Delta = \frac{\partial^2}{\partial r^2} + \frac{1}{r^2} \frac{\partial^2}{\partial \varphi^2} + \frac{1}{r} \frac{\partial}{\partial r} \qquad \text{(in polar coordinates)}.$$

2. Weighted residual. Let the function u be the solution of the Poisson equation $-\Delta u = f$ in Ω. Let $g_0 = -1/(2\pi) \ln r$ be the fundamental solution of the Laplacian. Start with the weighted residual formulation on the punctured domain $\Omega_\varepsilon = \Omega - N_\varepsilon(x)$, see Fig. 3.32a,

$$\int_{\Omega_\varepsilon} (-\Delta u - f) g_0[x] d\Omega_y = 0$$

and derive the integral representation (3.10) of u. Hint: apply integration by parts twice to the domain integral

$$\int_{\Omega_\varepsilon} -\Delta u \, g_0[x] d\Omega_y = 0$$

and then let the radius ε tend to zero. Recall that

$$\lim_{\varepsilon \to 0} \int_{\Gamma_{N_\varepsilon}} \frac{\partial}{\partial \nu} g_0 \, u(y) \, ds_y = c(x) u(x).$$

Why must you exclude an ε-neighborhood of the point x from the domain Ω?

3. Indirect method. Consider the boundary value problem

$$-\Delta u = 0 \quad \text{in} \quad \Omega \qquad u = 1 \quad \text{on} \quad \Gamma$$

on a square domain Ω. The solution is $u(x) = 1$. We want to obtain an approximate solution by placing a series of point sources on a circle outside of the problem domain, see Fig. 3.32b.

$$u(x) = \sum_{i=1}^{n} g_0(y_i, x) P_i \qquad g_0(y, x) = -\frac{1}{2\pi} \ln r .$$

a) How would you determine the size of the single point sources P_i?

b) Would they all have the same size?

c) Would the size of the point sources decrease or increase if you move the circle farther away?

148

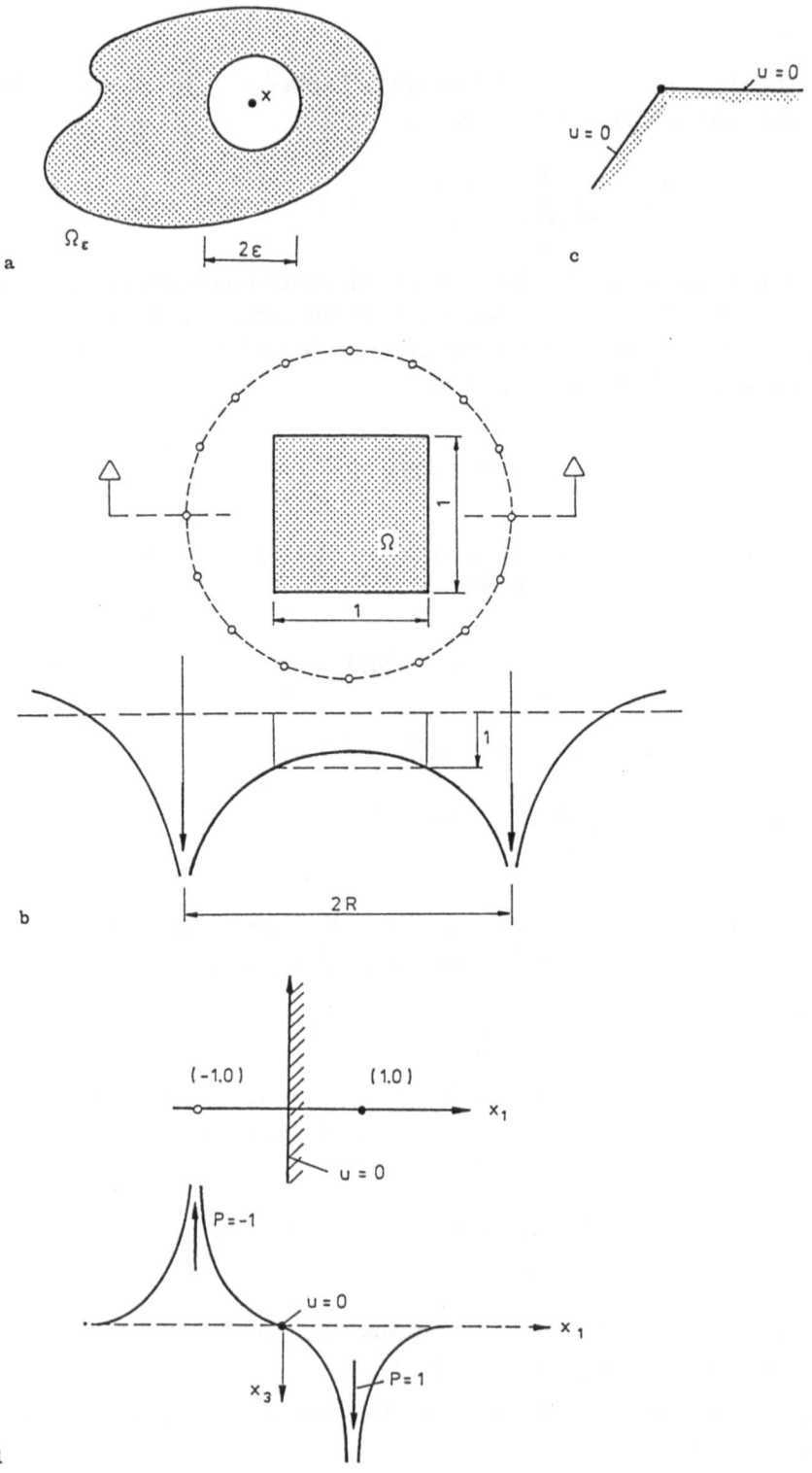

d) If all the point sources were of the same magnitude P what must be the size of P if the value of u at the centre of the rectangular domain is to be exact, $u = 1$.

4. If the potential u is zero at the two sides of a corner point, then the normal derivatives $\partial u/\partial n$ on both sides are zero too. Proof? (Hint: first show that the tangential derivatives $\partial u/\partial t$ are zero on both sides; this implies that the gradient $\nabla u = \{u_{,1}, u_{,2}\}$ is a zero vector (why?) and this in turn implies that the normal derivatives are zero on both sides, see Fig. 3.32c).

5. Mirror points. Think you have to find a potential field u in the right half-plane, $x_1 \geq 0$, which corresponds to a point source $P = 1$ located on the horizontal x_1-axis at $x_1 = 1$ and which vanishes on the vertical x_2-axis, $u = 0$. Show that you obtain a solution if you place a second point source of the same magnitude but with opposite sign, $P = -1$, at the 'mirror point' $x = (-1, 0)$, see Fig.3.32d. What sign should the auxiliary source have if the tractions (the normal derivative) were to vanish on the vertical axis? (Hint: $t = 0$ means that the slope is zero).

6. [102] p. 133. Using the program BE-LAPLACE solve the example in Fig. 3.33a with 12 quadratic elements and prescribing values of the potential u on the whole boundary. Can you guess the solution? (Hint: try a linear function $u(x) = a + bx_1 + cx_2$). If you have found the solution check that the results for the fluxes (normal derivatives) at the corners are very good.

7. A concrete slab is heated by three electric wires, see the cross-section in Fig. 3.33b. Model these three wires as three heat sources of magnitude, say, 10 and calculate the temperature distribution in the cross-section with the program BE-LAPLACE.

8. Calculate the temperature distribution in the living room, see Fig. 3.18, with the program BE-LAPLACE. Use different number of boundary elements and study how the solution is influenced by the discretisation. Distribute some light-bulbs over the room (heat sources) and calculate the temperature distribution.

9. Calculate the torsional rigidity I_t of the cross-section in Fig. 3.33c with the program BE-LAPLACE. First solve the boundary value problem

$$-\Delta u = 2 \qquad u = 0 \quad \text{on } \Gamma$$

Figure 3.32 a-d Exercises: **a** Punctured domain; **b** approximation of the unit deflection $u = 1$ in a square region Ω by the combined efforts of n point loads; **c** zero potential u at a corner implies zero slope; **d** two opposite point loads make that $u = 0$.

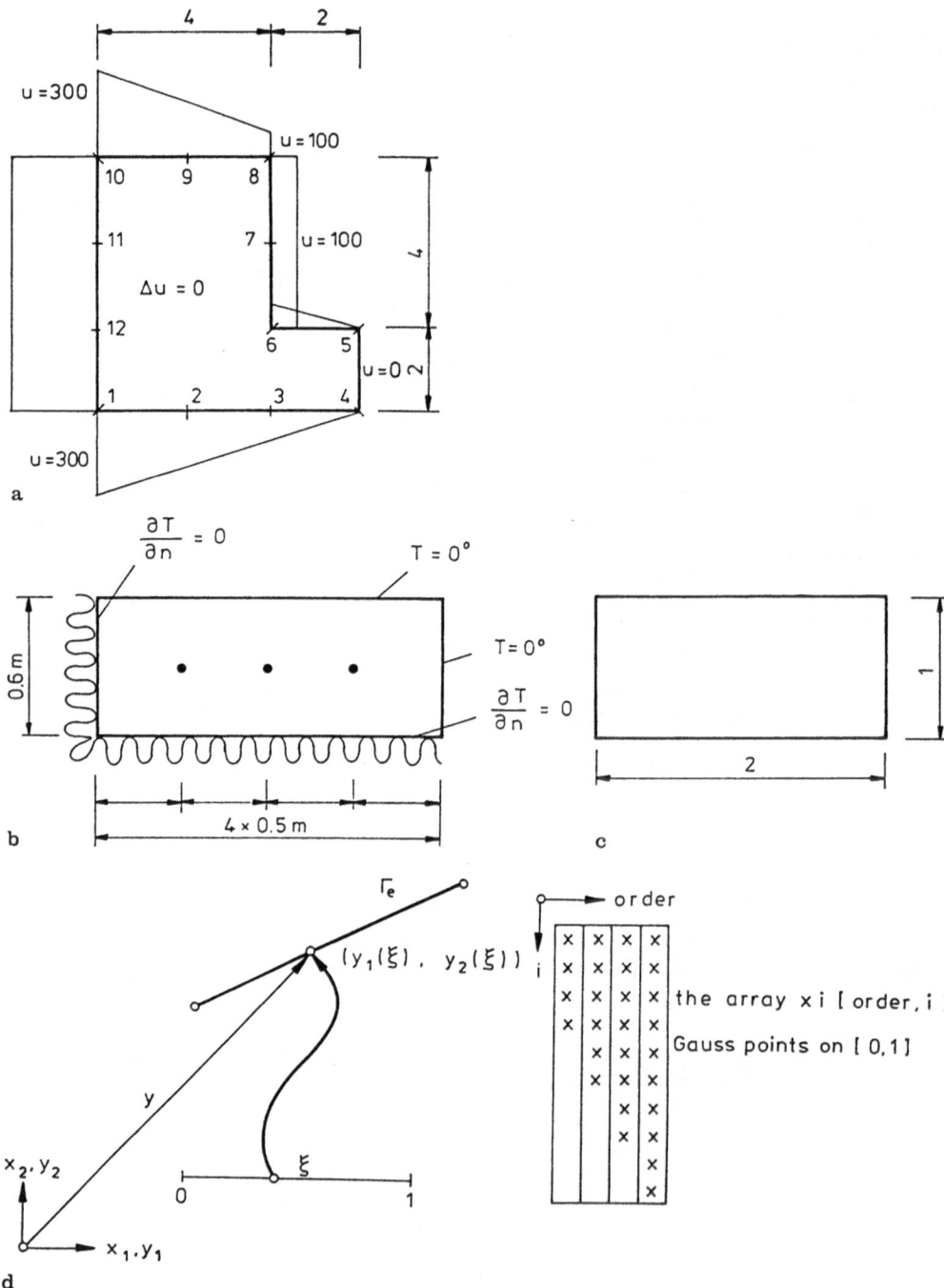

Figure 3.33 a-e Exercises: **a** distribution of the potential u along the edge; **b** cross-section of a concrete slab with heating cables; **c** cross-section of a beam; **d** master element and boundary element; **e** the array xi[order,i] of the Gauss points;

and then calculate the volume under the membrane.

$$I_t = 2 \int\limits_{\Omega} u d\Omega$$

The exact value is 0.458.

10. In the BEM the integrals are of the type

$$\int\limits_{\Gamma} k(\boldsymbol{y}, \boldsymbol{x}) \varphi(\boldsymbol{y}) ds\boldsymbol{y} \qquad (3.21)$$

All integrations are done on a master element, e.g. the interval [0,1]. The Gauss points and weights are usually given in the literature for the interval $[-1, 1]$. The transformation onto the unit-interval is achieved by

$$\xi_i = 0.5(1 + \xi_i) \qquad w_i = 0.5\, w_i$$

To be flexible we let the number of Gauss points vary with the distance of the source point $\boldsymbol{x}$ from Γ_e so that the points and weights of the different orders constitute two two-dimensional arrays xi[order, i] and w[order, i].

To transform an integral as (3.21) onto the master element we identify the points ξ of the master element with the points $\boldsymbol{y} = (y_1, y_2)$ of the boundary element Γ_e by two coordinate functions $y_1(\xi)$ and $y_2(\xi)$, see Fig. 3.33d. Write a subroutine which calculates for a straight boundary element given by its two end-points these two functions and which provides two two-dimensional arrays y_1[order, i] and y_2[order, i] which for any order of quadrature list the coordinates y_i of the Gauss-points at which the integrand must be evaluated on the element Γ_e. If the source point $\boldsymbol{x}$ is close to Γ_e then we shall use many points, say 10, and if it is not so close 8, then perhaps 6 and if it is far away 4 points. These are the different orders, see Fig. 3.33e.

Write a subroutine which provides the values of the three quadratic shape functions φ_i at the quadrature points of the master element in the form of three two-dimensional arrays phi_1[order, j], phi_2[order, j], phi_3[order, j].

Write a subroutine which you call with the coordinates of the source point $\boldsymbol{x}$ and the end points of the boundary element and which (i) determines the necessary order of quadrature and which (ii) provides two one-dimensional arrays G0[i] and G0NUE[i] that contain the values of the kernel functions g_0 and $\partial g_0 / \partial \nu$ at the Gauss points ξ_i of the master element. (Simply substitute the functions y_i into the kernel functions).

Finally write a subroutine which calculates the integral

```
integral := 0;
for i := 1 to number_of_Gauss_points do
```

```
integral := integral + G0[i]*phi_1[order,i]*weight[order,i]
integral := integral * length_of_element
```

You can save one multiplication if you introduce new weights

```
new_weight_1[order,i] := phi_1[order,i]*weight[order,i]
```

These three arrays new_weight_i[order,i] need only to be initialised once, at the program start. You can further speed things up if you add the terms 'by hand'

```
integral := G0[1]*new_weight_1[order,1] + G0[2]*new_weight_1[order,2]
          + G0[2]*new_weight_1[order,3] + ...
```

This you must do in your program naturally for each order as the number of Gauss-points is different for each order. Try this!

11. A point load $P = 1$ at the centre of a circular membrane of radius $R = 1$ will cause the deflection $u = -1/(2\pi N)\ln r$ (N is the tensile force within the membrane). Calculate the tensile forces (traction) $t = N\partial/\partial n\, u$ on the edge of the membrane and show that the tractions balance the exterior load

$$\int_\Gamma t\, ds = 1.$$

Show that the strain energy induced by the concentrated force

$$\frac{1}{2}\int_\Omega (u^2{}_{,1} + u^2{}_{,2})d\Omega = \frac{1}{2}\int_\Omega \frac{1}{r^2}d\Omega y$$

is infinite (Hint: $d\Omega = rdrd\varphi$). Note that the 'energy balance' *interior work = exterior work* still holds as also the exterior work of the concentrated force is infinite; its path extends to infinity because the deflection under the point force is infinite

$$\text{exterior work} = -\ln(r = 0) \times 1 = \infty \times 1.$$

Show that the deflection of the membrane is finite everywhere if a line load acts on the membrane. (Hint: for simplicity assume that the load is distributed along the interval $[0, R]$ of the x_1-axis and let the load be constant).

12. A membrane is stretched over the unit square and then blown up by a uniform pressure of magnitude 10, see Fig. 3.34a,

$$-\Delta u = 10 \qquad u = 0 \qquad \text{on} \quad \Gamma.$$

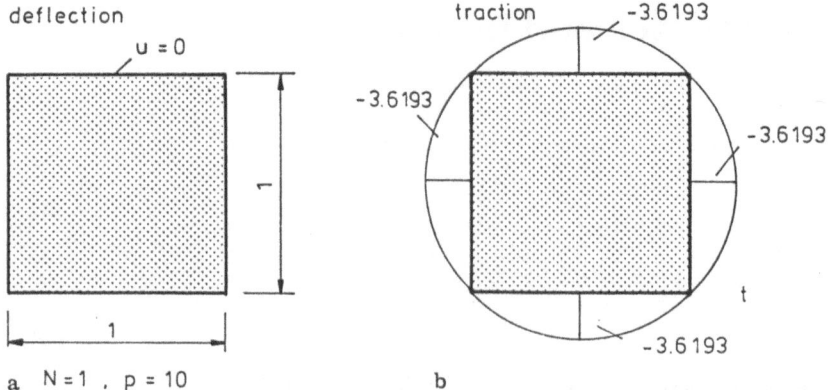

Figure 3.34 a-b Exercises: a square membrane with zero edge deflection; b distribution of tractions (slope) on the boundary;

An approximation with one quadratic element on each side has rendered the tractions in Fig. 3.34b. (The bubble function is the function φ_2^e). Substitute these tractions and the zero edge deflection, $u = 0$, into the influence function and calculate the deflection at the centre of the membrane

$$u(x) = \int_\Gamma g_0(y, x) t(y)\, ds_y + \int_\Omega g_0(y, x) p\, d\Omega_y$$

(Hint: transform the domain integral into an equivalent boundary integral and exploit the symmetry of the problem). A discretization with 10 quadratic elements on each side has rendered for the deflection u at the centre the (probably exact) value $u = 0.73671$. How close is your one-element result to this value?

4 Elastic plates and bodies

In this chapter we use the boundary element method to calculate the displacements and stresses within elastic plates and bodies.

4.1 Introduction

The points $\boldsymbol{x} = (x_1, x_2)$ of a plate have two degrees of freedom so that the displacement field of a plate

$$\boldsymbol{u}(\boldsymbol{x}) = \{u_1(\boldsymbol{x}), u_2(\boldsymbol{x})\}^T,$$

has a horizontal component, u_1, and a vertical component, u_2. The derivatives of this field determine the strains

$$\varepsilon_{11} = u_{1,1}, \qquad \varepsilon_{12} = \frac{1}{2}(u_{1,2} + u_{2,1}), \qquad \varepsilon_{22} = u_{2,2}$$

and the stresses

$$\sigma_{11} = 2\mu\,\varepsilon_{11} + \frac{2\mu\nu}{1 - 2\nu}(\varepsilon_{11} + \varepsilon_{22}), \qquad \sigma_{12} = 2\mu\,\varepsilon_{12},$$

$$\sigma_{22} = 2\mu\,\varepsilon_{22} + \frac{2\mu\nu}{1 - 2\nu}(\varepsilon_{11} + \varepsilon_{22})$$

within the material. The relation between the elastic constants

$$\nu = \text{Poisson's ratio},$$

$$\mu = G = \text{shear modulus},$$

and Young's modulus is

$$E = 2\mu(1 + \nu)$$

The term that is conjugated to the displacement vector u is the traction vector $t = \{t_1, t_2\}^T$. The traction vector is the product between the stress tensor S and the normal vector $n = \{n_1, n_2\}^T$ on the boundary

$$\sigma_{11}\, n_1 + \sigma_{12}\, n_2 = t_1\,,$$

$$\sigma_{21}\, n_1 + \sigma_{22}\, n_2 = t_2$$

The boundary conditions of a plate are formulated either in terms of displacements u_i or tractions t_i. Along the stress free edge of the cantilever plate in Fig. 4.1a the two components t_i of the traction vector must be zero and where the exterior shear forces are distributed it must hold that

$$t_1 = 0\,, \quad t_2 = -1000$$

The shear force is negative because its direction is opposite to the positive axis x_2.

These traction boundary conditions are supplemented by geometric boundary conditions

$$u_1 = u_2 = 0$$

at the support.

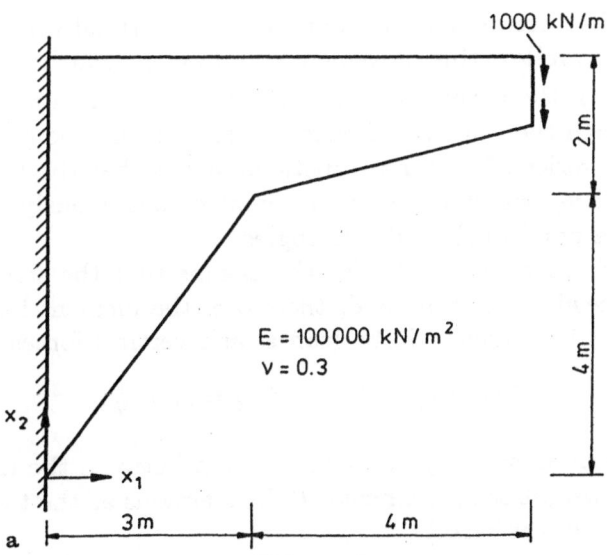

Figure 4.1a Elastic plate

Before we discuss the BE-approach let us recall how we determine the stresses within the cantilever plate with finite elements. First we subdivide the plate into, say, linear triangles. Each triangle is the base of three linear shape functions φ_i^e which in unison with the functions of the neighboring elements form a set of n hat-functions φ_i. The single hat-function has at the node x^i the value 1 and it is zero at all other nodes, see Fig. 3.2. With these n scalar-valued hat-functions φ_i we can now easily construct a set of $2n$ vector-valued displacement fields if we put the functions φ_i intermittently at the first and the second place

$$\boldsymbol{\Phi}_1 = \begin{bmatrix} \varphi_1 \\ 0 \end{bmatrix}, \quad \boldsymbol{\Phi}_2 = \begin{bmatrix} 0 \\ \varphi_1 \end{bmatrix}, \quad \boldsymbol{\Phi}_3 = \begin{bmatrix} \varphi_2 \\ 0 \end{bmatrix}, \quad \boldsymbol{\Phi}_4 = \begin{bmatrix} 0 \\ \varphi_2 \end{bmatrix}, \quad \cdots$$

Thus all odd coefficients in the FE-approximation

$$\boldsymbol{u}_h(\boldsymbol{x}) = u_i \boldsymbol{\Phi}_i(\boldsymbol{x}),$$

correspond to horizontal displacements and all even coefficients to vertical displacements. We, so to speak, expand the displacement field into a series of n purely horizontal and n purely vertical fields. A horizontal field is followed by a vertical field, this in turn by a horizontal field, etc.

The single displacement field $\boldsymbol{\Phi}_i$ resembles the shape of the plate if we push the node x^j

$$j = \begin{cases} (i+1)/2 & \text{if } i \text{ odd}, \\ i/2 & \text{if } i \text{ even} \end{cases}$$

by one unit of displacement into a horizontal or vertical direction and if the neighboring points follow this movement to an extent which is proportional to their distance from the active point x^j. Only those elements will be displaced which connect with the node x^j. Hence, approximating the displacement field of the plate by a series of such hat-functions means that the two displacement functions, u_1 and u_2, which originally resembled continuously curved surfaces, are replaced by a patchwork of flat triangles.

From a mechanical point of view this means that the stresses of the FE-solution are piecewise constant and, therefore, the element-loads zero. Only along the element boundaries do horizontal and vertical forces act

$$t_{\Delta 1} = t_1^l + t_1^r, \qquad t_{\Delta 2} = t_2^l + t_2^r,$$

which are responsible, so to speak, for the jump between the traction vectors, $\boldsymbol{t}^l$ and $\boldsymbol{t}^r$, of two neighboring elements. (The tractions at the two faces of a cut are equal if their sum is zero.)

The nodal displacements u_i of the FE-approximation are determined by minimizing the potential energy of the plate

$$\Pi_1(u_h) = \frac{1}{2}E(u_h, u_h) - \int_\Gamma \bar{t} \cdot u_h \, ds$$

The boundary integral in this expression is the work done by the exterior forces at the extreme end of the cantilever plate

$$\bar{t} = \{\bar{t}_1, \bar{t}_2\}^T = \{0, -1000\}^T$$

while the first integral, the *energy product*

$$E(u, u) = \int_\Omega S \cdot E \, d\Omega = \int_\Omega \sigma_{ij} \, \varepsilon_{ij} \, d\Omega,$$

is the scalar product between the stress tensor and the strain tensor of the field u. Because the energy product is a bilinear form we can expand the energy of the FE-approximation with respect to the energy products of its (vector-valued) basis functions

$$\frac{1}{2}E(u_h, u_h) = \frac{1}{2}E(u_i\Phi_i, u_j\Phi_j) = \frac{1}{2}u_i E(\Phi_i, \Phi_j)u_j = \frac{1}{2}u^T K u,$$

where

$$K_{ij} = E(\Phi_i, \Phi_j)$$

is the single element of the stiffness matrix.

The external work of the boundary forces is a linear form of the nodal displacements

$$\int_{\Gamma_L} \bar{t} \cdot u_h \, ds = \int_{\Gamma_L} \bar{t} \cdot u_i \Phi_i \, ds = f^T u$$

where

$$f_i = \int_{\Gamma_L} \bar{t} \cdot \Phi_i \, ds,$$

and therefore the potential energy of the FE-approximation can be expressed as

$$\Pi_1(u_h) = \frac{1}{2}u^T K u - f^T u$$

The requirement that the potential energy becomes a minimum leads to the system of equations

$$K_{ij} u_j = f_i, \qquad i = 1, 2, \ldots, 2n$$

158

and with regard to these equations it now holds that the *left side* of the single equation i is the energy product (= virtual strain energy) $E(\boldsymbol{\Phi}_i, \boldsymbol{u}_h) = \delta W_i$ between the FE-solution $\boldsymbol{u}_h$ and the displacement field $\boldsymbol{\Phi}_i$. Because the FE-solution is an equilibrium solution the virtual strain energy is equal to the virtual external work δW_e which is done by the FE-forces when they act through $\boldsymbol{\Phi}_i$.

The left side of equation i (up to now we only talked about this side) is also equal to the *right side*, that is the virtual external work f_i of the forces $\bar{t}$,

$$K_{ij}u_j = E(\boldsymbol{\Phi}_i, \boldsymbol{u}_h) = \delta W_e = f_i \,,$$

so that obviously the FE-solution is so tuned that the external work of its exterior forces is equal to the external work of the true exterior forces $\bar{t}$ with respect to all virtual displacements $\boldsymbol{\Phi}_i$,

$$\delta W_e(\boldsymbol{u}_h, \boldsymbol{\Phi}_i) = \delta W_e(\boldsymbol{u}, \boldsymbol{\Phi}_i)$$

The exterior forces of the FE-solution are the horizontal and vertical forces $t_{\Delta 1}$ and $t_{\Delta 2}$, which act along the element boundaries and the shear force at the end of the plate. As the interelement forces should tend to zero if the FE-solution is to converge the size of these forces can serve as an *error indicator*, see Fig. 4.1b.

How do we solve the same problem with boundary elements? The BE-solution is a solution in terms of the two influence functions for the horizontal and vertical displacements of a plate

$$u_i(\boldsymbol{x}) = \int_{\Gamma} [\, U_{ij}(\boldsymbol{y}, \boldsymbol{x}) t_j(\boldsymbol{y}) - T_{ij}(\boldsymbol{y}, \boldsymbol{x}) u_j(\boldsymbol{y})]\, ds_{\boldsymbol{y}} \,, \qquad (4.1)$$

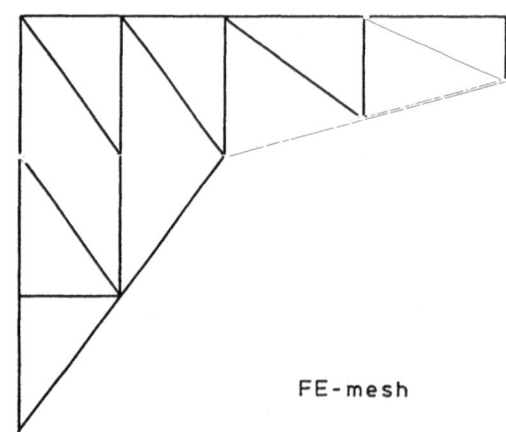

FE-mesh

b

Figure 4.1b FE-model of the plate

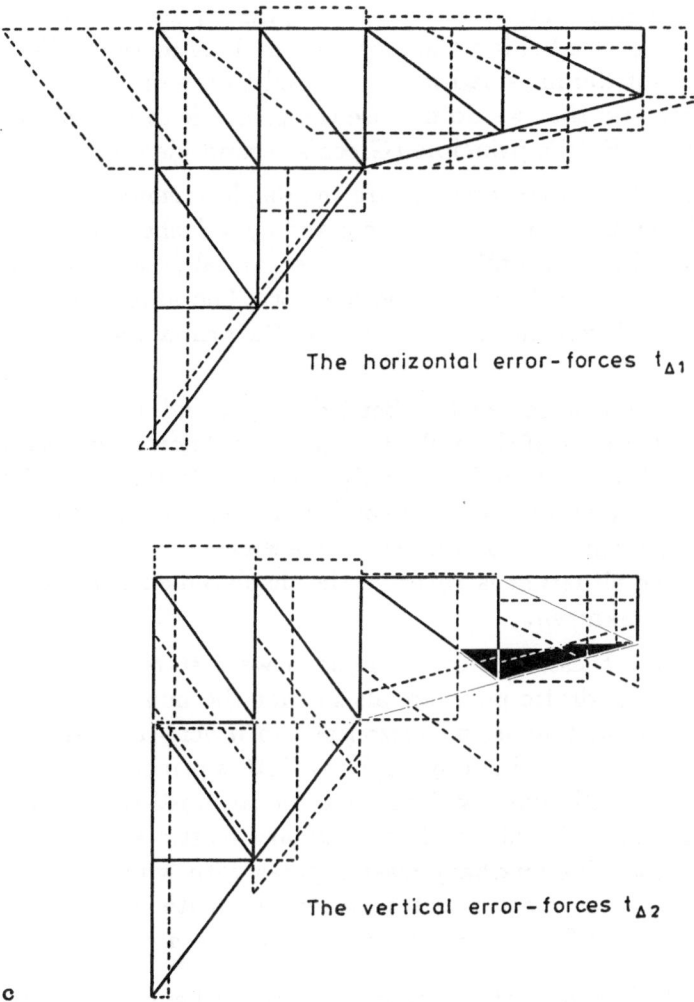

The horizontal error-forces $t_{\Delta 1}$

The vertical error-forces $t_{\Delta 2}$

c

Figure 4.1c The constant error forces are plotted parallel to the mesh lines they refer to. The magnitude of these forces is an indicator for the accuracy of the FE-solution. On the fixed edge the error forces are identical with the support reactions and at the free end they are identical with the exterior shear forces.

These functions are based on Betti's principle,

$$1 \times \text{displacement} = \text{displacement} \times \text{force} - \text{force} \times \text{displacement}$$

In mechanical terms Eq.(4.1) states that the displacement u_i at a point x can be obtained by calculating the reciprocal work, *singular loadcase × real loadcase*, of a singular elastic state and the primary state. The idea, basically, is the same as the idea behind the unit-dummy-load method. The difference is only that the unit-dummy-load is not applied to the original structure but to an infinite plate and that we do not use the principle of virtual forces (Green's first identity) but Betti's principle (Green's second identity).

The capital letters U_{ij} and T_{ij} are the displacements and the tractions at the edge of the embedded plate Ω, indicated by a dashed line, which belong to the elastic state in the infinite plate if a horizontal ($i = 1$) or vertical ($i = 2$) concentrated force $\hat{P} = 1$ acts at the point $\boldsymbol{x}$. The lower terms u_j and t_j are the boundary displacements and tractions of the primary system, the cantilever plate.

While the distribution of the Betti data U_{ij} and T_{ij} of the fundamental solutions along the edge of the embedded plate are known the four Betti data u_j and t_j of the real plate are only partially known; only two of the four terms are determined by support and loading conditions. At the free edge zero tractions t_j are prescribed but the conjugated displacements u_j are unknown. At the support level the situation is reverse: the displacements u_j are zero and the tractions t_j are unknown.

To determine these unknown functions we first replace all functions by, say, n piecewise quadratic polynomials so that the number of unknown nodal values shrinks to $2n$, that is, n horizontal and n vertical nodal values. These unknowns are then determined by applying Betti's principle or rather enforcing Betti's principle: the boundary data are to be so tuned that $2n$ tests of Betti's principle with the real state and $2n$ auxiliary elastic states render $2n$-times the correct result. The auxiliary elastic states with which we test the real or primary state are the elastic states an infinite plate is in if a horizontal or vertical concentrated force $\hat{P} = 1$ acts at one of the n collocation points.

Figure 4.2 illustrates the necessary steps we have to take if we want to determine the horizontal component, t_1^2, of the traction vector at node 2.

First we apply a concentrated horizontal force $\hat{P} = 1$ at node 2 of the infinite plate. This force causes displacements U_{1j} and tractions T_{1j} along the dashed boundary. Then we calculate the reciprocal work of these terms and the conjugated boundary terms of the real plate. Because the reciprocal work must be the same, $W_{1,2} = W_{2,1}$, we can solve this equation for t_1^2 and, therewith, t_1^2 is determined.

Naturally, the other nodal values t_i^j are unknown too and we therefore will have to repeat this procedure $2n$-times so that we finally obtain a whole system of equations. Note also that Fig. 4.2 depicts only the distribution of the horizontal components. The reciprocal work of the vertical components must also be considered. Only then is Betti's principle complete.

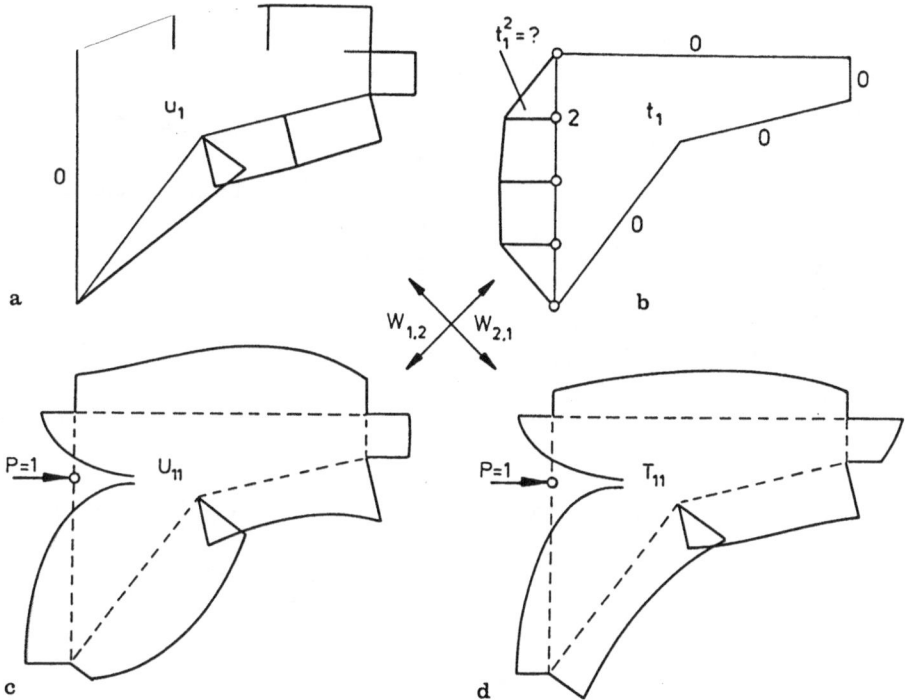

Figure 4.2 a-d. The cutout and the real plate: **a-b** the horizontal components on the edge of the real plate; **c-d** the same quantities on the edge of the cutout

But if we have done this, if we have determined all the $2n$ nodal values, then we can obtain with the influence function (4.1) the displacements and stresses in the interior at any point.

Summary: To calculate the displacement field with finite elements means to approximate the surfaces u_1 and u_2 by a patchwork of flat triangles. In mechanical terms such an approximation corresponds to a loadcase where, see Fig. 4.1b, horizontal and vertical line forces are acting along the element boundaries.

Unlike the FE-displacements, which are piecewise linear, the displacements u_1 and u_2 of the BE-solution are smooth functions. The same holds true for the stresses so that there is no reason to "to explain" the jumps in the stresses with the action of line forces. The BE-solution (4.1) also satisfies the homogeneous differential equation (4.4). But while the FE-solution satisfies the geometric boundary conditions exactly, the BE-solution violates these conditions and the statical boundary conditions as well. But according to St. Venant's principle we may hope that these errors will not propagate too far into the interior.

The stress distribution in the cantilever plate as calculated with the program BE-PLATES is shown in Fig. 4.3. It is interesting to observe that the

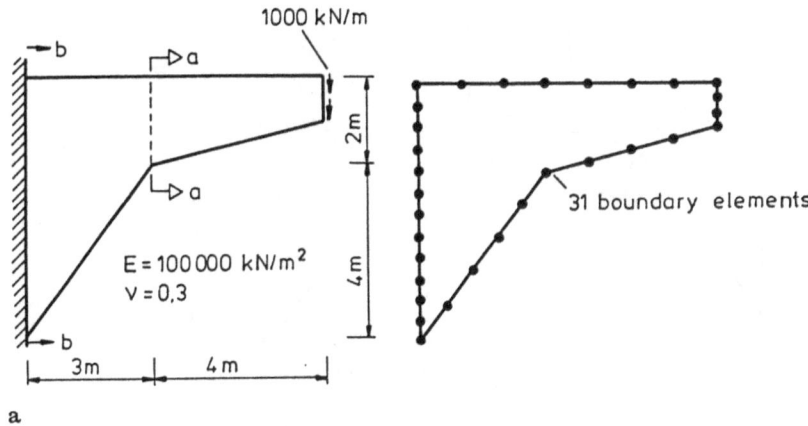

a

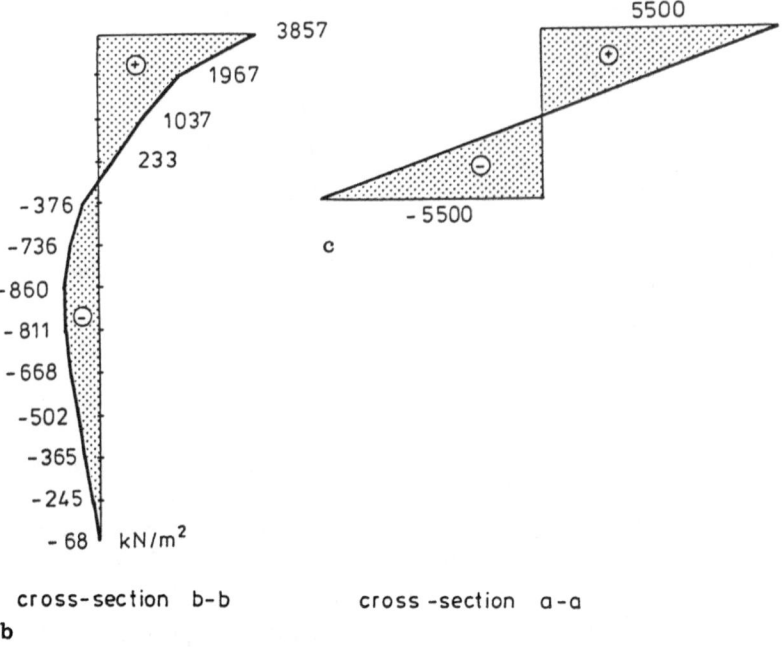

b

cross-section b-b   cross-section a-a

Figure 4.3 The BE-solution of the plate problem

stress distribution at the support, see Fig. 4.3b follows the plate theory while at the cross section b–b, see Fig. 4.3c, the stress distribution is linear as in the beam theory.

After these introductory remarks the underlying principles will be explained now in a more systematic fashion.

4.2 The influence functions

The elastic state of a body is characterized by its displacement field $u = \{u_i\}$, the strain tensor $E = [\varepsilon_{ij}]$ and the stress tensor $S = [\sigma_{ij}]$. These quantities satisfy the 15 equations

$$\frac{1}{2}(u_{i,j} + u_{j,i}) - \varepsilon_{ij} = 0, \qquad \text{(6 Eqs.)}, \qquad (4.2)$$

$$\frac{2\mu\nu}{1-2\nu}\varepsilon_{kk}\delta_{ij} + 2\mu\varepsilon_{ij} - \sigma_{ij} = 0, \qquad \text{(6 Eqs.)}, \qquad (4.3)$$

$$-\sigma_{ij,j} = p_i \qquad \text{(3 Eqs.)}. \qquad (4.4)$$

These equations are equivalent to a system of three differential equations for the displacement field $u = \{u_i\}$ alone,

$$-L_{ij}u_j := -\mu u_{i,jj} - \frac{\mu}{1-2\nu}u_{j,ji} = p_i, \qquad (4.5)$$

which is supplemented by geometric boundary conditions

$$u_i = \bar{u}_i$$

on a part Γ_1 of the boundary and statical boundary conditions

$$t_i = \sigma_{ij}n_j = \bar{t}_i$$

on the complementary part, Γ_2, of the boundary.

The derivation of the influence functions for the displacement field is an application of *Betti's principle*,

p: $\hat{u}, u \in C^2(\bar{\Omega})$,

q: $B(\hat{u}, u) = \int\limits_{\Omega} -L_{ij}\hat{u}_j u_i \, d\Omega + \int\limits_{\Gamma} [\hat{t}_i u_i - \hat{u}_i t_i] \, ds - \int\limits_{\Omega} \hat{u}_i(-L_{ij}u_j) \, d\Omega = 0$

In the following we shall substitute for $\hat{u}$ the fundamental solutions (*Kelvin solutions*) and for u the actual displacement field. The fundamental solutions represent the displacements at a point $y = (y_1, y_2, y_3)$ of the elastic continuum

$$U_{ij}(y, x) = \frac{1}{8\pi\mu(1-\nu)r}[(3 - 4\nu)\delta_{ij} + r_{,i}r_{,j}], \qquad (4.6)$$

if a concentrated force $\hat{P} = e_i$ is acting at some distant point $x = (x_1, x_2, x_3)$.

The first index i indicates the direction of the action at the source point x and the second index j the direction of the displacement at the observation point y.

The functions U_{ij} are symmetric with respect to x and y (this is *Maxwell's principle*) and they depend on the distance $r = |y - x|$ of the two points and on the spherical angles φ, ϑ under which we see the point y from the point x.

$$r_{,1} = \frac{y_1 - x_1}{r} = \cos\varphi\sin\vartheta, \quad r_{,2} = \frac{y_2 - x_2}{r} = \sin\varphi\sin\vartheta,$$

$$r_{,3} = \frac{y_3 - x_3}{r} = \cos\vartheta$$

In a state of plane strain the equations of the fundamental solution are

$$U_{ij}(y, x) = \frac{1}{8\pi\mu(1 - \nu)}[(3 - 4\nu)\ln\frac{1}{r}\delta_{ij} + r_{,i}\,r_{,j}]$$

where it is understood that the indices i, j run from 1 to 2. In 2-D the derivatives $r_{,i}$ of the distance r read

$$r_{,1} = \frac{y_1 - x_1}{r} = \cos\varphi, \quad r_{,2} = \frac{y_2 - x_2}{r} = \sin\varphi$$

The traction vector at a point y of a cross section (endowed with the normal vector $\nu = \{\nu_1, \nu_2, \nu_3\}^T$) of the loadcase $P = e_i$, has the three components

$$T_{ij}(y, x) = -\frac{1}{4\alpha\pi(1 - \nu)r^\alpha}[\frac{\partial r}{\partial\nu}((1 - 2\nu)\delta_{ij} + \beta r_{,i}\,r_{,j})$$

$$- (1 - 2\nu)\{r_{,i}\,\nu_j(y) - r_{,j}\,\nu_i(y)\}] \quad (4.7)$$

where $\alpha = 1$, $\beta = 2$ in 2-D and $\alpha = 2$, $\beta = 3$ in 3-D.

Figure 4.4 illustrates the situation that a concentrated horizontal force $P = e_1$ acts at the point x. This unit force will cause at some distant point y horizontal, U_{11}, and vertical, U_{12}, displacements and stresses T_{11} and T_{12}, see Fig. 4.4b. The latter depend on the direction of the cut at the point y, that is the normal vector ν at y.

To calculate with Betti's principle and the fundamental solutions the displacements at a point x we, first, formulate the second identity in the punctured domain Ω_ε and we then let the radius ε tend to zero, $B_\varepsilon \to B$, see [51].

In terms of mechanics the following happens:

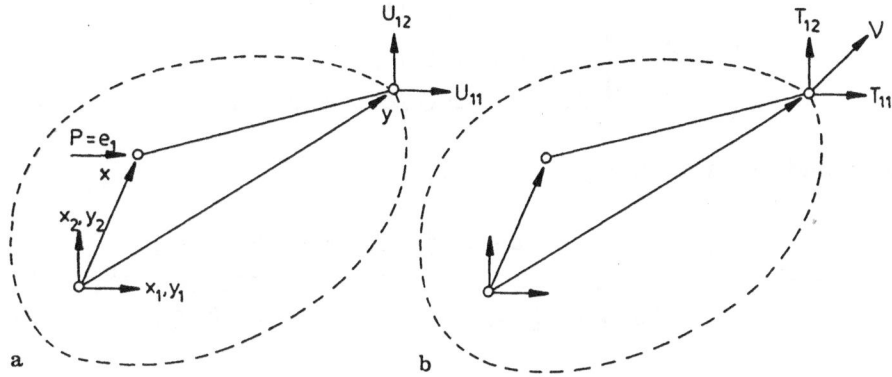

Figure 4.4 Attack of a concentrated horizontal force at the point x

a) We apply at the point x of the elastic continuum a concentrated force $\hat{P} = e_i$, see Fig. 4.5a.

b) The domain of the continuum that coincides in shape and extension with the real body is cut out of the continuum and so the prior internal tractions T_{ij} on the surface of the subdomain beome external tractions.

c) Analogously we separate the real body from its supports and let the support reactions become external tractions.

Both bodies are identical in shape and both are in a state of equilibrium so that according to Betti's principle we must have

$$W_{1,2} = C_{ij}(\boldsymbol{x})u_j(\boldsymbol{x}) + \int_\Gamma T_{ij}(\boldsymbol{y},\boldsymbol{x})u_j(\boldsymbol{y})\,ds\boldsymbol{y}$$

$$= \int_\Omega U_{ij}(\boldsymbol{y},\boldsymbol{x})p_j(\boldsymbol{y})\,d\Omega\boldsymbol{y} + \int_\Gamma U_{ij}(\boldsymbol{y},\boldsymbol{x})t_j(\boldsymbol{y})\,ds\boldsymbol{y} = W_{2,1}\ ,$$

or, if we solve this expression for $C_{ij}(\boldsymbol{x})u_j(\boldsymbol{x})$

$$C_{ij}(\boldsymbol{x})u_j(\boldsymbol{x}) = \int_\Gamma [U_{ij}(\boldsymbol{y},\boldsymbol{x})t_j(\boldsymbol{y}) - T_{ij}(\boldsymbol{y},\boldsymbol{x})u_j(\boldsymbol{y})]\,ds\boldsymbol{y}$$

$$+ \int_\Omega U_{ij}(\boldsymbol{y},\boldsymbol{x})p_j(\boldsymbol{y})\,d\Omega\boldsymbol{y} \tag{4.8}$$

This is the so-called *Somigliana identity*, the integral representation of the displacement field of an elastic body. The functions $C_{ij}(\boldsymbol{x})$ are the entries of the characteristic matrix

$$C_{ij}(\boldsymbol{x}) = \begin{cases} \delta_{ij}, & \boldsymbol{x} \in \Omega, \\ \dot{C}_{ij}(\boldsymbol{x}), & \boldsymbol{x} \in \Gamma, \\ 0, & \boldsymbol{x} \in \Omega^c \quad \text{(complement)} \end{cases}$$

whose boundary values are $(a = 1 - 2\nu)$,

$$\dot{C}(\boldsymbol{x}) = \frac{1}{8\pi(1-\nu)} \int\limits_{S_1(\boldsymbol{x},\Omega)} \begin{bmatrix} a + 3r_{,1}^2 & 3r_{,1}\,r_{,2} & 3r_{,1}\,r_{,3} \\ & a + 3r_{,2}^2 & 3r_{,2}\,r_{,3} \\ \text{sym.} & & a + 3r_{,3}^2 \end{bmatrix} dS_1 \qquad (4.9)$$

These integrals are to be taken over that section $S_1(\boldsymbol{x}, \Omega)$ of the unit sphere S_1 which lies inside the tangent cone, see section 1.1. If the surface is smooth at $\boldsymbol{x}$ then the value of the C-matrix is just $1/2 \times$ unit matrix

$$\dot{C}_{ij}(\boldsymbol{x}) = \frac{1}{2}\delta_{ij} \quad \text{(at a smooth point)}$$

In 2-D the functions $\dot{C}_{ij}$ are the elements of the matrix

$$\dot{C}(\boldsymbol{x}) = \frac{\Delta\varphi}{2\pi} I + \frac{1}{4\pi(1-\nu)}\{J(\varphi_1) - J(\varphi_2)\},$$

where $J(\varphi)$ is the matrix

$$J(\varphi) = \begin{bmatrix} \frac{1}{2}\sin 2\varphi & \sin^2\varphi \\ \text{sym.} & \frac{1}{2}\sin 2\varphi \end{bmatrix}$$

and where φ_1 and φ_2 denote the angles between the x_1-axis and the tangents t_1 and t_2 at the boundary point $\boldsymbol{x}$, see Fig. 3.6.

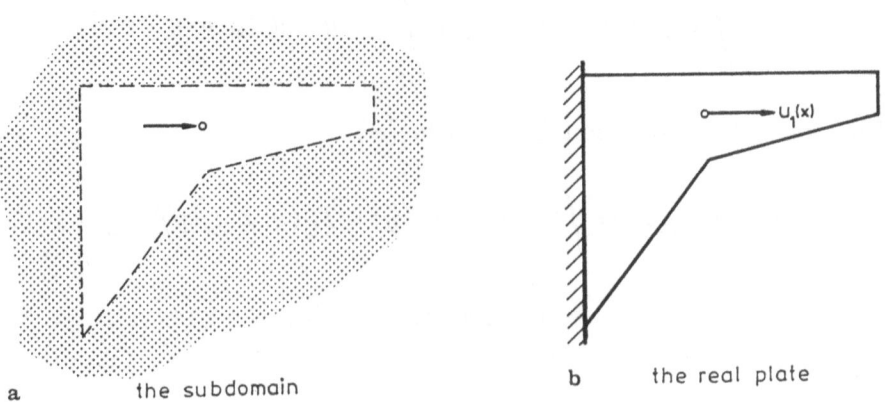

a   the subdomain   b   the real plate

Figure 4.5 a-b. Application of Betti's principle: **a** the cutout with the concentrated force; **b** the real plate and the unknown horizontal displacement

At smooth boundary points the expression simplifies to

$$\dot{C}(\boldsymbol{x}) = \frac{1}{2} \boldsymbol{I}$$

Because the Somigliana identity applies also if the displacement fields are simple translations such as $\boldsymbol{u} = \{1,0,0\}^T$, $\boldsymbol{u} = \{0,1,0\}^T$, etc., and because such translations leave the body stressfree we must have

$$C_{ij}(\boldsymbol{x}) = - \int\limits_{\Gamma} T_{ij}(\boldsymbol{y}, \boldsymbol{x}) \, ds_{\boldsymbol{y}} \tag{4.10}$$

so that the expression

$$\int\limits_{\Gamma} T_{ij}(\boldsymbol{y}, \boldsymbol{x})(u_j(\boldsymbol{y}) - u_j(\boldsymbol{x})) \, ds_{\boldsymbol{y}} = \int\limits_{\Gamma} U_{ij}(\boldsymbol{y}, \boldsymbol{x}) t_j(\boldsymbol{y}) \, ds_{\boldsymbol{y}}$$

$$+ \int\limits_{\Omega} U_{ij}(\boldsymbol{y}, \boldsymbol{x}) p_j(\boldsymbol{y}) \, d\Omega_{\boldsymbol{y}}$$

is identical with Eq.(4.8) the C-matrix has dropped out.

The Somigliana identity is the integral representation of the displacement field $\boldsymbol{u}(\boldsymbol{x})$ of an elastic body in terms of the displacements u_j and tractions t_j on the surface and the volume forces p_j within the body. If we place the source point $\boldsymbol{x}$ on the surface then the Somigliana identity formulates a coupling condition between the boundary values u_j and t_j which allows us to determine the unknown boundary terms.

4.3 Discretization

The coupling condition (4.8) represents a system of three integral equations for three unknown boundary functions and as the surface consists of infinitely many points $\boldsymbol{x}$ it represents a system of $3 \times$ infinitely many linear equations.

To solve this coupling condition approximately we interpolate the Betti data by piecewise constant, linear or quadratic functions, see Fig. 3.13 and determine their $3n$ unknown nodal values by satisfying the integral equations at n collocation points.

The global basis functions $\varphi_i(\boldsymbol{x})$ and $\psi_i(\boldsymbol{x})$, which interpolate the Betti data

$$u_j(\boldsymbol{x}) = u_j^i \varphi_i(\boldsymbol{x}), \quad t_j(\boldsymbol{x}) = t_j^i \psi_i(\boldsymbol{x}),$$

have finite support, that is only on a small part of the surface they are different from zero. The typical function $\varphi_i(\boldsymbol{x})$ has at the node $\boldsymbol{x}^i$ the value 1 and it is zero at all other nodes.

$$\varphi_i(\boldsymbol{x}^{\,j}) = \delta_{ij}$$

A boundary displacement field with three such identical components represents a "unit displacement" of the surface: the node $\boldsymbol{x}^{\,i}$ slides along the diagonal of the unit cube while all the other nodes stay in place remain fixed.

Similarly, the functions $\psi_i(\boldsymbol{x})$ represent unit tractions. Unlike the displacements φ_i these functions can be discontinuous because we must be prepared to interpolate the traction vector also at the edges of a body. This we do with two functions, ψ_j and ψ_{j+1}. If the approximation is linear then the two functions ψ_j and ψ_{j+1} are the two halves of a hat-function, see Fig. 4.6.

Because we interpolate all the three displacements $u_j(\boldsymbol{x})$ and all the three tractions $t_j(\boldsymbol{x})$ at a node by the same scalar-valued basis function φ_i and ψ_i, respectively, (if t_1 is discontinuous then so too are t_2 and t_3), the interpolation of the boundary functions is of the form

$$\boldsymbol{u}(\boldsymbol{x}) = \varphi_i(\boldsymbol{x})\,\boldsymbol{u}^{\,i}, \qquad \boldsymbol{t}(\boldsymbol{x}) = \psi_i(\boldsymbol{x})\,\boldsymbol{t}^{\,i} \qquad \text{(scalar} \times \text{vector)}$$

where

$$\boldsymbol{u}^{\,i} = \{u_1^i, u_2^i, u_3^i\}^T, \qquad \boldsymbol{t}^{\,i} = \{t_1^i, t_2^i, t_3^i\}^T$$

are the vectors of nodal values. On substituting these expressions into the Somigliana identity it follows

$$C_{mn}(\boldsymbol{x}^k)u_n^k + \int_\Gamma T_{mn}(\boldsymbol{y}, \boldsymbol{x}^k)\varphi_i(\boldsymbol{y})\,ds_{\boldsymbol{y}}\,u_n^i = \int_\Gamma U_{mn}(\boldsymbol{y}, \boldsymbol{x}^k)\psi_i(\boldsymbol{y})\,ds_{\boldsymbol{y}}\,t_n^i$$

$$+ \int_\Omega U_{mn}(\boldsymbol{y}, \boldsymbol{x}^k)p_n(\boldsymbol{y})\,d\Omega_{\boldsymbol{y}}\,.$$

(no sum over k in the first term)

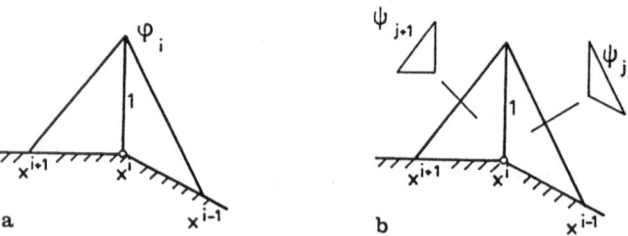

Figure 4.6 a-b. Interpolation at a corner: a continuous interpolation of the displacements; b discontinuous interpolation of the traction with the two halves of a hat function

where summation is done over i and n. The index k is the index of the collocation point and the index m is the number of the integral equation, the first, second, or third equation.

In a matrix notation these three equations read

$$C(x^k)u^k + \int_\Gamma T(y,x^k)\varphi_i(y)\,ds_y\,u^i = \int_\Gamma U(y,x^k)\psi_i(y)\,ds_y\,t^i$$

$$+ \int_\Omega U(y,x^k)p(y)\,d\Omega_y \quad (4.11)$$

If we introduce the matrices (3×3)

$$H_{ki} = C(x^k)\delta_{ki} + \int_\Gamma T(y,x^k)\varphi_i(y)\,ds_y\,,$$

$$G_{ki} = \int_\Gamma U(y,x^k)\psi_i(y)\,ds_y$$

and the vector (3 components)

$$p_k = \int_\Omega U(y,x^k)p(y)\,d\Omega_y$$

then we can write for these equations

$$H_{ki}\,u^i = G_{ki}\,t^i + p_k$$

This coupling condition must be formulated at all n nodes, $k = 1, 2\ldots,n$, so that we finally obtain a whole system of equations

$$Hu = Gt + p$$

where H, G, u, t and p denote appropriate *hyper-matrices* and *hyper-vectors*. (A hyper-matrix is a matrix whose entries are matrices and a hyper-vector is a vector whose components are vectors.)

If the displacement field $u(x)$ represents a translation of the kind $u = \{1,1,1\}^T$, then the hyper-vectors p and t are zero vectors and it follows

$$Hu = o$$

Hence, it must hold

$$H_{kk} = -\sum_{\substack{i=1 \\ i \neq k}} H_{ki}$$

This means that the element H_{kk} on the diagonal of the hyper-matrix H can be calculated by summing over the off-diagonal terms. This equation corresponds to Eq.(4.10).

4.4 Element matrices for plates

To the construction of the global basis functions φ_i and ψ_i from local basis functions φ_i^e corresponds the assemblage of the global matrices H and G from element matrices H^e and G^e.

If we use quadratic elements then the element matrices G^e have three columns each of which contains K blocks of size 2×2, see Fig. 4.7a. The first block in the upper left corner contains the four integrals

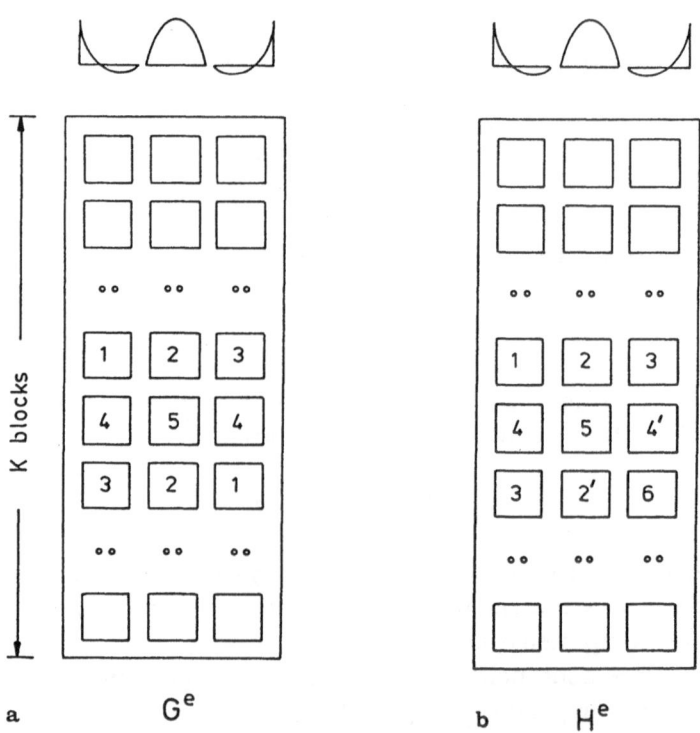

Figure 4.7 The element matrices G^e and H^e

$$\int_{\Gamma_e} U_{ij}(\boldsymbol{y}, \boldsymbol{x}^k) \varphi_1^e(\boldsymbol{y}) \, ds_y, \quad i,j = 1,2,$$

which represent the influence which the element layer φ_1^e exerts on the collocation point $\boldsymbol{x}^k = \boldsymbol{x}^1$ according to the weight of the four kernels U_{ij}. The neighboring block to the right represents the influence which the second element layer, φ_2^e, exerts on the point $\boldsymbol{x}^1$ and the third block represents the influence of the third element layer, φ_3^e. If we move one block down in $\boldsymbol{G}^e$ then we move to the next collocation point $\boldsymbol{x}^k$.

The numbered blocks contain the singular integrals: if the source point $\boldsymbol{x}$ is either the first point, $\boldsymbol{x} = \boldsymbol{x}^a$, the mid-point, $\boldsymbol{x} = \boldsymbol{x}^m$, or the end point $\boldsymbol{x} = \boldsymbol{x}^e$ of the element itself. In the notations of Fig. 4.8 and with the abbreviations

$$a = (3 - 4\nu), \qquad b = 8\pi\mu(1-\nu), \qquad d = 4\pi(1-\nu),$$

$$c_e = \cos\varphi_e = \frac{y_1^B - y_1^A}{l_e}, \qquad s_e = \sin\varphi_e = \frac{y_2^B - y_2^A}{l_e},$$

$$l_e = \text{length of the element}, \qquad \lambda_e = \ln l_e,$$

we obtain

Block 1

$$\frac{1}{b} \begin{bmatrix} -al_e/36\,(6\lambda_e - 17) + l_e/6\,c_e^2 & l_e/6\,c_e s_e \\ \text{sym.} & -al_e/36\,(6\lambda_e - 17) + l_e/6\,s_e^2 \end{bmatrix},$$

Block 2

$$\frac{1}{b} \begin{bmatrix} -al_e/9\,(6\lambda_e - 5) + 2/3\,l_e c_e^2 & 2/3\,l_e c_e s_e \\ \text{sym.} & -al_e/9\,(6\lambda_e - 5) + 2/3\,l_e s_e^2 \end{bmatrix},$$

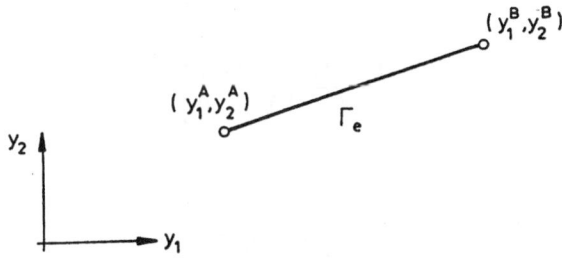

Figure 4.8 The orientation of an element with regard to the global system of coordinates

Block 3

$$\frac{1}{b}\begin{bmatrix} -al_e/36\,(6\lambda_e+1)+l_e/6\,c_e^2 & l_e/6\,c_e s_e \\ \text{sym.} & -al_e/36\,(6\lambda_e+1)+l_e/6\,s_e^2 \end{bmatrix},$$

Block 4

$$\frac{1}{b}\begin{bmatrix} -al_e\,(1/6\,\lambda_e-0.17108)+l_e/6\,c_e^2 & l_e/6\,c_e s_e \\ \text{sym.} & -al_e\,(1/6\lambda_e-0.17108)+l_e/6\,s_e^2 \end{bmatrix},$$

Block 5

$$\frac{1}{b}\begin{bmatrix} -a8l_e(1/12\,\lambda_e-z)+2/3\,l_e c_e^2 & 2/3\,l_e c_e s_e \\ \text{sym.} & -a8l_e(1/12\,\lambda_e-z)+2/3\,l_e s_e^2 \end{bmatrix},$$

$z = 0.1688733$

All fractions

$$2/3\,l_e c_e s_e = (2/3)\,l_e c_e s_e \qquad \text{etc.}$$

are to be interpreted in the strict sense.

The singular blocks of the matrix $\boldsymbol{H}^e$ are:

Block 1

$$0.5\begin{bmatrix} \dot{c}_{11} & \dot{c}_{12} \\ \dot{c}_{21} & \dot{c}_{22} \end{bmatrix}_{\text{node left}} + \frac{\lambda_e(1-2\nu)}{d}\begin{bmatrix} 0 & -1 \\ 1 & 0 \end{bmatrix},$$

Block 2

$$\frac{2(1-2\nu)}{d}\begin{bmatrix} 0 & -1 \\ 1 & 0 \end{bmatrix},$$

Block 4

$$\frac{1-2\nu}{d}\begin{bmatrix} 0 & -1 \\ 1 & 0 \end{bmatrix},$$

Block 5

$$\begin{bmatrix} 0.5 & 0 \\ 0 & 0.5 \end{bmatrix},$$

Block 6

$$0.5 \begin{bmatrix} \dot{c}_{11} & \dot{c}_{12} \\ \dot{c}_{21} & \dot{c}_{22} \end{bmatrix}_{\text{node right}} + \frac{\lambda_e(1-2\nu)}{d} \begin{bmatrix} 0 & 1 \\ -1 & 0 \end{bmatrix}$$

Blocks 2' and 4' are the blocks 2 and 4 multiplied with (-1). Block 3 is zero. Blocks 1, 5 and 6 are essentially the C-matrix. The factor 0.5 within the blocks 1 and 6 is introduced because the element matrices $\boldsymbol{H^e}$ overlap and so the C-matrix is evenhandedly split into two parts.

$$\dot{c}_{ij} = 0.5\,\dot{c}_{ij} + 0.5\,\dot{c}_{ij}$$

The factor 0.5 in block 5 is the 1/2 from $\dot{c}_{ij} = 1/2\,\delta_{ij}$ (the mid-node is always a smooth node). The two matrices that accompany the c-terms in block 1 and 6 will drop out during the assemblage (block 6 of the left element is added to block 1 of the right element), if the two neighboring elements have the same length, because it then holds

$$\lambda_{e,\text{element left}} = \lambda_{e,\text{element right}}$$

The entries of these two matrices are the Cauchy principal values of the four integrals

$$\int_\Gamma T_{ij}(\boldsymbol{y}, \boldsymbol{x}^k)\varphi_i^e(\boldsymbol{y})\,ds_{\boldsymbol{y}}$$

If the collocation point $\boldsymbol{x}^k$ however has the same distance from its two neighbors, then these principal values are zero, because "what is minus on the left is positive on the right", but if the distance between the collocation points $\boldsymbol{x}^k$ is different then only the principal value of the symmetric part is zero and the integral over that part Γ^k of Γ^e, which has no counterpart with respect to the mirror point $\boldsymbol{x}^k$, renders a non-vanishing contribution, see Fig. 4.9.

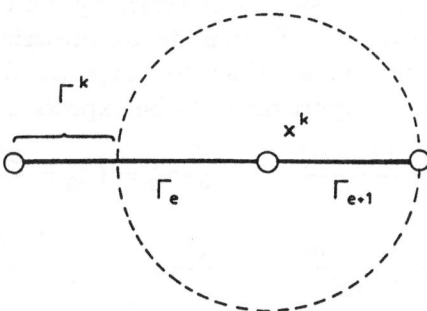

Figure 4.9 Neighboring elements with varying lengths

4.5 Boundary conditions

The storage of the element matrices G^e and H^e in the global matrices G and H reflects the continuity of the functions u_i and t_i. The matrices overlap if the functions are continuous at the nodes. Otherwise they lie side by side as the element matrices G^e of the two elements that enclose a corner point. Due to the discontinuity of the tractions at corner points the number of the t degrees-of-freedom (t-$dofs$) is always greater than the number of the u degrees-of-freedom (u-$dofs$), so that the assembled system of equations has the form

$$\begin{bmatrix} & H & \end{bmatrix} \begin{bmatrix} u \end{bmatrix} = \begin{bmatrix} & G & \end{bmatrix} \begin{bmatrix} t \end{bmatrix} + \begin{bmatrix} p \end{bmatrix}.$$

The form $Ax = b$ is obtained, if we arrange the columns of G and H according to known and unknown nodal values.

At a smooth node two of the four nodal values

$$u_1^i = u_1(x^i), \qquad u_2^i = u_2(x^i),$$

$$t_1^j = t_1(x^i), \qquad t_2^j = t_2(x^i)$$

are prescribed. The corresponding columns are multiplied with the prescribed values and added to the vector b on the right-hand side.

At a corner point the traction vector t has four degrees-of-freedom

$$t_1(x_-^i), \qquad t_2(x_-^i), \qquad (\text{left}),$$

$$t_1(x_+^i), \qquad t_2(x_+^i), \qquad (\text{right})$$

so that, adding the two degrees-of-freedom of the displacements, the number of degrees-of-freedom raises to six. But in general four of the six nodal values sometimes even all six are determined by boundary conditions. If the displacement field u is prescribed on both sides of the corner point then the two traction vectors t^l and t^r on the two sides of the node are prescribed as well.

This follows simply from the fact that the tangential derivatives of the two displacement components can approximately be expressed as, see Fig. 4.10,

$$\frac{\partial}{\partial \tau} u_1^l = u_{\tau 1}^l = \frac{u_1^c - u_1^a}{\Delta a}, \qquad \frac{\partial}{\partial \tau} u_2^l = u_{\tau 2}^l = \frac{u_2^c - u_2^a}{\Delta a},$$

$$\frac{\partial}{\partial \tau} u_1^r = u_{\tau 1}^r = \frac{u_1^b - u_1^c}{\Delta b}, \qquad \frac{\partial}{\partial \tau} u_2^r = u_{\tau 2}^r = \frac{u_2^b - u_2^c}{\Delta b},$$

and therefore the two gradients as well

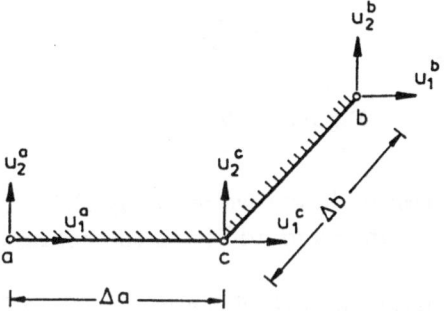

Figure 4.10 Position of the three collocation points

$$u_{1,1} = \frac{1}{D}(\nu_1^r u_{\tau 1}^l - \nu_1^l u_{\tau 1}^r), \qquad D = \nu_1^l \nu_2^r - \nu_2^l \nu_1^r,$$

$$u_{1,2} = \frac{1}{D}(\nu_2^r u_{\tau 1}^l - \nu_2^l u_{\tau 1}^r), \qquad l = \text{left}, \quad r = \text{right},$$

$$u_{2,1} = \frac{1}{D}(\nu_1^r u_{\tau 2}^l - \nu_1^l u_{\tau 2}^r),$$

$$u_{2,2} = \frac{1}{D}(\nu_2^r u_{\tau 2}^l - \nu_2^l u_{\tau 2}^r)$$

The gradients determine the strains ε_{ij} and the stresses σ_{ij} at the corner point and therefore also the traction vectors t^l and t^r.

At a roller support, see Fig. 4.11, the horizontal and vertical degrees-of-freedom are coupled according to

$$u_2 = u_1 \tan \alpha, \qquad t_2 = -\frac{1}{\tan \alpha} t_1$$

We treat the horizontal components u_1 and t_1 as the independent degrees-of-freedom (*masters*) and the vertical components as the dependent degrees-of-

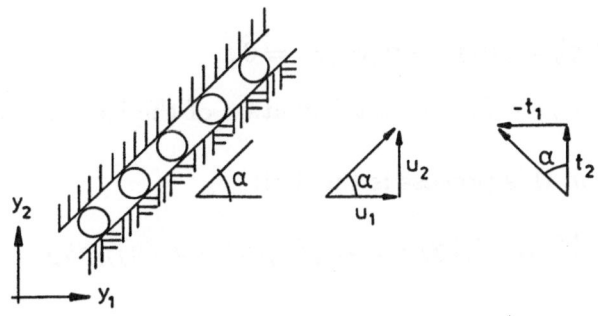

Figure 4.11 Roller support

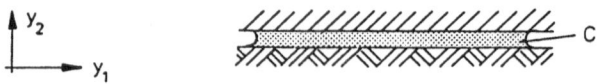

Figure 4.12 Elastic support

freedom (*slaves*). Correspondingly we multiply the column that is associated with u_2 with the factor $\tan \alpha$ and add it to the column u_1. The same is done with the column of t_2.

At a horizontal elastic support with stiffness c, see Fig. 4.12, the boundary conditions are

$$t_1 = 0, \qquad cu_2 - t_2 = 0$$

If we declare u_1 and t_2 to be the independent degrees-of-freedom then we must multiply the column of u_2 (the slave) with c^{-1} and add it to the column of t_2.

4.6 Stresses

Stresses depend on the strain in the material, that is, the derivatives of the displacement field. The influence functions are

$$u_{i,d}(\boldsymbol{x}) = \int\limits_{\Gamma} [U_{ij,x_d} t_j - T_{ij,x_d}(u_j(\boldsymbol{y}) - u_j(\boldsymbol{x}))] ds_{\boldsymbol{y}}$$

$$+ \int\limits_{\Omega} U_{ij,x_d} p_j d\Omega_{\boldsymbol{y}} \tag{4.12}$$

These formulas look, as if we had simply differentiated the influence functions for the displacements u_i with respect to x_d. However, the point $\boldsymbol{x}$ lies in the domain of integration and so we did the following: we formulated Betti's principle first in the punctured domain Ω_ε, differentiated with respect to x_d and we then let the radius ε of the hole shrink to zero,

$$\lim_{\varepsilon \to 0} B(g_{0,x_d}^i, \boldsymbol{u}(\boldsymbol{y}) - \boldsymbol{u}(\boldsymbol{x}))_{\Omega_\varepsilon} = 0,$$

$$g_0^i = \{U_{ij}\} = \text{fundamental solution of the loadcase } \hat{\boldsymbol{P}} = \boldsymbol{e}_i$$

The main result in this process is the limit

$$\lim_{\varepsilon \to 0} \int\limits_{\Gamma_{N_\varepsilon}(\boldsymbol{x})} [U_{ij,x_d} t_j(\boldsymbol{y}) - T_{ij,x_d}(u_j(\boldsymbol{y}) - u_j(\boldsymbol{x}))] ds_{\boldsymbol{y}} = -u_{i,d}(\boldsymbol{x})$$

$$\boldsymbol{x} = \text{internal point} \tag{4.13}$$

Considering that the limit of the domain integral is free of integral-free terms

$$\lim_{\varepsilon \to 0} \int_{\Omega_\varepsilon} U_{ij,x_d}\, p_j\, d\Omega_y = \int_\Omega U_{ij,x_d}\, p_j\, d\Omega_y \,,$$

the limit (4.12) is then obvious. For the stresses it follows then easily

$$\sigma_{ij}(\boldsymbol{x}) = \int_\Gamma [D_{kij} t_k - S_{kij} u_k]\, ds_y + \int_\Omega D_{kij} p_k\, d\Omega_y$$

where

$$D_{kij} = \frac{1}{r^\alpha}\{(1-2\nu)[\delta_{ki}r_{,j} + \delta_{kj}r_{,i} - \delta_{ij}r_{,k}] + \beta r_{,i}\, r_{,j}\, r_{,k}\}\frac{1}{4\alpha\pi(1-\nu)}$$

and

$$S_{kij} = \frac{2\mu}{r^\beta}\{\beta \frac{\partial r}{\partial \nu}[(1-2\nu)\delta_{ij}r_{,k} + \nu(\delta_{ik}r_{,j} + \delta_{jk}r_{,i}) - \gamma r_{,i}\, r_{,j}\, r_{,k}]$$

$$+ \beta\nu(\nu_i r_{,j}\, r_{,k} + \nu_j r_{,i}\, r_{,k}) + (1-2\nu)(\beta\nu_k r_{,i}\, r_{,j} + \nu_j \delta_{ik}$$

$$+ \nu_i \delta_{jk}) - (1-4\nu)\nu_k \delta_{ij}\}\frac{1}{4\alpha\pi(1-\nu)}$$

The parameters α, β, γ assume the values

$$\text{2-D} \quad \alpha = 1, \quad \beta = 2, \quad \gamma = 4\,,$$

$$\text{3-D} \quad \alpha = 2, \quad \beta = 3, \quad \gamma = 5$$

The calculation of the stresses σ_{ij} on the edge of a plate is done in the program BE-PLATES as follows: first the stresses are calculated with regard to the local system of coordinates ($\tilde{\ }$), this is the system whose axis $\tilde{x}_1$ runs parallel to the boundary, see Fig. 4.13, and they are then transformed into the global system.

In the local system the horizontal displacements of the node and its two neighbors are

$$\tilde{u}_1^a = \cos\alpha\, u_1^a + \sin\alpha\, u_2^a, \qquad \text{analogously for } \tilde{u}_1^b \text{ and } \tilde{u}_1^c$$

Hence, the strain parallel to the edge is approximately

$$\tilde{\varepsilon}_{11} = \frac{1}{2}\left(\frac{\tilde{u}_1^c - \tilde{u}_1^a}{\Delta a} + \frac{\tilde{u}_1^b - \tilde{u}_1^c}{\Delta b}\right),$$

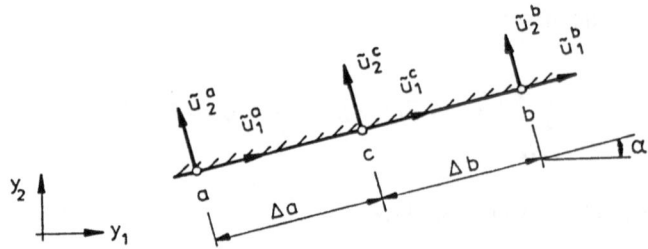

Figure 4.13 Local and global system of coordinates

and, therefore, the stresses in the local system are

$$\tilde{\sigma}_{12} = -\tilde{t}_1 = \cos\alpha\, t_1 + \sin\alpha\, t_2\,,$$

$$\tilde{\sigma}_{22} = -\tilde{t}_2 = -\sin\alpha\, t_1 + \cos\alpha\, t_2\,,$$

$$\tilde{\sigma}_{11} = \frac{\nu}{(1-\nu^2)}[\tilde{\sigma}_{22}(1+\nu) - \frac{E\nu}{(1-2\nu)}\tilde{\varepsilon}_{11}] + E\,\tilde{\varepsilon}_{11}\frac{(1-\nu)}{(1-2\nu)}(1+\nu)\,,$$

and the stresses in the global system become

$$\sigma_{11} = 0.5\Delta p + 0.5\Delta m\,\cos(-2\alpha) + \tilde{\sigma}_{12}\sin(-2\alpha)\,,$$

$$\sigma_{22} = 0.5\Delta p - 0.5\Delta m\,\cos(-2\alpha) - \tilde{\sigma}_{12}\sin(-2\alpha)\,,$$

$$\sigma_{12} = -\Delta m\,\sin(-2\alpha) + \sigma_{12}\cos(-2\alpha)\,,$$

$$\Delta p = \tilde{\sigma}_{11} + \tilde{\sigma}_{22}\,, \qquad \Delta m = \tilde{\sigma}_{11} - \tilde{\sigma}_{22}$$

4.7 The domain integrals

A domain integral as

$$\int_\Omega f(\boldsymbol{x})\,d\Omega$$

can be transformed into a boundary integral, if $f(\boldsymbol{x})$ is the right-hand side of a differential equation as $DF = f$ where the operator D satisfies an integral identity as

$$\int_\Omega DF\,1\,d\Omega = \int_\Gamma r(F)\,1\,ds \qquad \text{(first identity with } \hat{u} = 1)$$

To apply this idea to the domain integrals in the Somigliana identity,

$$\int_\Omega U_{ij}(\boldsymbol{y}, \boldsymbol{x}) p_j(\boldsymbol{y}) \, d\Omega_{\boldsymbol{y}}.,$$

we have to find a system of differential equations which has the fundamental solutions U_{ij} as its right-hand side, $DF = U_{ij}$, [52].

It now holds if a vector field $\boldsymbol{g} = \{g_j\}$ satisfies the three equations

$$-\mu \Delta \Delta g_j = p_j, \tag{4.14}$$

then the so-called *Boussinesq-Somigliana-Galerkin solution*, see [53],

$$u_j = g_{j,kk} - \frac{1}{2(1-\nu)} g_{k,kj} \tag{4.15}$$

satisfies the equation

$$-L_{ij} u_j = p_i$$

To the right-hand side $\boldsymbol{p} = \delta(\boldsymbol{x} - \boldsymbol{y}) e_i$ of Eq.(4.14) belongs the solution

$$G_{ij} = \frac{1}{8\pi\mu} r \, \delta_{ij}$$

Hence the fundamental solution is the *B-S-G solution* of the field G_{ij},

$$U_{ij} = G_{ij,kk} - \frac{1}{2(1-\nu)} G_{ik,kj} \tag{4.16}$$

To the system (4.15) belongs the integral identity

$$\int_\Omega \left(g_{j,kk} - \frac{1}{2(1-\nu)} g_{k,kj} \right) d\Omega = \int_\Gamma \left(g_{j,k} - \frac{1}{2(1-\nu)} g_{k,j} \right) n_k \, ds,$$

and after taking the limit, $\Omega_\varepsilon \to \Omega$, it therefore follows

$$\int_\Omega U_{ij} p_j \, d\Omega_{\boldsymbol{y}} = \int_\Gamma \left(G_{ij,k} - \frac{1}{2(1-\nu)} G_{ik,j} \right) \nu_k p_j \, ds_{\boldsymbol{y}}$$

$$= \frac{1}{8\pi\mu} \int_\Gamma \left(p_i \nu_k r_{,k} - \frac{1}{2(1-\nu)} p_k r_{,k} \nu_i \right) ds_{\boldsymbol{y}}$$

In 2-D the solution of the two-dimensional problem (4.14) is

$$G_{ij} = -\frac{1}{8\pi\mu} r^2 \ln r \, \delta_{ij}$$

and it belongs to the *B-S-G solution*

$$\tilde{U}_{ij} = \frac{1}{8\pi\mu(1-\nu)}\left\{(3-4\nu)\ln\frac{1}{r}\,\delta_{ij} - r_{,i}\,r_{,j} - \left(\frac{7-8\nu}{2}\right)\delta_{ij}\right\}$$

$$= U_{ij} - \left(\frac{7-8\nu}{16\pi\mu(1-\nu)}\right)\delta_{ij}$$

This solution differs by a rigid-body motion from the fundamental solution U_{ij}. Naturally the Somigliana identity remains correct if we replace U_{ij} by $\tilde{U}_{ij}$. We, so to speak, only add the equilibrium condition

$$\left(\frac{7-8\nu}{16\pi\mu(1-\nu)}\right)\left[\int_\Gamma \delta_{ij}t_j\,ds + \int_\Omega \delta_{ij}p_j d\Omega\right] = 0$$

and the matrix $\tilde{T}_{ij}$ is of course the same matrix as T_{ij}. If the program BE-PLATES must solve a loadcase with distributed domain forces then it applies this technique and replaces the kernel U_{ij} by the kernel $\tilde{U}_{ij}$ in all relevant expressions and the domain integral by the following boundary integral

$$\int_\Omega \tilde{U}_{ij}p_j\,d\Omega y = \frac{1}{8\pi\mu}\int_\Gamma r[(2\ln r + 1)(\frac{1}{2(1-\nu)}p_k r_{,k}\,\nu_i - p_i r_\nu)]\,ds y$$

The same technique can be extended to centrifugal forces, see [54, p. 220]. If the load satisfies a Poisson equation (as constant, linear and quadratic loads do) then the more general method of Rizzo and Shippy is applicable, [55].

The contributions of the boundary integrals to the calculation of the stresses

$$\sigma_{ij} = \int_\Gamma [D_{kij}t_k - S_{kij}u_k]\,ds y + \int_\Gamma \hat{S}_{ij}\,ds y$$

are in 3-D, see [54, p. 220],

$$\hat{S}_{ij} = \frac{1}{8\pi r}[\nu_m r_{,m}\,(p_i r_{,j} + p_j r_{,i}) + \frac{1}{1-\nu}\{\nu\delta_{ij}(\nu_m r_{,m}\,p_s r_{,s} - p_m \nu_m)$$

$$- \frac{1}{2}(p_m r_{,m}\,[\nu_i r_{,j} + \nu_j r_{,i}] + (1-2\nu)(p_i\nu_j + p_j\nu_i))\}]$$

and in 2-D

$$\hat{S}_{ij} = \frac{1}{8\pi}[2\nu_m r_{,m}\,(p_i r_{,j} + p_j r_{,i}) + \frac{1}{1-\nu}\{\nu\delta_{ij}(2\nu_m r_{,m}\,p_s r_{,s}$$

$$+ (1 + 2 \ln r) p_m \nu_m) - p_m r_{,m} (\nu_i r_{,j} + \nu_j r_{,i})$$

$$+ \frac{1 - 2\nu}{2} (1 + 2 \ln r)(p_i \nu_j + p_j \nu_i)\}]$$

4.8 Double nodes

Double nodes were invented to interpolate discontinuous tractions with continuous layers. Points on the boundary where the traction is discontiuous are split into two nodes i and j and the displacement and the traction vectors at the two nodes are treated as independent quantities, see Fig. 4.14. In agreement with this the collocation equations are formulated at the point i

$$C(\boldsymbol{x}^i)\boldsymbol{u}(\boldsymbol{x}^i) + \int_\Gamma \boldsymbol{T}(\boldsymbol{y}, \boldsymbol{x}^i)\boldsymbol{u}(\boldsymbol{y})\, ds_{\boldsymbol{y}} = \int_\Gamma \boldsymbol{U}(\boldsymbol{y}, \boldsymbol{x}^i)\boldsymbol{t}(\boldsymbol{y})\, ds_{\boldsymbol{y}},$$

and at the point j

$$C(\boldsymbol{x}^j)\boldsymbol{u}(\boldsymbol{x}^j) + \int_\Gamma \boldsymbol{T}(\boldsymbol{y}, \boldsymbol{x}^j)\boldsymbol{u}(\boldsymbol{y})\, ds_{\boldsymbol{y}} = \int_\Gamma \boldsymbol{U}(\boldsymbol{y}, \boldsymbol{x}^j)\boldsymbol{t}(\boldsymbol{y})\, ds_{\boldsymbol{y}}$$

Because the position of the two nodes is the same, $\boldsymbol{x}^i = \boldsymbol{x}^j$, the two rows i and j are up to the terms on the diagonal identical, see Fig. 4.15.

Figure 4.14 Application of double nodes

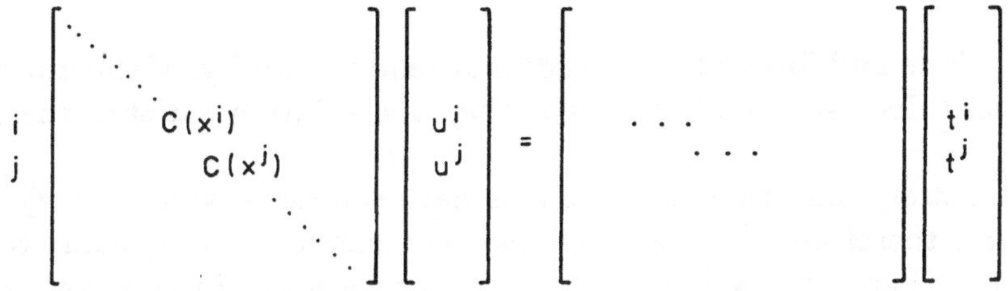

$$\begin{bmatrix} \ddots \\ & C(x^i) \\ & & C(x^j) \\ & & & \ddots \end{bmatrix} \begin{bmatrix} \vdots \\ u^i \\ u^j \\ \vdots \end{bmatrix} = \begin{bmatrix} & \cdots \\ \cdots & & \cdots \\ & \cdots \end{bmatrix} \begin{bmatrix} \vdots \\ t^i \\ t^j \\ \vdots \end{bmatrix}$$

Figure 4.15 The inner structure of the matrices

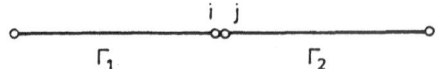

Figure 4.16 Double nodes

Therefore, the two displacement vectors, $u(x^i)$ and $u(x^j)$, may not be prescribed at the same time; otherwise the system matrix would contain two identical rows. The singular nature of the kernels requires that integration over element 1 is done analytically if the collocation point is the point j and, vice versa, see Fig. 4.16. The singular integral over the element itself is calculated indirectly by rigid-body motions.

The author avoids double nodes if possible. He rather works with discontinuous boundary layers because, in his opinion, these are a more natural approach. According to the rule: *the worse the regularity of a boundary term, the better the regularity of the conjugated kernel*, force terms may be discontinuous at the collocation points because the adjoint kernels, being of displacement type, are well behaved regular functions and therefore the integrals exist even if the boundary layers are discontinuous at the collocation points.

Closely connected with the concept of double nodes is the idea to place additional collocation points outside of the problem domain where $c(x) = 0$. But the experience of the author and most of his colleagues, though not all, see [56], is that this technique produces singular matrices. Each position of x produces different results.

4.9 Infinite domains

The Somigliana identity is applicable to the solutions of exterior problems if they satisfy the so-called *radiation condition*:

Pick an arbitrary fixed point x as the centre of a circle with radius R, see Fig. 4.17, and, next, calculate the integral

$$\int_\Gamma [U(y,x)t(y) - T(y,x)u(y)]\, ds_y$$

over this circle and finally let the radius R tend to infinity. If the integral tends to zero and does so for all points x then the solution satisfies the radiation condition.

Fortunately, all practical problems have solutions which satisfy this condition and, therefore, the boundary element method is applicable to exterior problems. There are even some relaxations because, in contrast to interior problems, the equilibrium condition

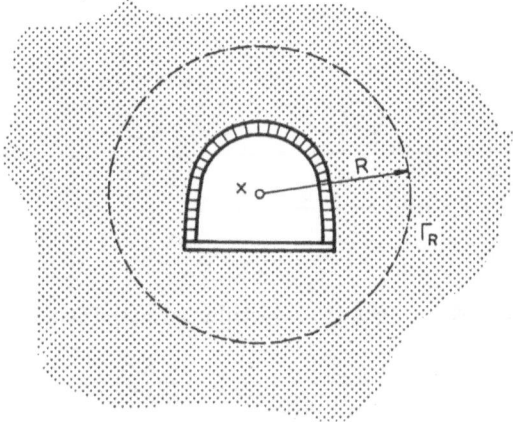

Figure 4.17 Tunnel

$$\int_\Gamma t \cdot (a + b \times x)\, ds = 0, \qquad \text{for all rigid-body motions } a + b \times x,$$

is no necessary condition, that is the tubes of a tunnel may be loaded with arbitrary forces. Such a loading does not even violate the radiation condition: according to St. Venant's principle the displacement field at infinity will behave as the displacement field generated by the statically equivalent group of concentrated forces (the resultant of the distributed forces that act on the tube), that is the displacement field would behave as the fundamental solutions. But the fundamental solutions satisfy the radiation condition and therefore the load may be arbitrary.

4.10 Examples

4.10.1 Beam with openings

The stresses in the beam in Fig. 4.18 were calculated with boundary elements. The boundary was subdivided into 134 linear elements. Because of the symmetry of the problem it sufficed to discretize only one half of the beam. The openings were approximated with 20 straight elements each. Figure 4.19 shows the distribution of the main stresses.

4.10.2 Spring element

The second example is a curved elastic spring, see Fig. 4.20. Such spring elements connect rotating shafts as in a ship's diesel engine. They let the

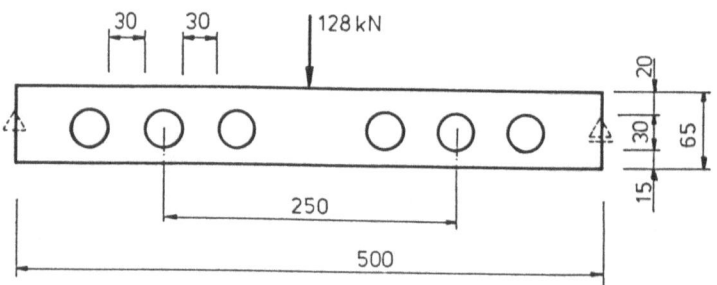

Figure 4.18 Beam with circular cutouts (Dallmann)

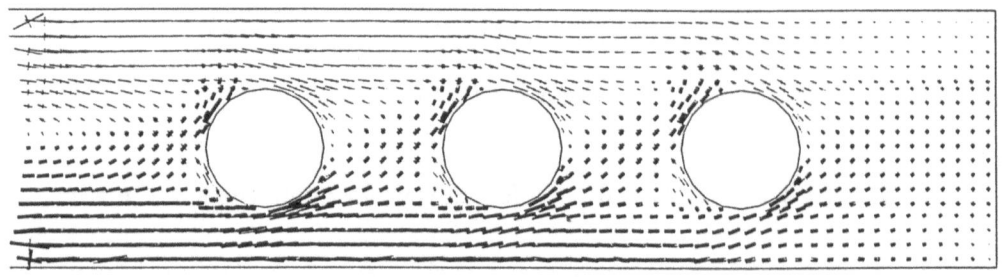

Figure 4.19 Main stresses

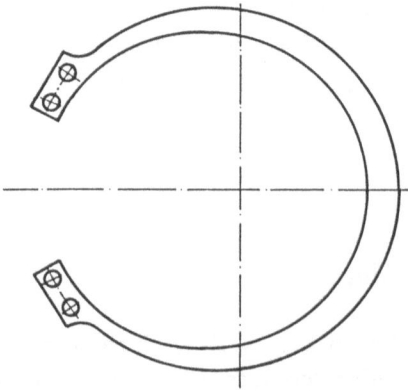

Figure 4.20 Spring element (Dallmann)

torque pass through but they dampen sudden, impulsive accelerations. The spring has a diameter of approximately 40 cm (15.7 inches).

The spring was analysed with finite elements and with boundary elements. Both methods utilized the symmetry of the problem. The finite element mesh

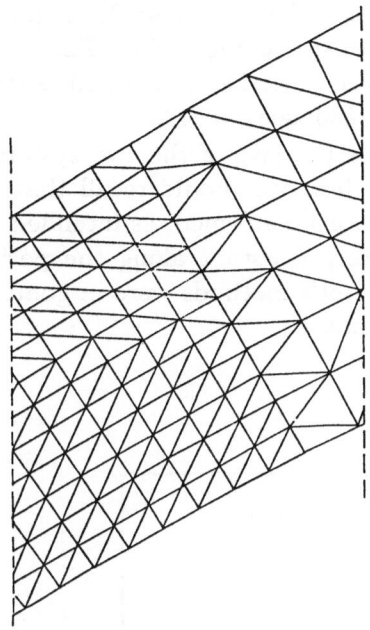

Figure 4.21 Detail of the FE-discretization

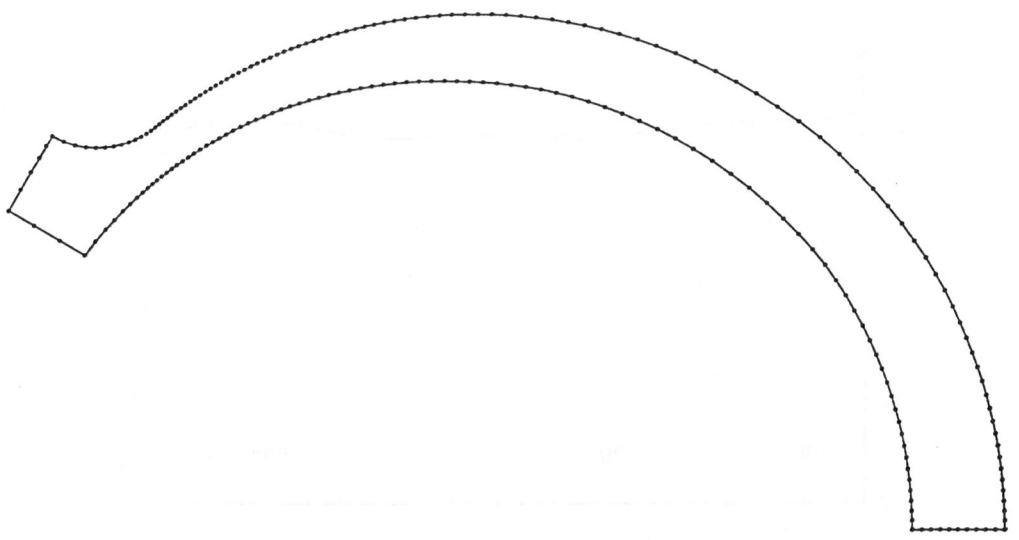

Figure 4.22 BE-discretization of the spring element

consisted of 1520 quadratic elements and 3255 nodes which corresponds to $2 \times 3255 = 6510$ unknowns, see Fig. 4.21. The boundary element subdivision consisted of 107 straight elements and 215 nodes so that the number of unknowns was $2 \times 215 = 430$, see Fig. 4.22.

Figure 4.23 compares the size of the two system matrices, of David and Goliath. However, to be fair, we should recall that the FE-matrix is sparse. The result of the calculations confirmed, as intended by the designer, that the stresses were approximately constant along the perimeter of the spring, see Fig. 4.24. The results of both methods were in good agreement and were also confirmed by measurements.

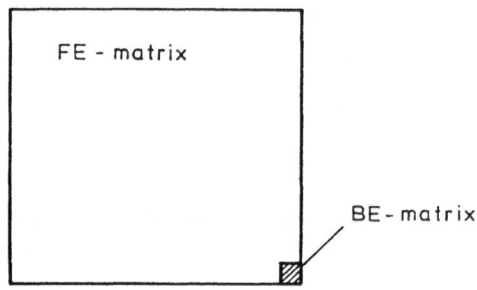

Figure 4.23 A comparison between the two matrices

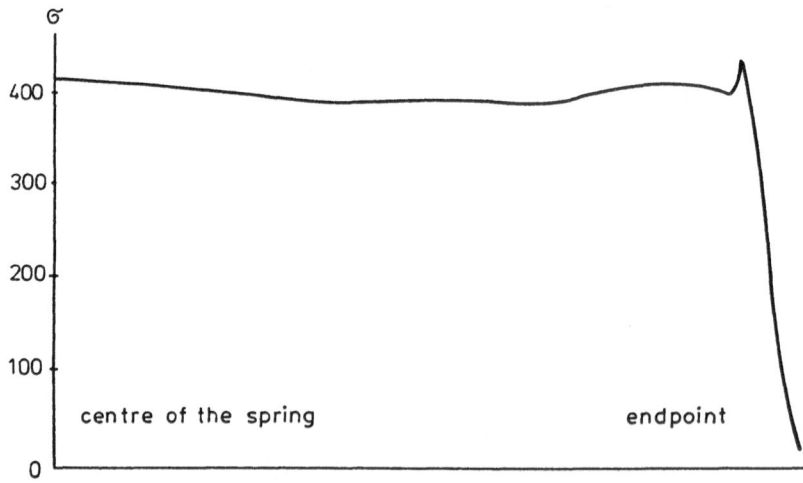

Figure 4.24 Stress distribution along the perimeter of the spring element

4.10.3 The search for cavities an inverse problem

In the absence of volume forces the elastic state of a body is determined by the displacements and tractions on the surface alone. Hence, it should be possible to make guesses about the size and the location of cavities within a body by studying the displacements and tractions on the surface. Whether this is possible was investigated by Kroener, [46], who tried to find holes in elastic plates by investigating the displacements and stresses on the edge of a plate.

So let us assume that we know the Betti data u_i and t_i on the boundary precisely, but the size and location of a hole in the interior of a plate only approximately. To find the centre point of the hole and its shape, we solve the plate problem and we compare the calculated boundary displacements u_{hi} with the measured displacements at n boundary points x^i. If the assumptions about the size and position of the hole were correct then the error function (Kroener only compared the horizontal displacements)

$$F = \left(\sum_{i=1}^{n} [u_1(x^i) - u_{h1}(x^i)]^2 / n \right)^{1/2}$$

should be zero. Otherwise the assumptions about the size and position of the hole must be improved. This is an iterative process with the aim to minimize the error function. This function depends on the parameter of the hole. In the case of an elliptic hole the parameters are the coordinates of the centre point x_c, y_c, the two half-axis a, b and the angle α between the main axis and the horizontal direction.

As an example we cite the search for an elliptic hole in a cantilever plate, see Fig. 4.25.

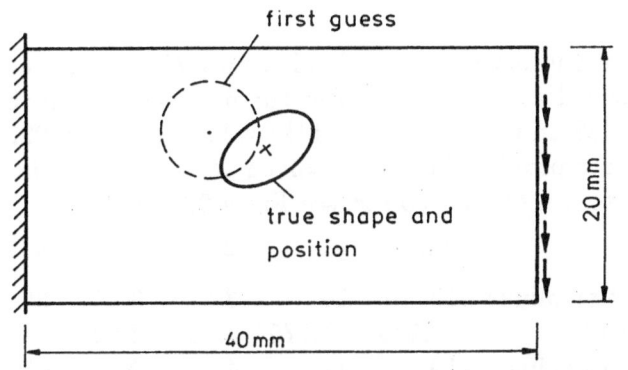

Figure 4.25 A cantilever plate with an elliptic hole (Kroener)

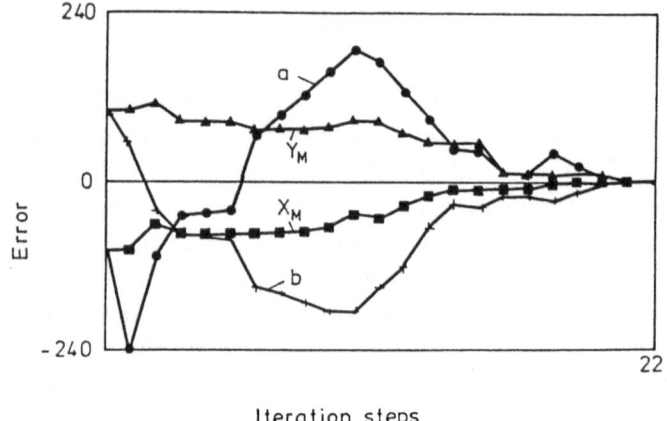

Figure 4.26 The convergence of the system parameters

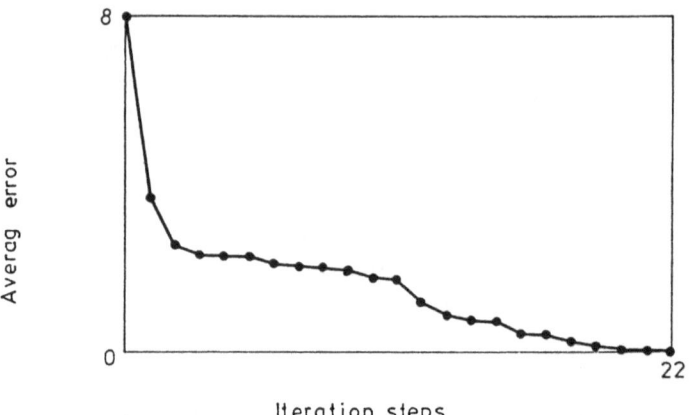

Figure 4.27 The behavior of the error F

The edge of the plate was subdivided into 24 constant elements and the error was measured at 10 points. After the first calculation the square of the error was $F = 7.88$ units and at the end the value was $F = 0.02$. To achieve this accuracy the plate problem had to be solved 148-times and the total computer time added up to 3 h 45 min 03 sec, see Fig. 4.26 and 4.27.

	x_M	y_M	a	b	α
initial values	15	14	3.5	3.5	35
result	18.973	11.975	3.985	2.503	35
exact value	19	12	4	2.5	35
	mm	mm	mm	mm	degree

4.11 Singularities

At re-entrant corners and points where the boundary conditions change the stresses become singular. To illustrate the effect of such singularities on a BE-solution we consider the cantilever plate in Fig. 4.28.

At the transition point *free edge, fixed edge* the boundary conditions read

$$\text{free edge} \qquad t_1 = 0, \qquad t_2 = 0,$$

$$\text{fixed edge} \qquad u_1 = 0, \qquad u_2 = 0$$

If we pass the node in a counter clockwise direction then we must first satisfy homogeneous traction boundary conditions and then homogeneous displacement boundary conditions. The first set of conditions

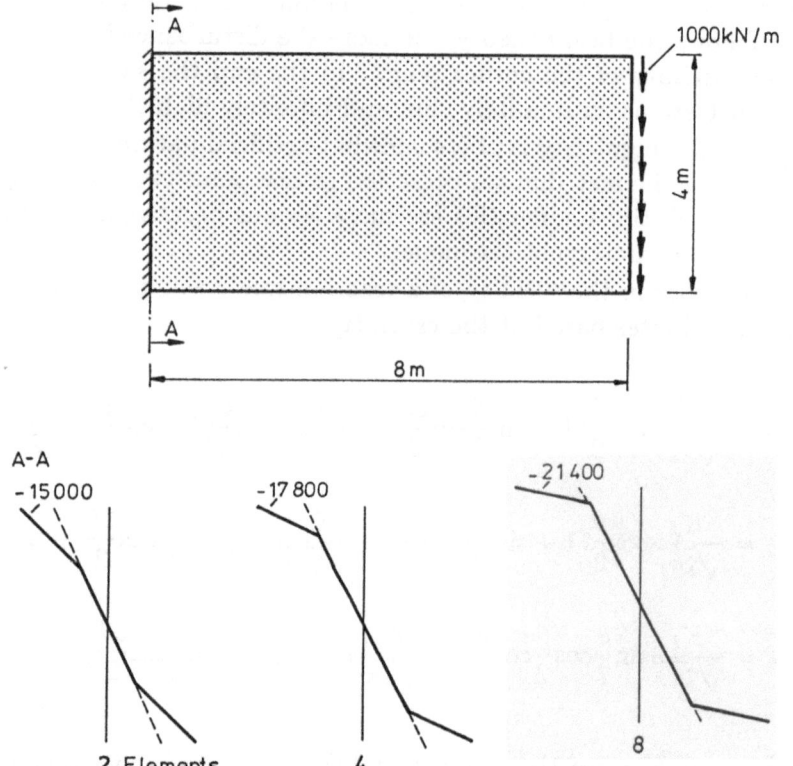

Figure 4.28 Stress distribution at the support for different element subdivisions

$$\begin{bmatrix} \sigma_{11} & \sigma_{12} \\ \sigma_{21} & \sigma_{22} \end{bmatrix} \begin{bmatrix} 0 \\ 1 \end{bmatrix} = \begin{bmatrix} 0 \\ 0 \end{bmatrix}, \qquad \begin{bmatrix} 0 \\ 1 \end{bmatrix} = \text{normal vector at node 1},$$

implies that

$$\sigma_{12} = \sigma_{22} = 0$$

Because the vertical displacement $u_2 = 0$ is zero along the fixed edge, its derivative in this direction is zero too, $u_{2,2} = 0$, and, therefore also the strain $\varepsilon_{22} = 0$. Because of

$$\sigma_{22} = \frac{E}{1+\nu}[\varepsilon_{22} + \frac{\nu}{1-2\nu}(\varepsilon_{11} + \varepsilon_{22})] = 0 \qquad \text{and} \qquad \varepsilon_{22} = 0,$$

this implies that $\varepsilon_{11} = 0$ also must be zero and, due to $\varepsilon_{22} = 0$, the stress σ_{11} as well. But this result is in contrast to the beam theory. According to the beam theory the stresses σ_{11} on the flanks of the plate should be the maximum stresses. Obviously the change in the boundary condition, free edge fixed edge, leads to a contradiction between the plate theory and the beam theory. Figure 4.28b contains for different boundary subdivisions the distribution of the traction $t_1 = -\sigma_{11}$ along the fixed edge. The finer the subdivision the more the solution follows the beam theory, the more the disturbance is restricted to the immediate vicinity of the corner point. In the end the singularity seemingly will demonstrate as an infinitely thin and infinitely high stress peak.

While this singularity is of a somewhat artificial nature it is more or less a consequence of the fact that we prescribe at one point two opposing boundary conditions crack tip singularities are more serious because they do not vanish if we modify the mathematical model.

The stresses in the vicinity of a traction-free crack can be given in terms of polar coordinates based at the crack tip

$$\sigma_{11} = \frac{K_I}{\sqrt{2\pi r}}\cos\frac{\theta}{2}(1 - \sin\frac{\theta}{2}\sin\frac{3\theta}{2}) - \frac{K_{II}}{\sqrt{2\pi r}}\sin\frac{\theta}{2}(2 + \cos\frac{\theta}{2}\cos\frac{3\theta}{2}) + \cdots$$

$$\sigma_{22} = \frac{K_I}{\sqrt{2\pi r}}\cos\frac{\theta}{2}(1 + \sin\frac{\theta}{2}\sin\frac{3\theta}{2}) + \frac{K_{II}}{\sqrt{2\pi r}}\sin\frac{\theta}{2}\cos\frac{\theta}{2}\cos\frac{3\theta}{2} + \cdots$$

$$\sigma_{12} = \frac{K_I}{\sqrt{2\pi r}}\sin\frac{\theta}{2}\cos\frac{\theta}{2}\cos\frac{3\theta}{2} + \frac{K_{II}}{\sqrt{2\pi r}}\cos\frac{\theta}{2}(1 - \sin\frac{\theta}{2}\sin\frac{3\theta}{2}) + \cdots$$

where K_I and K_{II} are the mode I and mode II stress intensity factors, see Fig. 4.29. The corresponding displacements are

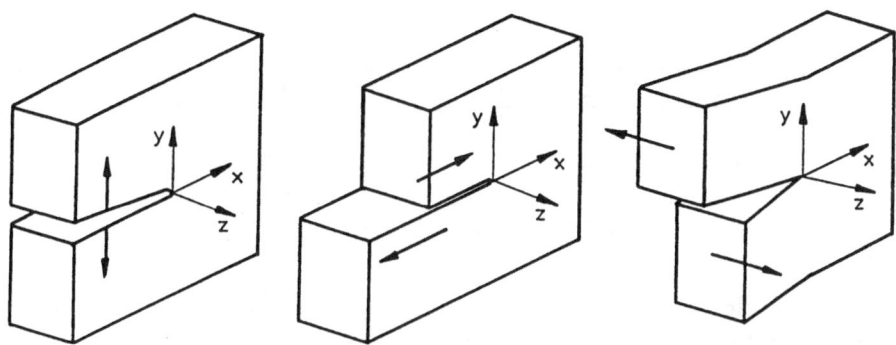

Figure 4.29 Crack opening modes

$$u_1 = \frac{K_I}{4\mu}\sqrt{\frac{r}{2\pi}}\left((2\gamma - 1)\cos\frac{\theta}{2} - \cos\frac{3\theta}{2}\right)$$

$$+ \frac{K_{II}}{4\mu}\sqrt{\frac{r}{2\pi}}\left((2\gamma + 3)\sin\frac{\theta}{2} + \sin\frac{3\theta}{2}\right) + \ldots$$

$$u_2 = \frac{K_I}{4\mu}\sqrt{\frac{r}{2\pi}}\left((2\gamma - 1)\sin\frac{\theta}{2} - \sin\frac{3\theta}{2}\right)$$

(4.17)

$$- \frac{K_{II}}{4\mu}\sqrt{\frac{r}{2\pi}}((2\gamma - 3)\cos\frac{\theta}{2} + \cos\frac{3\theta}{2}) + \ldots$$

where

$$\gamma = (3 - 4\nu) \qquad \text{plane strain}, \qquad \gamma = \frac{3 - \nu}{1 + \nu} \qquad \text{plane stress}.$$

To cope with these stress singularities many different strategies have been adopted: use of special Green's functions, flat crack modelling, symmetric crack modelling, singular elements and the idealization of the crack as an open notch. Cruse reports, [58], that the latter approach cannot be recommended. If the notch is too thick the surfaces of the crack are modelled too far apart and if they are too close the system becomes ill-conditioned. In flat crack modelling the two crack surfaces collapse into one plane and the unknowns are the relative displacements between the two cracks and the displacement of the collapsed crack. This model, too, has its deficiencies. The most elegant approach is the use of a special fundamental solution that satisfies the boundary conditions at the crack surface. However, this technique is as is so often the case in fracture mechanics mainly limited to two-dimensional problems. Blandford et al. [59] adopted an FE-technique; they introduced special crack elements

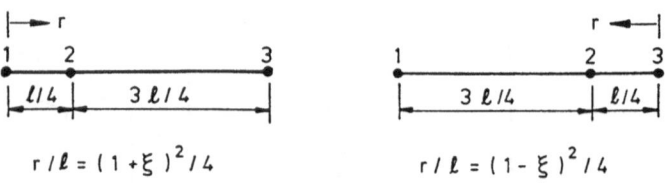

Figure 4.30 Quarter point element

(quarter-point elements) and they simultaneously split the problem domain into two separate domains, so that the they can model the displacements and tractions on the crack faces independently. By placing the midpoint node of an isoparameteric quadratic element at the quarter-point, see Fig. 4.30, the displacement functions exhibit the $\sqrt{r}$ variation

$$u_i = a_i^1 + a_i^2 \sqrt{r} + a_i^3 \tag{4.18}$$

which corresponds to a $1/\sqrt{r}$ singularity in the tractions. However, boundary displacements and tractions are independent quantities in the BEM so that Eq.(4.18) must be multiplied with $\sqrt{l/r}$ to exhibit a variation

$$t_i = \sqrt{\frac{l}{r}}(a_i^1 + a_i^2 \sqrt{r} + a_i^3) = \frac{b_i^1}{\sqrt{r}} + b_i^2 + b_i^3 \sqrt{r}$$

that corresponds to the stress singularity in the vicinity of the crack tip. (The length l is a normalizing term.)

By evaluating the displacements (4.18) at the three nodes of the element and solving these equations for the coefficients a_i^j we can express the variation of the displacements in terms of the nodal values

$$u = u_A + (-3u_a + 4u_B - u_C)\sqrt{\frac{r}{l}} + (2u_A - 4u_B + 2u_C)\frac{r}{l}$$

$$v = v_A + (-3v_a + 4v_B - v_C)\sqrt{\frac{r}{l}} + (2v_A - 4v_B + 2v_C)\frac{r}{l}$$

where $u = u_1$ and $v = u_2$ are the displacements along the crack axis and normal to the crack axis, respectively. Equating these expansions termwise with William's expansion (4.17) the stress intensity factors become

$$K_I = \frac{\mu}{\gamma + 1}\sqrt{\frac{2\pi}{l}}[4(v_B - v_D) + v_E - v_C]$$

$$K_{II} = \frac{\mu}{\gamma + 1}\sqrt{\frac{2\pi}{l}}[4(u_B - u_D) + u_E - u_C]$$

where the points D and E lie on the opposite face of the crack, see Fig. 4.31.

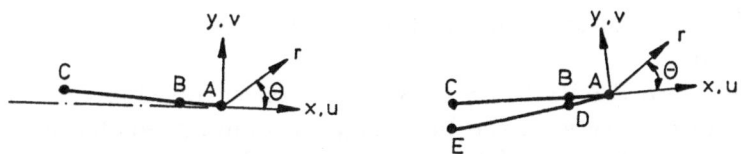

Figure 4.31 Nodal points in the vicinity of the crack tip

Special transition elements can be appended to the crack element to ease the transition between the zone of high stress gradients and to extend the distance "over which the crack is sensed".

4.12 Concentrated forces

If a concentrated force $P = (P_1, P_2)$ acts at a point $\boldsymbol{\xi} = (\xi_1, \xi_2)$ then the integral of the traction vector $\boldsymbol{Sn} = \boldsymbol{t}$ taken over smaller and smaller circles centred at $\boldsymbol{\xi}$ must tend to $\boldsymbol{P}$,

$$\lim_{\varepsilon \to 0} \int_{\Gamma_{N\varepsilon}} \boldsymbol{Sn}\, ds_{\boldsymbol{x}} = \boldsymbol{P}$$

Because the perimeter p of the circle tends to zero as $p = 2\pi\varepsilon$ the components σ_{ij} of the stress tensor must counter-balance this action by tending to infinity as $1/2\pi\varepsilon$.

This is exactly the behavior of a fundamental solution near the source point and, therefore, the displacement field of a plate in the presence of a concentrated force

$$\boldsymbol{u}(\boldsymbol{x}) = \boldsymbol{u}_r(\boldsymbol{x}) + g_0^1(\boldsymbol{\xi}, \boldsymbol{x})P_1 + g_0^2(\boldsymbol{\xi}, \boldsymbol{x})P_2$$

consists of the two fundamental solutions g_0^i (one for each component of the force) and a regular displacement field $\boldsymbol{u}_r$ which cares that the boundary conditions are satisfied.

The integral representation of this compound field is the Somigliana identity plus the two fundamental solutions

$$C(\boldsymbol{x})\boldsymbol{u}(\boldsymbol{x}) = \cdots + g_0^1(\boldsymbol{\xi}, \boldsymbol{x})P_1 + g_0^2(\boldsymbol{\xi}, \boldsymbol{x})P_2$$

This can also be explained by assuming that the volume forces $\boldsymbol{p}$ have shrunk to the point $\boldsymbol{\xi}$. This explains why the point $\boldsymbol{\xi}$ has taken over the place of the integration variable $\boldsymbol{y}$.

While the FE-method would have to introduce special singular elements to handle such a loadcase the correct singularity is built into the BE-solution and so only the boundary data of the regular part u_r of the solution must be approximated.

The program BE-PLATES allows only concentrated forces in the interior. For concentrated forces on the boundary we would need the fundamental solutions g_0^i of the elastic halfplane. Only these functions satisfy the boundary conditions of a free boundary. But we can easily move the point of attack a little bit inside the plate with no grave consequences for the accuracy of the results.

4.13 Three-dimensional problems

The simplest approximation of the surface of an elastic body is a facet-like approximation with plane triangles, that is, linear shape functions. If the shape functions are also the local basis functions then we speak of isoparametric elements. For such elements analytical expressions for the singular integrals exist. In the following we cite from the work of Li et al. [60].

Let us assume that the source point is node 1 of the triangle. If we introduce polar coordinates with centre at the source point then, because of

$$ds = r\, dr\, d\varphi\,,$$

the order of the singularity reduces by a factor of one. Weakly singular integrals become regular integrals and strongly singular integrals become weakly singular integrals. This reduction is now coupled with a mapping of the triangle D onto the unit square D', see Fig. 4.32. This transformation maps the point P onto the left edge of the square, the line $P_1' P_4'$. The inverse mapping, the mapping from the square back onto the triangle, is

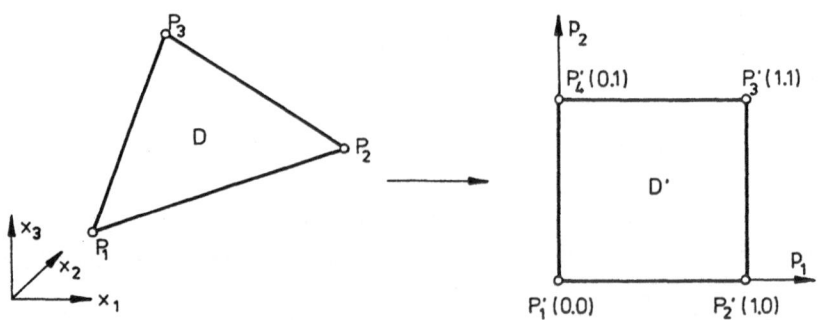

Figure 4.32 The triangular element is mapped onto the unit square

$$x_i = (1 - \rho_1)x_i^{(1)} + \rho_1(1 - \rho_2)x_i^{(2)} + \rho_1\rho_2 x_i^{(3)}, \qquad i = 1, 2, 3$$

where $x_i^{(k)}$ are the coordinates of the three vertices

$$P_k = (x_1^{(k)}, x_2^{(k)}, x_3^{(k)}), \qquad k = 1, 2, 3$$

The distance of a point y in D from the source point P_1 measures in local coordinates

$$r = (\sum_{i=1}^{3}(y_i - x_i^{(1)})^2)^{1/2} = \rho_1(a_1 + a_2\rho_2 + a_3\rho_2^2)$$

where

$$a_1 = \sum_{i=1}^{3}(x_i^{(21)})^2, \quad a_2 = 2\sum_{i=1}^{3}(x_i^{(21)}x_i^{(32)}), \quad a_3 = \sum_{i=1}^{3}(x_i^{(32)})^2,$$

$$x_i^{(21)} = x_i^{(2)} - x_i^{(1)}, \qquad x_i^{(32)} = x_i^{(3)} - x_i^{(2)}$$

The derivatives are

$$r_{,i} = \frac{y_i - x_i^{(1)}}{r} = \frac{x_i^{(21)} + \rho_2 x_i^{(32)}}{(a_1 + a_2\rho_2 + a_3\rho_2^2)^{1/2}}$$

Due to the change *triangle square* the Jacobian J is no longer constant, $J = 2\rho_1 A$, so that the area element depends on the coordinate ρ_1

$$ds = 2\rho_1 A \, d\rho_1 \, d\rho_2, \qquad A = \text{area of the triangle}$$

Using the notations

$$u_j^{(k)} = \text{value of } u_j \text{ at the point } P_k, \quad t_j^{(k)} = \text{value of } t_j \text{ at the point } P_k,$$

we obtain for the singular integrals the expressions

$$\int_{\Gamma_e} U_{ij}t_j \, ds_y = \frac{A}{16\pi(1 - \nu)\mu}\{[(3 - 4\nu)\delta_{ij}I_0 + x_i^{(21)}x_j^{(21)}I_2$$

$$+ (x_j^{(32)}x_i^{(21)} + x_i^{(32)}x_j^{(21)})I_3 + x_i^{(32)}x_j^{(32)}I_4]t_j^{(1)}$$

$$+ [(3 - 4\nu)\delta_{ij}(I_0 - I_1) + x_i^{(21)}x_j^{(21)}(I_2 - I_3)$$

$$+ (x_j^{(32)}x_i^{(21)} + x_i^{(32)}x_j^{(21)})(I_3 - I_4)$$

$$+ x_i^{(32)}x_j^{(32)}(I_4 - I_5)]t_j^{(2)}$$

$$+ [(3 - 4\nu)\delta_{ij}I_1 + x_i^{(21)}x_j^{(21)}I_3 + (x_j^{(32)}x_i^{(21)} + x_i^{(32)}x_j^{(21)})I_4$$

$$+ x_i^{(32)}x_j^{(32)}I_5]t_j^{(3)}\}$$

and

$$\int_{\Gamma_e} T_{ij}u_j\, ds\boldsymbol{y} = \frac{1 - 2\nu}{8\pi(1 - \nu)}[\varepsilon_{jik}x_k^{(32)}I_0 + 2A(-\nu_i x_j^{(21)} + \nu_j x_i^{(21)})I_2$$

$$+ 2A(-\nu_i x_j^{(32)} + \nu_j x_i^{(32)})I_3]u_j^{(1)} + \frac{(1 - 2\nu)A}{4\pi(1 - \nu)}\{(\nu_i x_j^{(21)} - \nu_j x_i^{(21)})I_2$$

$$+ [(\nu_i x_j^{(32)} - \nu_j x_i^{(32)}) + (-\nu_i x_j^{(21)} + \nu_j x_i^{(21)})]I_3 + (-\nu_i x_j^{(32)} + \nu_j x_i^{(32)})I_4\}u_j^{(2)}$$

$$+ \frac{(1 - 2\nu)A}{4\pi(1 - \nu)}[(\nu_i x_j^{(21)} - \nu_j x_i^{(21)})I_3 + (\nu_i x_j^{(32)} - \nu_j x_i^{(32)})I_4]u_j^{(3)}$$

$$+ \left[\frac{(1 - 2\nu)}{8\pi(1 - \nu)}\varepsilon_{jik}\int_{\bar{2}1+\bar{1}3}\frac{1}{r}\, dy_k\right]u_j^{(1)} \tag{4.19}$$

The last integral in Eq.(4.19), the integral over the two sides $\bar{2}1$ and $\bar{1}3$, must not be calculated because it drops out when the element contributions are assembled, see Fig. 4.33, because we integrate over each line that ends at the source point two times in opposite directions.

The terms I_i mean[1]

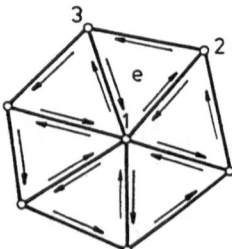

Figure 4.33 The integrals over the interior mesh lines cancel

---

[1] The integrals I_1 and I_5 were calculated by R.Dallmann. The original values, [60], are misprinted.

$$I_0 = \int_0^1 p^{-1/2}\, d\rho = \frac{1}{\sqrt{a_3}} \{ \ln\left[2\sqrt{sa_3} + a_2 + 2a_3\right] - \ln(2\sqrt{a_1 a_3} + a_2) \},$$

$$I_1 = \int_0^1 \rho\, p^{-1/2}\, d\rho = \frac{w}{a_3} - \frac{\sqrt{a_1}}{a_3} - \frac{a_2}{2a_3} I_0$$

$$I_2 = \int_0^1 p^{-3/2}\, d\rho = \frac{2(2a_2 + 2a_3)}{-qw} + \frac{2a_2}{\sqrt{a_1}\, q},$$

$$I_3 = \int_0^1 \rho\, p^{-3/2}\, d\rho = \frac{2(2a_1 + a_2)}{qw} - \frac{4a_1}{\sqrt{a_1}\, q},$$

$$I_4 = \int_0^1 \rho^2 p^{-3/2}\, d\rho = -\frac{2a_2^2 - 4a_1 a_3 + 2a_1 a_2}{a_3 qw} + \frac{2a_1 a_2}{a_3 q \sqrt{a_1}} + \frac{1}{a_3} I_0,$$

$$I_5 = \int_0^1 \rho^3 p^{-3/2}\, d\rho = \frac{-a_3 q + a_2(10a_1 a_3 - 3a_2^2) + a_1(8a_1 a_3 - 3a_2^2)}{-a_3^2 qw}$$

$$+ \frac{a_1(8a_1 a_3 - 3a_2^2)}{a_3^2 q \sqrt{a_1}} - \frac{1,5a_2}{a_3^2} I_0,$$

where

$$\rho = \rho_2, \quad p = a_1 + a_2 \rho + a_3 \rho^2, \quad s = a_1 + a_2 + a_3, \quad w = \sqrt{s}$$

$$q = a_2^2 - 4a_1 a_3$$

With these analytical expressions the element matrices G^e and H^e are complete. The element matrix G^e consists of three columns, each of which contains K blocks, where K is the number of collocation points. Block k in column m consists of the 3×3 elements (i, j)

$$\int_{\Gamma_e} U_{ij}(y, x^k)\varphi_m^e(y)\, ds_y \quad k = \text{block number}, \quad m = \text{column number}$$

The functions $\varphi_m^e, m = 1, 2, 3$ are the local basis functions, which take on the value 1 at the node m and zero at the other two nodes, as indicated in Fig. 4.34.

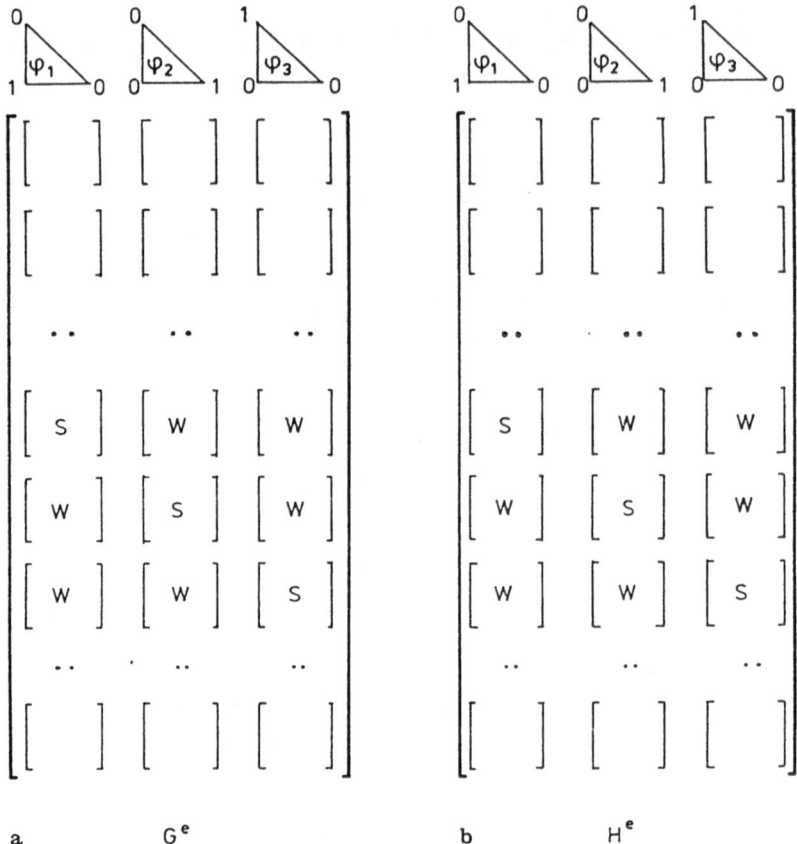

Figure 4.34 Element matrices G^e and H^e

For two reasons the matrix H^e requires yet some additional considerations. One is of a mathematical nature and the other one has to do with the partition of the boundary into elements. Let us start with the latter. The singular part of the matrix H^e consists of three rows that each contain three blocks, see Fig. 4.34b. The blocks S on the "main diagonal" represent the influence of the local basis function associated with the collocation point on the point itself. The neighboring blocks W represent the influences of the local basis functions of the neighboring nodes on the collocation point. To these blocks we would have to add on the main diagonal a matrix $1/n \times$ the C-matrix where n is the number of elements which enclose the collocation point. After the assemblage, the n matrices $1/n \times$ C-matrix would add up to the full C-matrix. It seems, therefore, simpler to delay this manoeuvre till after the assemblage and then to simply add the full C-matrix to the main diagonal of the matrix H.

1. Smooth point

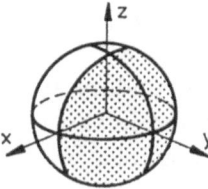

$$C = \frac{1}{8\pi} \begin{bmatrix} 4\pi & 0 & 0 \\ 0 & 4\pi & 0 \\ 0 & 0 & 4\pi \end{bmatrix}$$

2. Edge

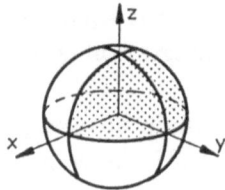

$$C = \frac{1}{8\pi} \begin{bmatrix} 2\pi & 0 & 0 \\ 0 & 2\pi & \frac{2}{1-v} \\ 0 & \frac{2}{1-v} & 2\pi \end{bmatrix}$$

3. Corner

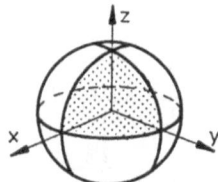

$$C = \frac{1}{8\pi} \begin{bmatrix} \pi & \frac{1}{1-v} & \frac{1}{1-v} \\ \frac{1}{1-v} & \pi & \frac{1}{1-v} \\ \frac{1}{1-v} & \frac{1}{1-v} & \pi \end{bmatrix}$$

Figure 4.35 The area $S_1(x, \Omega)$ on the unit sphere and the associated C-matrix

Now to the mathematical problem. The C-matrix is the integral of trigonometric functions over that portion of the unit sphere $S_1(x, \Omega)$, which lies inside the tangent cone. As long as this portion has a simple shape, i.e. as long as its edges run parallel to the coordinate lines $\varphi = $ const. and $\vartheta = $ const. the C-matrix can be calculated by integration, see Fig. 4.35.

At "skew" angles though, angles whose tangent cone cuts the sphere S_1 in an irregular curve, integration in general is no longer possible. These two difficulties are both circumvented if we calculate the block on the main diagonal

$$H_{kk} = C(x^k) + \int_\Gamma T(y, x^k)\varphi_k(y) \, ds_y$$

of the hyper-matrix H by summing over the off-diagonal terms.

$$H_{kk} = -\sum_{\substack{i=1 \\ i \neq k}} H_{ki}$$

On element level this means that we neglect the singular blocks, marked with S in Fig. 4.34, and calculate only the weakly singular blocks, marked W, analytically.

Better approximations of the surface as of the boundary functions are achieved with isoparametric quadratic elements. However, in this case an analytical integration of the singular integrals is no longer possible. For that reason we adopt the following approach, see Fig. 4.36. The element Γ_e is mapped onto the master element Γ_e' and this element depending on the position of the source point is split into two or three triangles, respectively, which themselves in turn are mapped onto the unit square. In the last step the unit square is mapped finally onto the square with corner points $(\pm 1, \mp 1)$. Integration is done on this element. All the necessary formulas are given in [60].

In 3-D elasticity the BEM can fully play out its main advantage: the reduction of the dimensionality of a problem. Figure 4.37 shows a crankshaft that was analysed with the program DBETSY-3D, [61]. The mesh generator required approximately only 20 parameters as input. The mesh consisted of

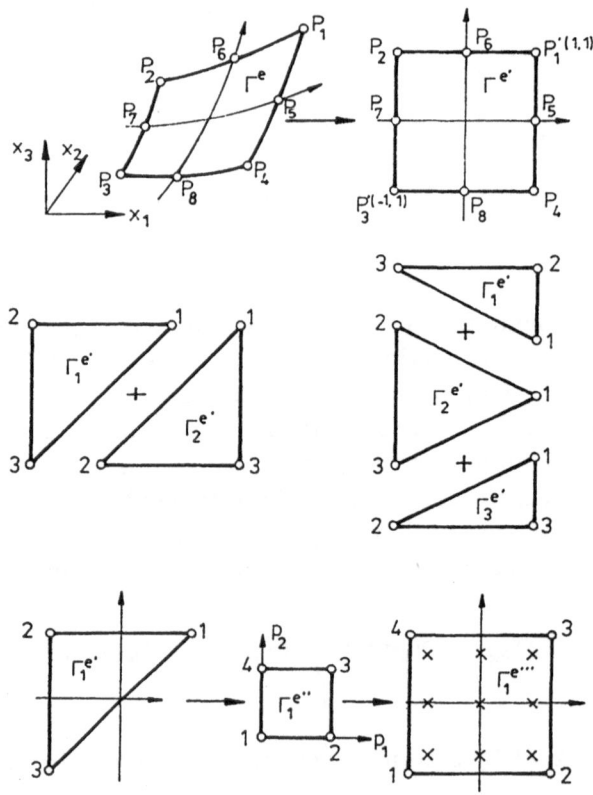

Figure 4.36 The curved element is mapped stepwise onto simple elements

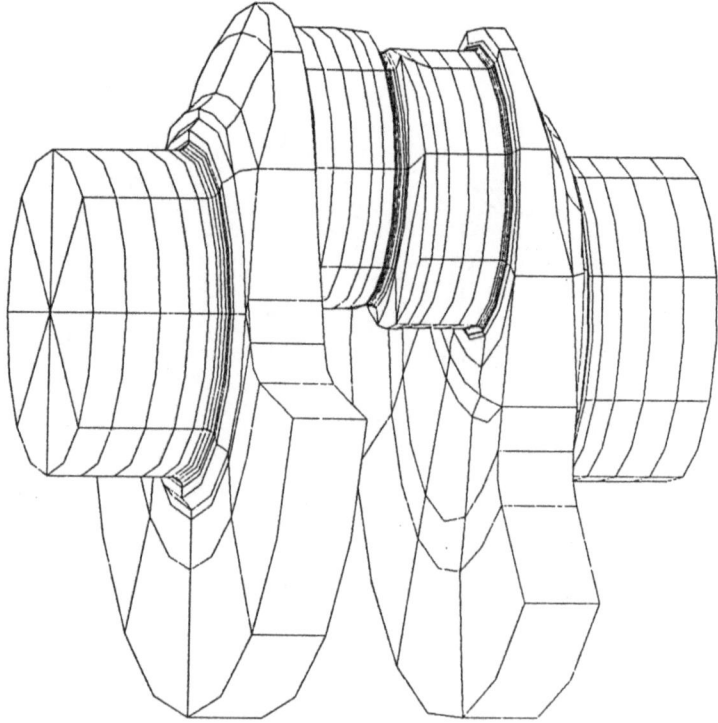

Figure 4.37 Double throw of a crossed crankshaft which was subdivided into five substructures with a total of 636 elements and 1714 nodes

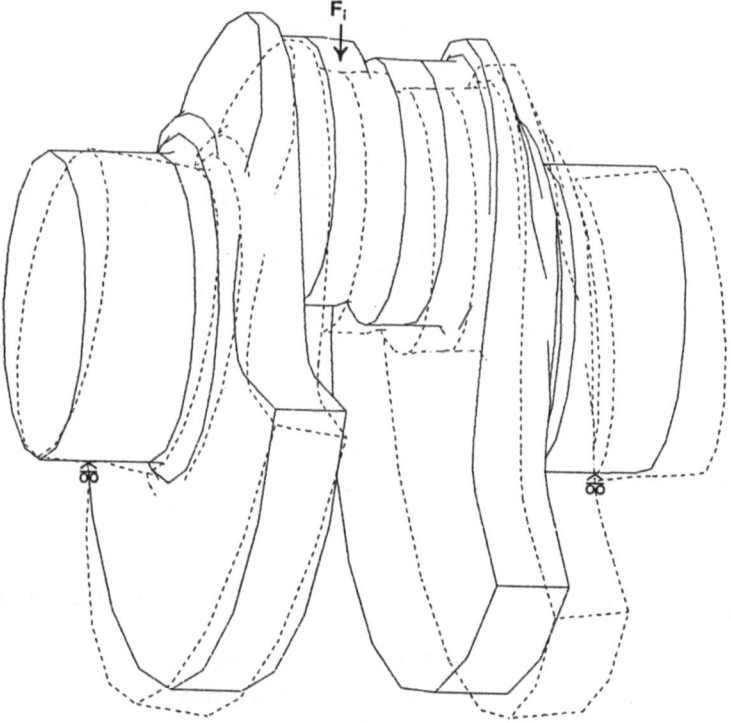

Figure 4.38 Deformed (dashed-line) and undeformed structure

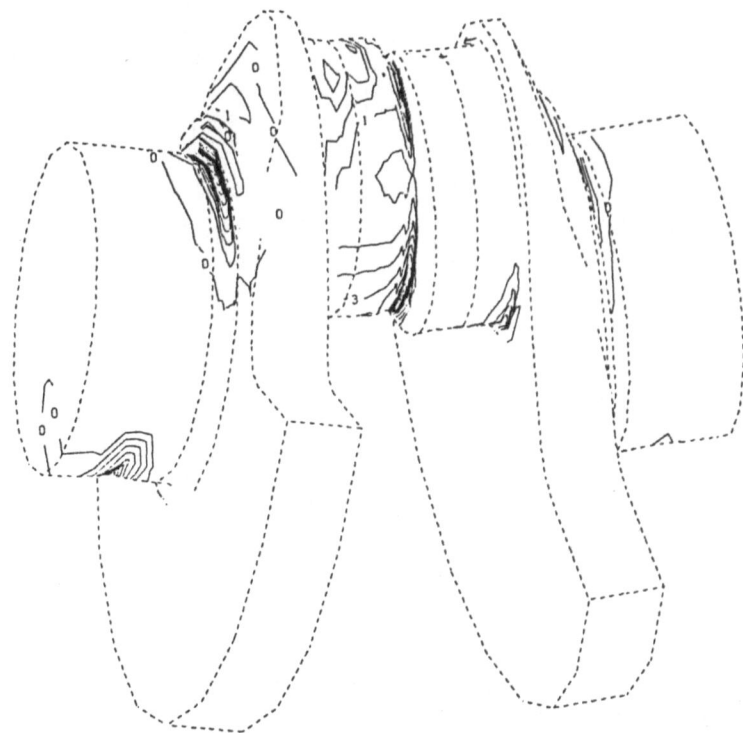

Figure 4.39 Distribution of surface-stresses (von Mises)

1714 nodes, 636 elements and the interior was subdivided into 5 substructures. Figure 4.38 shows the deformed structure (dotted lines) over the original structure shortly after ignition has occurred. The crankshaft is supported at the centres of the main journals. Figure 4.39 shows the lines of constant stresses. The dense distribution near the notches signals the occurrence of high stress gradients at these points.

4.14 Axisymmetric problems

by K. Kremer

The cylindrical coordinates of the source point x and the field point y are denoted by

$$
\begin{aligned}
x_1 &= r\cos\varphi & y_1 &= \rho\cos\psi \\
x_2 &= r\sin\varphi & y_2 &= \rho\sin\psi \\
x_3 &= z & y_3 &= \zeta
\end{aligned}
$$

The fundamental solutions are the response of the continuum due to the action of radial, axial and tangential ringloads, see Fig. 4.40. The components of these fields are obtained by adding the influence of the single point sources, that is by integrating over the circle. The combined action of the radial forces will cause at some distant point y the displacements

$$U_{rr}(\boldsymbol{y}, \boldsymbol{x}) = \frac{1}{2\pi} \int_0^{2\pi} [U_{11}\cos\varphi + U_{21}\sin\varphi]d\varphi$$

$$U_{rz}(\boldsymbol{y}, \boldsymbol{x}) = \frac{1}{2\pi} \int_0^{2\pi} [U_{13}\cos\varphi + U_{23}\sin\varphi]d\varphi$$

the axial forces will cause the displacements

$$U_{zr}(\boldsymbol{y}, \boldsymbol{x}) = \frac{1}{2\pi} \int_0^{2\pi} U_{31}d\varphi$$

$$U_{zz}(\boldsymbol{y}, \boldsymbol{x}) = \frac{1}{2\pi} \int_0^{2\pi} U_{33}d\varphi$$

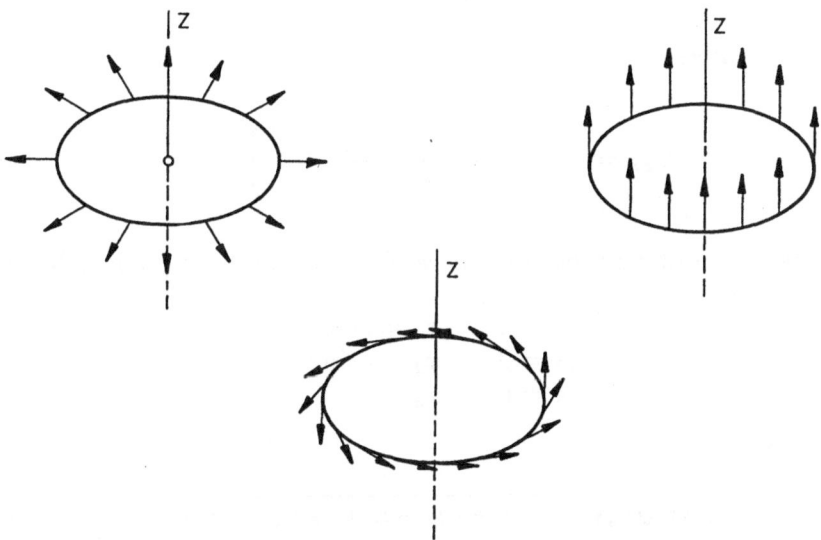

Figure 4.40 Radial, axial and tangential ringloads

and the tangential forces the displacements

$$U_{\varphi\varphi}(\boldsymbol{y}, \boldsymbol{x}) = \frac{1}{2\pi} \int_0^{2\pi} [-U_{12} \sin\varphi + U_{22} \cos\varphi] d\varphi$$

Note that radial and axial effects influence each other but no such coupling exists between circumferential (tangential) effects and the other two effects.

The kernel in the integral U_{zz} is

$$U_{33} = \frac{1}{16\pi\mu(1-\nu)r}[(3-4\nu) + r_{,3}^2]$$

Because of

$$r_{,3} = \frac{y_3 - x_3}{r} = \frac{\zeta - z}{r}$$

its equation is equivalent to

$$U_{33} = \frac{1}{16\pi\mu(1-\nu)}\left[\frac{3-4\nu}{r} + \frac{(\zeta-z)^2}{r^3}\right]$$

so that, if we let

$$A = \frac{1}{16\pi\mu(1-\nu)}$$

the integral becomes

$$U_{zz} = \frac{A}{2\pi} \int_0^{2\pi} \left[\frac{3-4\nu}{r} + \frac{(\zeta-z)^2}{r^3}\right] d\varphi \qquad (4.20)$$

If we assume, without any loss of generality, that the field point $\boldsymbol{y}$ lies in the y_1, y_3-plane

$$\begin{aligned} y_1 &= \rho & x_1 &= r\cos\varphi \\ y_2 &= 0 & x_2 &= r\sin\varphi \\ y_3 &= \zeta & x_3 &= z \end{aligned}$$

then the distance

$$r = \sqrt{(y_1 - x_1)^2 + (y_2 - x_2)^2 + (y_3 - x_3)^2}$$

becomes

$$r = \sqrt{(\rho - r\cos\varphi)^2 + (r\sin\varphi)^2 + (\zeta - z)^2}$$

$$= \sqrt{\rho^2 - 2\rho r\cos\varphi + r^2\cos^2\varphi + r^2\sin^2\varphi + (\zeta - z)^2}$$

$$= \sqrt{\rho^2 - 2\rho r\cos\varphi + r^2 + (\zeta - z)^2}$$

$$= \sqrt{(\rho + r)^2 - 2\rho r(1 + \cos\varphi) + (\zeta - z)^2}$$

If we let

$$C = \sqrt{(\rho + r)^2 + (\zeta - z)^2}$$

then this simplifies to :

$$r = \sqrt{C^2 - 2\rho r(1 + \cos\varphi)} = C\sqrt{1 - \frac{2\rho r}{C^2}(1 + \cos\varphi)}$$

or

$$r = C\sqrt{1 - \frac{2\rho r}{C^2}2\cos^2\frac{\varphi}{2}} \tag{4.21}$$

Introducing $\varphi = 2\alpha - \pi$ we obtain

$$r = C\sqrt{1 - \frac{4\rho r}{C^2}\cos^2(\alpha - \frac{\pi}{2})}$$

or

$$r = C\sqrt{1 - \frac{4\rho r}{C^2}\sin^2\alpha} = C\sqrt{1 - m^2\sin^2\alpha}$$

where

$$m = \frac{2\sqrt{\rho r}}{C}$$

and

$$\tilde{z} = \zeta - z$$

If we substitute Eq.(4.21) into Eq.(4.20) then we obtain

$$U_{zz} = \frac{2A}{\pi}\left[\int_0^{\frac{\pi}{2}} \frac{3 - 4\nu}{C\sqrt{1 - m^2\sin^2\alpha}}d\alpha + \int_0^{\frac{\pi}{2}} \frac{(\zeta - z)^2}{C^3(1 - m^2\sin^2\alpha)^{3/2}}d\alpha\right]$$

which in terms of elliptic integrals can be written as

$$U_{zz} = \frac{2A}{\pi}\left[\frac{3 - 4\nu}{C}K(\frac{\pi}{2}, m) + \frac{(\zeta - z)^2}{C^3}\frac{1}{1 - m^2}E(\frac{\pi}{2}, m)\right]$$

$$=\frac{2A}{C\pi}\left[(3-4\nu)K(\frac{\pi}{2},m)+\frac{\tilde{z}^2}{D}E(\frac{\pi}{2},m)\right]$$

where

$$D=(\rho-r)^2+(\zeta-z)^2$$

The same treatment can be applied to the other integrals as well. Considering the equations

$$\int_0^{\frac{\pi}{2}}\frac{\cos^2\vartheta}{(1-m^2\sin^2\vartheta)^{\frac{1}{2}}}d\vartheta=\frac{1}{m^2}[E(\frac{\pi}{2},m)-(1-m^2)K(\frac{\pi}{2},m)]$$

$$\int_0^{\frac{\pi}{2}}\frac{1}{(1-m^2\sin^2\vartheta)^{\frac{3}{2}}}d\vartheta=\frac{1}{1-m^2}E(\frac{\pi}{2},m)$$

$$\int_0^{\frac{\pi}{2}}\frac{\cos^2\vartheta}{(1-m^2\sin^2\vartheta)^{\frac{3}{2}}}d\vartheta=\frac{1}{m^2}[K(\frac{\pi}{2},m)-E(\frac{\pi}{2},m)]$$

$$\int_0^{\frac{\pi}{2}}\frac{\cos^4\vartheta}{(1-m^2\sin^2\vartheta)^{\frac{3}{2}}}d\vartheta=\left(\frac{2}{m^4}-\frac{1}{m^2}\right)E(\frac{\pi}{2},m)-\left(\frac{2}{m^4}-\frac{2}{m^2}\right)K(\frac{\pi}{2},m)$$

we obtain

$$U_{rr}=\frac{A}{C\pi\rho r}\left\{\left[-4(1-\nu)C^2+\frac{(\rho^2-r^2)^2+\tilde{z}^2(\rho^2+r^2)}{D}\right]E(\frac{\pi}{2},m)\right.$$

$$\left.+\left[4(1-\nu)(\rho^2+r^2+\tilde{z}^2)-(\rho^2+r^2)\right]K(\frac{\pi}{2},m)\right\}$$

$$U_{rz}=\frac{A\tilde{z}}{C\pi r}\left\{\frac{\rho^2-r^2+\tilde{z}^2}{D}E(\frac{\pi}{2},m)-K(\frac{\pi}{2},m)\right\}$$

$$U_{zr}=\frac{A\tilde{z}}{C\pi\rho}\left\{\frac{\rho^2-r^2-\tilde{z}^2}{D}E(\frac{\pi}{2},m)+K(\frac{\pi}{2},m)\right\}$$

$$U_{\varphi\varphi}=\frac{1}{2\pi\mu r\rho C}[-C^2E(\frac{\pi}{2},m)+(r^2+\rho^2+\tilde{z}^2)K(\frac{\pi}{2},m)]$$

The tractions of the three fundamental solutions are the integrals

$$T_{rr}(\boldsymbol{y}, \boldsymbol{x}) = \frac{1}{2\pi} \int_0^{2\pi} [T_{11}\cos\varphi + T_{21}\sin\varphi] d\varphi$$

$$T_{rz}(\boldsymbol{y}, \boldsymbol{x}) = \frac{1}{2\pi} \int_0^{2\pi} [T_{13}\cos\varphi + T_{23}\sin\varphi] d\varphi$$

$$T_{zr}(\boldsymbol{y}, \boldsymbol{x}) = \frac{1}{2\pi} \int_0^{2\pi} T_{31} d\varphi$$

$$T_{zz}(\boldsymbol{y}, \boldsymbol{x}) = \frac{1}{2\pi} \int_0^{2\pi} T_{33} d\varphi$$

$$T_{\varphi\varphi}(\boldsymbol{y}, \boldsymbol{x}) = \frac{1}{2\pi} \int_0^{2\pi} [-T_{12}\sin\varphi + T_{22}\cos\varphi] d\varphi$$

and although the calculation of these integrals, in principle, can be done as before we prefer to calculate the stresses by differentiating the newly derived fundamental solutions directly.

The stresses are defined as

$$\sigma_{rr} = \frac{2\mu\nu}{1-2\nu}\left[\frac{u_r}{r} + \frac{\partial u_z}{\partial z} + \left(\frac{1-\nu}{\nu}\right)\frac{\partial u_r}{\partial r}\right]$$

$$\sigma_{rz} = \mu\left[\frac{\partial u_z}{\partial r} + \frac{\partial u_r}{\partial z}\right]$$

$$\sigma_{zz} = \frac{2\mu\nu}{1-2\nu}\left[\frac{u_r}{r} + \frac{\partial u_r}{\partial r} + \left(\frac{1-\nu}{\nu}\right)\frac{\partial u_z}{\partial z}\right]$$

$$\sigma_{r\varphi} = \mu\left[\frac{\partial u_\varphi}{\partial r} - \frac{u_\varphi}{r}\right]$$

$$\sigma_{rz} = \mu\left[\frac{\partial u_\varphi}{\partial z}\right]$$

so that the stresses caused by the radial ringload are

$$\sigma^r_{rr} = \frac{2\mu\nu}{1-2\nu}\left[\frac{U_{rr}}{r} + \frac{\partial U_{rz}}{\partial z} + \left(\frac{1-\nu}{\nu}\right)\frac{\partial U_{rr}}{\partial r}\right]$$

$$\sigma^r_{rz} = \mu\left[\frac{\partial U_{rz}}{\partial r} + \frac{\partial U_{rr}}{\partial z}\right]$$

$$\sigma^r_{zz} = \frac{2\mu\nu}{1-2\nu}\left[\frac{U_{rr}}{r} + \frac{\partial U_{rr}}{\partial r} + \left(\frac{1-\nu}{\nu}\right)\frac{\partial U_{rz}}{\partial z}\right]$$

the stresses caused by the axial ringload

$$\sigma^z_{rr} = \frac{2\mu\nu}{1-2\nu}\left[\frac{U_{zr}}{r} + \frac{\partial U_{zz}}{\partial z} + \left(\frac{1-\nu}{\nu}\right)\frac{\partial U_{zr}}{\partial r}\right]$$

$$\sigma^z_{rz} = \mu\left[\frac{\partial U_{rz}}{\partial r} + \frac{\partial U_{zr}}{\partial z}\right]$$

$$\sigma^z_{zz} = \frac{2\mu\nu}{1-2\nu}\left[\frac{U_{zr}}{r} + \frac{\partial U_{zr}}{\partial r} + \left(\frac{1-\nu}{\nu}\right)\frac{\partial U_{zz}}{\partial z}\right]$$

and the stresses caused by the tangential ringload

$$\sigma^\varphi_{r\varphi} = \mu\left[\frac{\partial U_{\varphi\varphi}}{\partial r} - \frac{U_{\varphi\varphi}}{r}\right]$$

$$\sigma^\varphi_{r\varphi} = \mu\frac{\partial U_{\varphi\varphi}}{\partial z}$$

The derivatives of the elliptic integrals are

$$\frac{dK(\frac{\pi}{2},m)}{dm} = \frac{E(\frac{\pi}{2},m) - (1-m^2)K(\frac{\pi}{2},m)}{m(1-m^2)}$$

$$\frac{dE(\frac{\pi}{2},m)}{dm} = \frac{E(\frac{\pi}{2},m) - K(\frac{\pi}{2},m)}{m}$$

so that we obtain, after some rather technical steps, for the stresses which are caused by the radial load the expressions

$$\sigma_{rr}^r = \frac{A\mu}{2\pi\rho^2 rC}\left\{\left[(7-8\nu)C^2 + \frac{4(1-2\nu)[-r^2 B + \tilde{z}^2(2r^2 + F)]}{D}\right.\right.$$

$$+\frac{-3B^2 - 6\tilde{z}^2 B + 9\tilde{z}^4}{D}$$

$$+\left.\frac{4[(r^2-\rho^2)^3 - \tilde{z}^6 + \tilde{z}^2 B(\tilde{z}^2 + B)]L}{C^2 D^2}\right] E$$

$$+\left[3B - 9\tilde{z}^2 - 4(1-2\nu)(\rho^2 + 2M)\right.$$

$$\left.\left.-\frac{(r^2-\rho^2)^3 - \tilde{z}^6 + \tilde{z}^2 B(\tilde{z}^2 + B)}{C^2 D}\right] K\right\}$$

$$\sigma_{rz}^r = \frac{A\mu\tilde{z}}{\pi\rho rC}\left\{\left[-\frac{6\tilde{z}^2 + 2(1-2\nu)L}{D} + \frac{4(\tilde{z}^4 - B^2)L}{C^2 D^2}\right] E\right.$$

$$\left.+\left[3 + 2(1-2\nu) - \frac{\tilde{z}^4 - B^2}{C^2 D}\right] K\right\}$$

$$\sigma_{zz}^r = \frac{2A\mu}{\pi rC}\left\{\left[\frac{(B+\tilde{z}^2)(1-2\nu) + 3\tilde{z}^2}{D} - \frac{4\tilde{z}^2 L(B+\tilde{z}^2)}{C^2 D^2}\right] E\right.$$

$$\left.-\left[1-2\nu - \frac{\tilde{z}^2(B+\tilde{z}^2)}{C^2 D}\right] K\right\}$$

for the stresses caused by the axial load the expressions

$$\sigma_{rr}^z = \frac{A\mu\tilde{z}}{\pi\rho^2 C}\left\{\left[\frac{4(1-2\nu)\rho^2 - 6(B-\tilde{z}^2)}{D} - \frac{4L(B-\tilde{z}^2)^2}{C^2 D^2}\right] E\right.$$

$$\left.+\left[-3 + \frac{(B-\tilde{z}^2)^2}{C^2 D}\right] K\right\}$$

$$\sigma_{rz}^z = \frac{-2A\mu}{\pi\rho C}\left\{\left[\frac{(B-\tilde{z}^2)(1-2\nu) + 3\tilde{z}^2}{D} + \frac{4\tilde{z}^2 L(B-\tilde{z}^2)}{C^2 D^2}\right] E\right.$$

$$\left.+\left[1-2\nu - \frac{\tilde{z}^2(B-\tilde{z}^2)}{C^2 D}\right] K\right\}$$

$$\sigma_{zz}^z = \frac{4A\mu\tilde{z}}{\pi C}\left\{\left[-\frac{1-2\nu}{D} - \frac{4\tilde{z}^2 L}{C^2 D^2}\right] E + \left[\frac{\tilde{z}^2}{C^2 D}\right] K\right\}$$

and by the tangential load the expressions

$$\sigma^\varphi_{r\varphi} = \frac{1}{2\pi\rho}\left[-\frac{2r^2 + 4\rho^2 + 4\tilde{z}^2}{2r^2C}K + \frac{\rho^4 - r^4 + z^4 + 2\rho^2\tilde{z}^2 + 3C^2D}{2r^2CD}\right]$$

$$\sigma^\varphi_{\varphi z} = \frac{\tilde{z}}{2\pi r\rho C}\left[K - \frac{r^2 + \rho^2 + \tilde{z}^2}{D}E\right]$$

where the capital letters have the following meaning

$$A = \frac{1}{16\pi G(1 - \nu)}$$

$$B = \rho^2 - r^2$$

$$C = \sqrt{(\rho + r)^2 + (\zeta - z)^2}$$

$$D = (\rho - r)^2 + (\zeta - z)^2$$

$$E = E(\frac{\pi}{2}, m)$$

$$F = \rho^2 + (\zeta - z)^2$$

$$K = K(\frac{\pi}{2}, m)$$

$$L = \rho^2 + r^2 + (\zeta - z)^2$$

$$M = r^2 + (\zeta - z)^2$$

The tractions are then simply the product between the stress tensor and the normal vector on the boundary

$$T_{ij} = \sigma^i_{kj}n_k$$

By formulating Betti's principle with these two fundamental solutions we obtain two influence functions for the radial ($i = 1$) and vertical ($i = 2$) displacement of an axisymmetric displacement field

$$c_{ij}u_i(\boldsymbol{x}) + \int_\Gamma T_{ij}(\boldsymbol{x}, \boldsymbol{y})u_j(\boldsymbol{y})ds_{\boldsymbol{y}} = \int_\Gamma U_{ij}(\boldsymbol{x}, \boldsymbol{y})t_j(\boldsymbol{y})ds_{\boldsymbol{y}}$$

The integrals are to be taken over the contour Γ of the axisymmetric body. The characteristic functions c_{ij} are the same functions as in 2-D elasticity and also the numerical technique, naturally, is identical with the technique developed in the foregoing chapters. The contour is divided into boundary elements and the $2n$ nodal values of the boundary functions are determined by formulating $2n$ equations at n collocation points.

$$Hu = Gt$$

The rigid-body-motion trick which furnishes the diagonal terms of the H-matrix is only applicable in the z-direction where it renders

$$\begin{bmatrix} 0 & h_{rz} \\ 0 & h_{zz} \end{bmatrix}_{ii} = -\sum_{j=1}^{n} \begin{bmatrix} 0 & h_{rz} \\ 0 & h_{zz} \end{bmatrix}_{ij} \quad (j \neq i)$$

and in the tangential φ-direction ($h = h_{\varphi\varphi}$)

$$h_{ii} = -\frac{1}{r_i} \sum_{j=1}^{n} h_{ij} r_j$$

To calculate the other main-diagonal terms of the H-matrix we can use the boundary data of one of the following three simple stress states.

Hydrostatic pressure

In this case we have

$$\sigma_{ij} = \delta_{ij}$$

so that the tractions are

$$t_r = n_r \qquad t_z = n_z$$

The displacements follow from Hooke's law

$$u_r = \frac{1 - 2\nu}{E} r \qquad u_z = \frac{1 - 2\nu}{E} z$$

State of plane strain

Here the boundary conditions are

$$\sigma_{rr} = \sigma_{\varphi\varphi} = 1 \qquad e_{zz} = 0$$

and the corresponding boundary values are

$$t_r = n_r \qquad t_z = 2\nu n_z$$

$$u_r = \frac{(1 - 2\nu)(1 + \nu)}{E} r \qquad u_z = 0$$

State of plane stress

The conditions

$$\sigma_{rr} = \sigma_{\varphi\varphi} = 1 \qquad \sigma_{zz} = 0$$

render the following boundary data

$$t_r = n_r \qquad t_z = 0$$

$$u_r = \frac{1-\nu}{E}\, r \qquad u_z = \frac{-2\nu}{E}\, z$$

With either one of these three stress states the missing terms of the H-matrix can be calculated

$$\begin{bmatrix} h_{rr} & 0 \\ h_{zr} & 0 \end{bmatrix}_{ii} \begin{bmatrix} u_r \\ 0 \end{bmatrix}_i = \sum_{j=1}^{n} \begin{bmatrix} g_{rr} & g_{rz} \\ g_{zr} & g_{zz} \end{bmatrix}_{ij} \begin{bmatrix} t_r \\ t_z \end{bmatrix}_j$$

$$- \sum_{j=1}^{n} \begin{bmatrix} h_{rr} & h_{rz} \\ h_{zr} & h_{zz} \end{bmatrix}_{ij} \begin{bmatrix} u_r \\ u_z \end{bmatrix}_j \qquad (j \neq i)$$

$$- \begin{bmatrix} 0 & h_{rz} \\ 0 & h_{zz} \end{bmatrix}_{ii} \begin{bmatrix} 0 \\ u_z \end{bmatrix}_i$$

To calculate the stresses

$$\sigma_{rr}, \sigma_{\varphi\varphi}, \sigma_{zz}, \sigma_{rz}, \sigma_{r\varphi}, \sigma_{z\varphi}$$

on the boundary we define a local system of coordinates with directions tangential (1) and normal (2) to the element. For the strains in radial and axial direction it then follows

$$e_{\varphi\varphi} = \frac{u_r}{r}$$

$$u_1(\xi) = \left[\sum_{i=1}^{3} \Phi_i(\xi)(u_r)_i \right] m_{1r} + \left[\sum_{i=1}^{3} \Phi_i(\xi)(u_z)_i \right] m_{1z}$$

$$e_{11} = \frac{1}{l} \left\{ \left[\sum_{i=1}^{3} \frac{\partial \Phi_i(\xi)}{\partial \xi}(u_r)_i \right] m_{1r} + \left[\sum_{i=1}^{3} \frac{\partial \Phi_i(\xi)}{\partial \xi}(u_z)_i \right] m_{1z} \right\}$$

where m_{1i} is the component of the tangent vector with regard to the direction i. In this local system of coordinates the tractions are

$$t_1 = t_z n_r - t_r n_z \qquad t_2 = t_r n_r + t_z n_z$$

and, therefore, the local stresses

$$\sigma_{11} = \frac{E}{1 - \nu^2}(e_{11} + \nu e_{\varphi\varphi}) + \frac{\nu}{1 - \nu}t_2$$

$$\sigma_{22} = t_2 \qquad\qquad \sigma_{12} = \sigma_{21} = t_1$$

In terms of β, the angle between the normal vector on the boundary and the horizontal direction, the global stresses can be expressed as

$$\sigma_{rr} = \sigma_{11} \sin^2 \beta + \sigma_{22} \cos^2 \beta - 2\sigma_{12} \sin \beta \cos \beta$$

$$\sigma_{zz} = \sigma_{11} \cos^2 \beta + \sigma_{22} \sin^2 \beta + 2\sigma_{12} \sin \beta \cos \beta$$

$$\sigma_{rz} = -\sigma_{11} \sin \beta \cos \beta + \sigma_{22} \sin \beta \cos \beta + \sigma_{12}(\cos^2 \beta - \sin^2 \beta)$$

$$\sigma_{\varphi\varphi} = E e_\varphi + \nu(\sigma_{11} + \sigma_{22})$$

$$\sigma_{r\varphi} = \left[(\frac{t_\varphi}{\mu} + \frac{u_\varphi}{r} \cos \beta) \cos \beta - u_{\varphi,s} \sin \beta - \frac{u_\varphi}{r} \right] \mu$$

$$\sigma_{\varphi z} = \left[(\frac{t_\varphi}{\mu} + \frac{u_\varphi}{r} \cos \beta) \sin \beta + u_{\varphi,s} \cos \beta \right] \mu$$

If the contour of the axisymmetric body has its endpoint on the axis of rotation then at some instances the radial distance of the source point is zero, $r = 0$, therefore $m = 0$ as well so that

$$K(\frac{\pi}{2}, 0) = E(\frac{\pi}{2}, 0) = \frac{\pi}{2}$$

It seems that in such a case the fundamental solutions become infinite. But a radial load at the source point cannot cause displacements and stresses and therefore we must have $U_{ri} = \sigma_{ij}^r = 0$. The other terms do not contain the polar coordinate r in the denominator so that the fundamental solutions become

$$U_{rr}(P, Q) = 0$$

$$U_{rz}(P, Q) = 0$$

$$U_{zr}(P, Q) = -\frac{A r_Q \tilde{z}}{C^3}$$

$$U_{zz}(P, Q) = \frac{A}{C} \left[3 - 4\nu + \frac{1}{C^2}\tilde{z}^2 \right]$$

$$U_{\varphi\varphi}(P, Q) = 0$$

$$T_{rr}(P, Q) = 0$$

$$T_{rz}(P,Q) = 0$$

$$T_{zr}(P,Q) = 2\mu \left\{ \frac{A\tilde{z}}{C^3} \left[2(1+\nu) - \frac{3\tilde{z}^2}{C^2} \right] n_r - \frac{Ar_Q}{C^3} \left[1 - 2\nu + \frac{3\tilde{z}^2}{C^2} \right] n_z \right\}$$

$$T_{zz}(P,Q) = 2\mu \left\{ \frac{A\tilde{z}}{C^3} \left[1 - 2\nu - \frac{3\tilde{z}^2}{C^2} \right] n_z - \frac{Ar_Q}{C^3} \left[1 - 2\nu + \frac{3\tilde{z}^2}{C^2} \right] n_r \right\}$$

$$\sigma^\varphi_{r\varphi}(P,Q) = 0 \qquad \sigma^\varphi_{\varphi z}(P,Q) = 0$$

This case requires no particular precautions in a boundary element code. It is only that we have to take care that the program does not calculate the zero terms but leaves them as such. Because the corresponding row of the system matrix then only contains zeros we only have to add a 1 on the main diagonal to make the system regular and to obtain the correct r-displacement. Note also that the upper row of the corresponding C-matrix must not to be evaluated in this case.

The elliptic integrals are best evaluated by the formulas, [62],

$$K(\frac{\pi}{2},m) = \ln 4 + \sum_{i=1}^{n} a_i(1 - m^2)^i + \ln(\frac{1}{1-m^2}) \left[\frac{1}{2} \sum_{i=1}^{n} b_i(1 - m^2)^i \right]$$

$$E(\frac{\pi}{2},m) = 1 + \sum_{i=1}^{n} c_i(1 - m^2)^i + \ln(\frac{1}{1-m^2}) \left[\sum_{i=1}^{n} d_i(1 - m^2)^i \right]$$

These series converge very rapidly. For $n = 5$ terms the error in the series is less than 10^{-8} %.

The first five coefficients a_i, b_i, c_i and d_i in these series are

$a_1 = 0.096578619622$	$b_1 = 0.12499929597$
$a_2 = 0.031559431627$	$b_2 = 0.070148757782$
$a_3 = 0.023761224857$	$b_3 = 0.044983875539$
$a_4 = 0.025962888452$	$b_4 = 0.018751660276$
$a_5 = 0.0066398011146$	$b_5 = 0.0018472341632$
$c_1 = 0.44315287472$	$d_1 = 0.24999920273$
$c_2 = 0.057566998484$	$d_2 = 0.093564907830$
$c_3 = 0.031761145524$	$d_3 = 0.054260524448$
$c_4 = 0.031662347457$	$d_4 = 0.021836021169$
$c_5 = 0.0076529606032$	$d_5 = 0.0021247918284$

4.15 Examples

All the following BE-results were obtained with straight elements and quadratic basis functions.

The first example is a cylinder with a circular cavity, see Fig. 4.41, which is stressed uniformly in the z-direction. Peterson [54] obtained his solution with a 3D–BE–program and a mesh as in Fig. 4.41a. Figure 4.42 contains a comparison of the BE-results with Peterson's 3-D results and the analytic results. The deviations of the values ◇ result from the approximation of the circular cavity with straight elements. The artificially introduced edges effect an increase in the stress intensity factor.

The second example is a conical shell, which was chosen to assess how good the BE-code can handle such thin-walled structures. In Fig. 4.43 is plotted the radial displacement at the upper edge versus the ratio thickness : element length, l/t.

The third example is a pressurized tank, see Fig. 4.44. This too is a thin-walled structure. The discretization of such structures with boundary elements requires nearly as much effort as with finite elements. For results see table 4.1. The results of Márkus [63], though based on a series expansion, are only of semi-analytical character because the stiffening effect of the welding between the cylindrical tube and the circular main tank could only be approximately cared for.

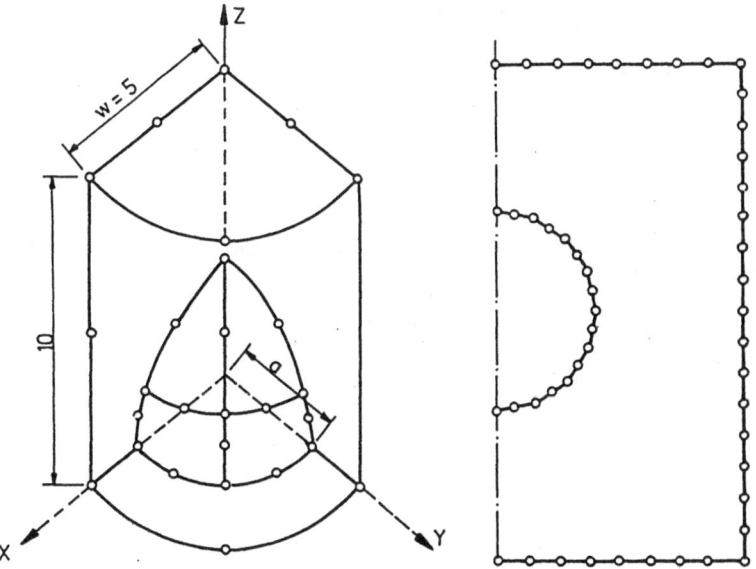

Figure 4.41 3-D BE-model of a shaft with a circular cavity and the same structure modelled with axisymmetric elements

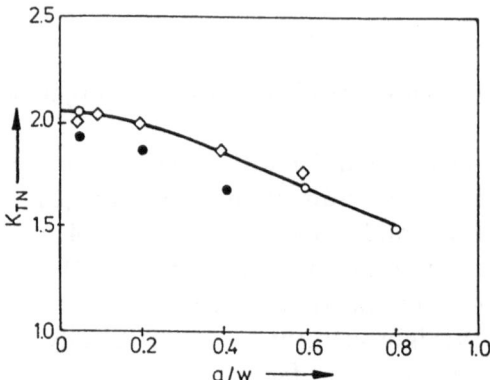

Figure 4.42 The stress intensity factors

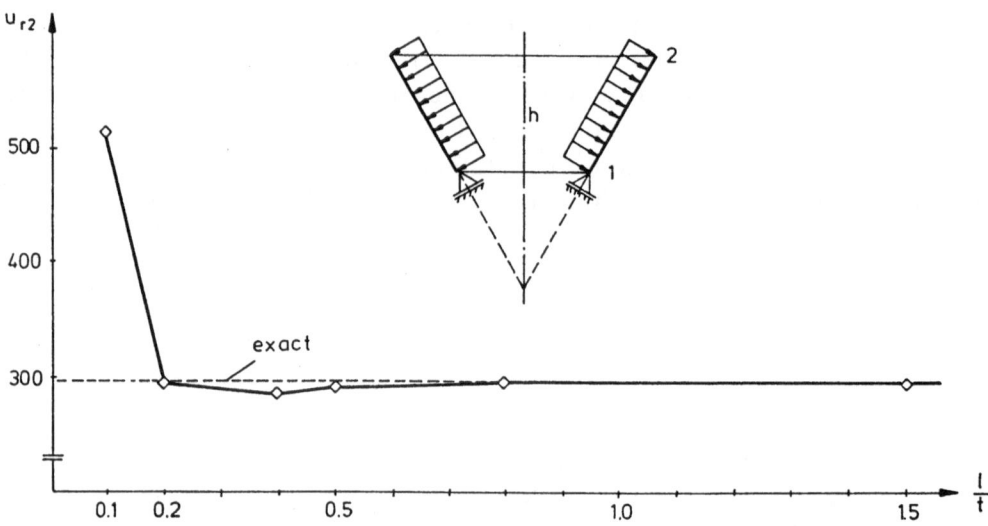

Figure 4.43 Conical shell. Plotted is the radial displacement at the upper edge versus the ratio l/t = elementlength/thickness

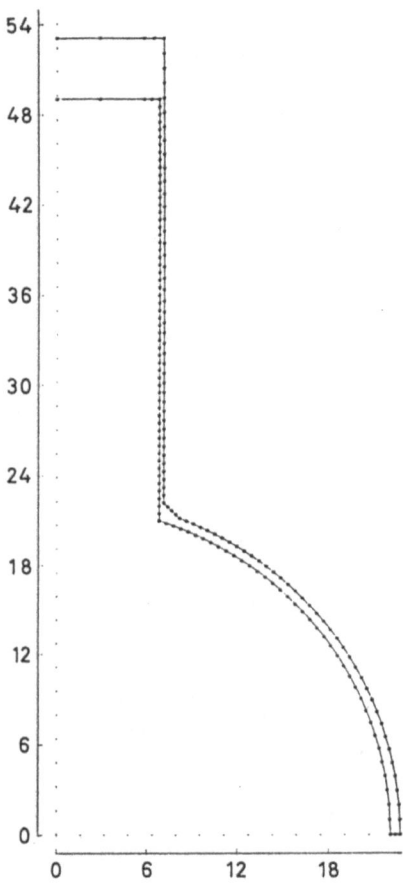

Figure 4.44 Axisymmetric model of a thin-walled tank

Table 4.1

		Márkus	BEM	BEM	FEM
nodes			193	237	231
unit		$[10^{-3}cm]$	$[10^{-3}cm]$	$[10^{-3}cm]$	$[10^{-3}cm]$
base	u_r	1.387	1.406	1.380	1.370
cylinder base	u_r	1.437	1.438	1.449	1.435
cylinder base	u_z	–	3.012	3.049	2.989
cylinder	u_r	0.655	≈ 0.7	0.650	0.650
head(inside)	u_z	–	3.606	3.625	3.553
head(outside)	u_z	–	3.459	3.487	3.434

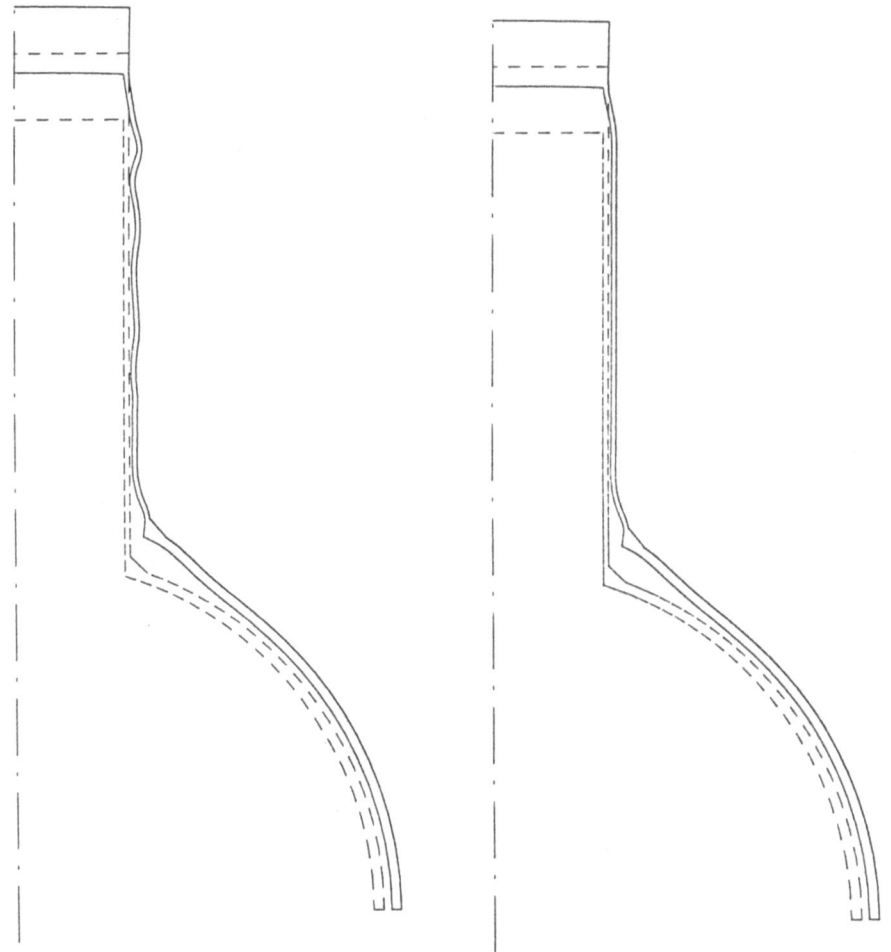

Figure 4.45 The wavy deformation pattern is an indication that a discretization with 193 nodes for this thin-walled structure is not sufficient

Figure 4.46 Deformation pattern obtained with 237 nodes

The wavy deformation pattern of the upper part of the tank, see Fig. 4.45, is an indication that a discretization with 193 nodes is not sufficient. A discretization with 237 nodes straightened this out, see Fig. 4.46. A control calculation with 347 nodes showed no further gain in accuracy.

Exercises

1. Let a concentrated force $P = e_i$ be located at a point x of the infinite plane elastic continuum and let Γ be a circle centred at x, see Fig. 4.47a,b. Show

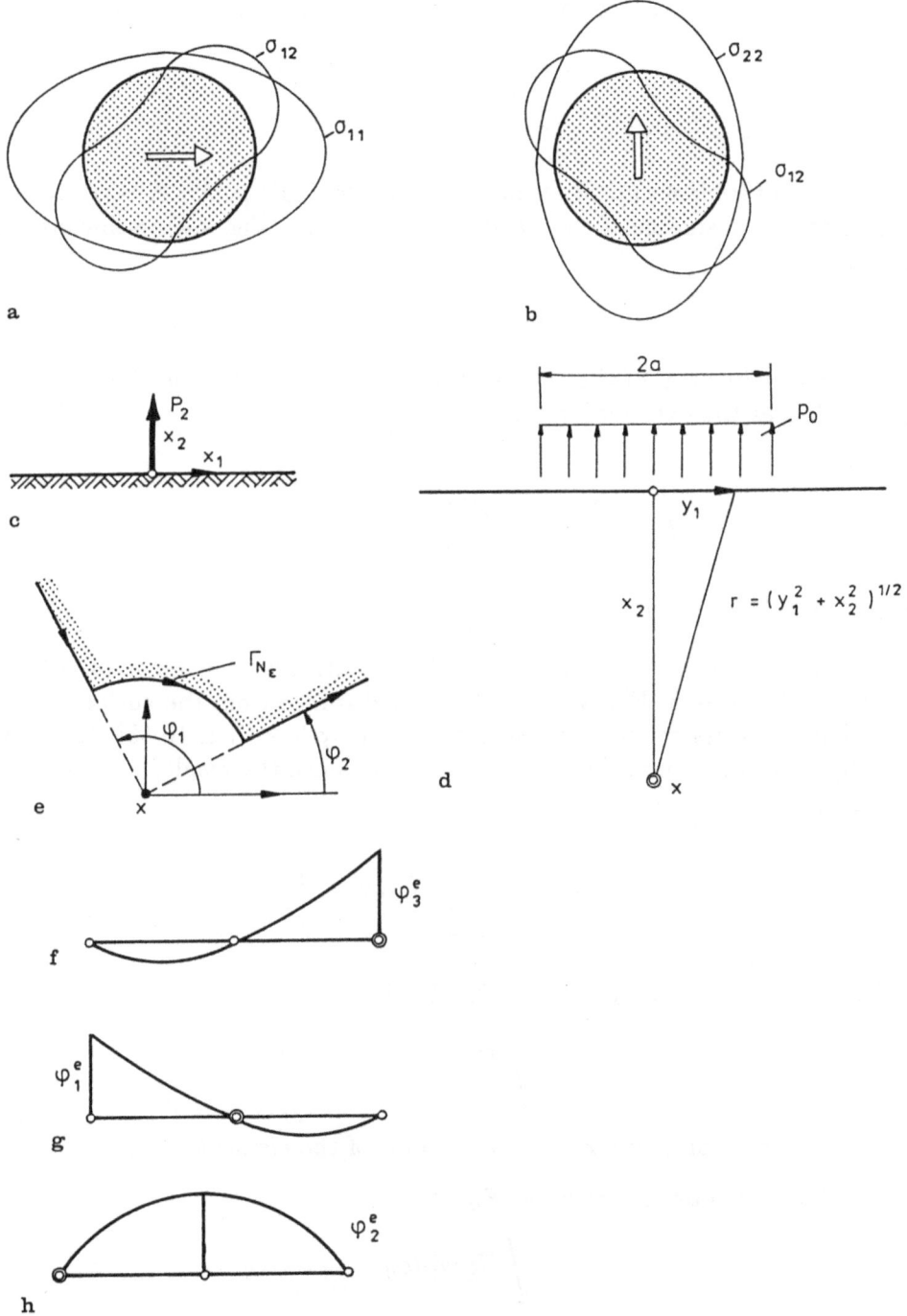

Figure 4.47 a-h Exercises: **a-b** stress distribution; **c** point load on a halfspace; **d** statically equivalent distributed load; **e** shape of the punctured domain near the source point x; **f-h** quadratic basis functions

that the integral of the tractions satisfies the equilibrium condition

$$\int_{\Gamma} T_{ij} ds_y = \delta_{ij}.$$

2. Flamant's problem and St. Venant's principle, [103] p. 25. A vertical force P_2 applied to the surface of the elastic half-space as in Fig. 4.47c will cause the stresses

$$\sigma_{ij} = -\frac{2P_2}{\pi} \frac{r_{,i} r_{,j} r_{,2}}{r}.$$

Show that the integral of the stresses taken over any half-circle centred at the origin balances the exterior load, i.e. show that

$$\int_{\Gamma} t_i ds_y = \int_{\Gamma} \sigma_{ij} \nu_j ds_y = \begin{cases} 0 & i = 1 \\ 1 & i = 2 \end{cases}$$

(Hint: use that $\nu_j = r_{,j}$ on Γ).

Replace the concentrated force P_2 by a statically equivalent distributed load $p_0 = P_2/2a$ as in Fig. 4.47d. At which distance from the surface do the vertical stresses σ_{22} of the two loadcases differ by less than 10 %? How does the length $2a$ over which the load is spread influence the result?

3. Calculate, see Fig. 4.47e,

$$C_{ij}(\boldsymbol{x}) = \lim_{\varepsilon \to 0} \int_{\Gamma_{N_\varepsilon}} T_{ij}(\boldsymbol{y}, \boldsymbol{x}) \, ds_y$$

4. Calculate the four integrals, see Fig. 4.47f,

$$\int_{\Gamma} U_{ij} \varphi_3^e ds_y$$

when the collocation point $\boldsymbol{x}$ is the end-point of the element.

5. Calculate the four integrals, see Fig. 4.47g,

$$\int_{\Gamma} U_{ij} \varphi_1^e ds_y$$

when the collocation point $\boldsymbol{x}$ is the centre-point of the element.

6. In 2-D the traction kernels are defined as

$$T_{ij}(y,x) = -\frac{1}{4\pi(1-\nu)r}[\frac{\partial r}{\partial \nu}((1-2\nu)\delta_{ij} + 2r_{,i}\, r_{,j})$$

$$- (1-2\nu)\{r_{,i}\,\nu_j(y) - r_{,j}\,\nu_i(y)\}]\,.$$

Show that if the integration point y and the collocation point x lie on the same (straight) element then holds

$$T_{11} = T_{22} = 0 \qquad T_{12} = \frac{-1}{4\pi(1-\nu)r}[(1-2\nu)r_\tau] = -T_{21}\,.$$

(Hint: use $\nu_1 = \tau_2$, $\nu_2 = -\tau_1$). In preparation for the following exercise recall that the tangential derivative r_τ has the value - 1 if y lies to the left of x and the value + 1 if y lies to the right of x.

7. The kernels T_{ij} are strongly singular. Their integrals are calculated by first removing a small symmetric ε-neighborhood of the source point and then letting ε tend to zero (Cauchy principal value)

$$\int_\Gamma T_{ij}\varphi ds_y = \lim_{\varepsilon\to 0} \int_{\Gamma_{N_\varepsilon}} T_{ij}\varphi ds_y\,.$$

The kernels have a positive and a negative infinite peak at the source point x (the switch in sign is due to r_τ) which makes that on approaching this singularity from both sides the singularity annihilates itself, see Fig. 4.48a. Would these integrals also exist if the accompanying function (the boundary layer) φ would be discontinuous at x, say $\varphi = 0$ on the left-hand side and $\varphi = 1$ on the right-hand side? Explain then why we cannot consider the singular integrals of the traction kernels elementwise — as we could in the case of the displacement kernels U_{ij} — but that for to have a continuous continuation at the end nodes we must consider the neighboring element as well, see Fig. 4.48b. Apply this technique to calculate the Cauchy principal values

$$\int_\Gamma T_{ij}\varphi ds_y = \lim_{\varepsilon\to 0} \int_{\Gamma_\varepsilon} T_{ij}\varphi ds_y$$

where Γ_ε is the boundary Γ minus an ε-neighborhood of the source point x and φ is the global basis function centred at x. Assume that both elements have the same length. Would the result also be so simple if the length of the two elements were different? Would you obtain the same results if the two elements would form a corner?

8. Calculate the four integrals, see Fig. 4.47h,

$$\int_\Gamma T_{ij}\varphi_2^e ds_y$$

222

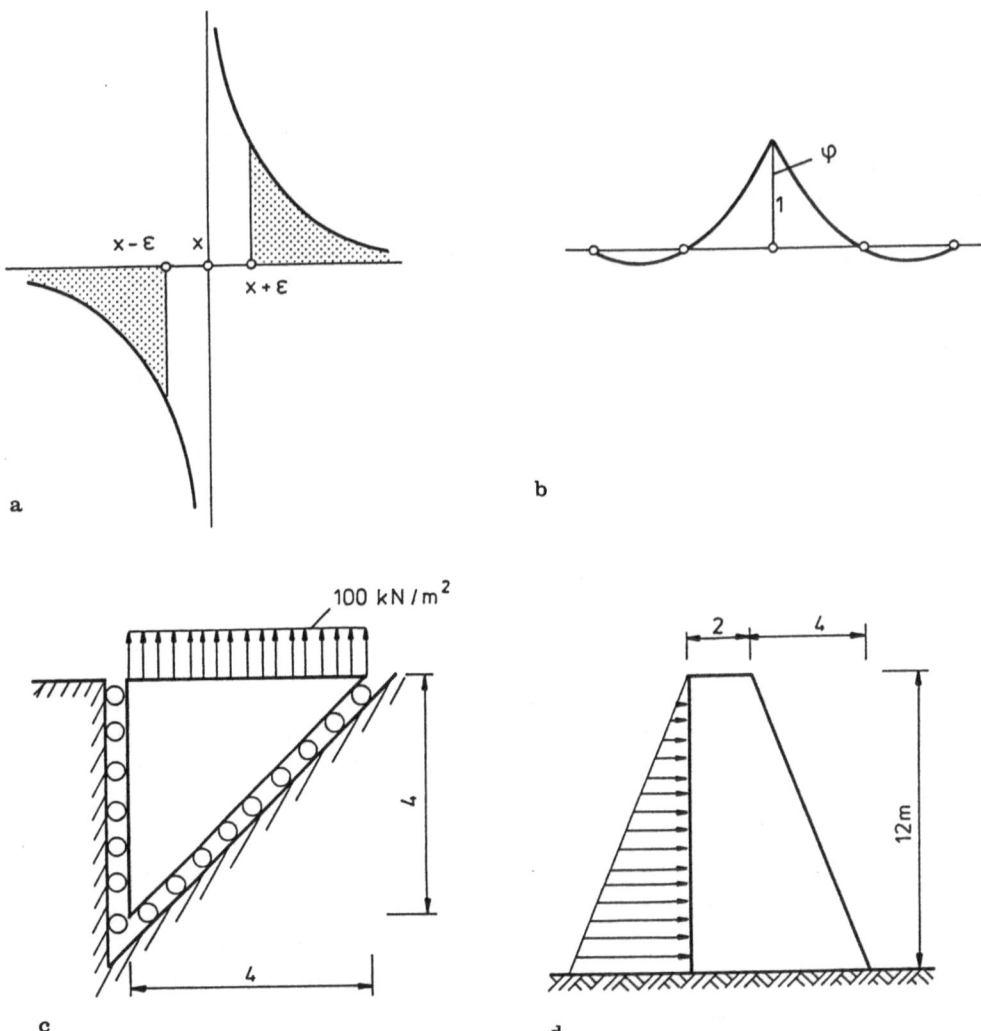

Figure 4.48 a-d Exercises: **a** the singularity of the traction kernel T_{12}; **b** the global basis function is continuous at the node; **c** triangular plate on roller supports; **d** water dam

when the collocation point x is the first point of the element. Note that the shape function φ_2^e has a zero at the collocation point so that in this case the integrand is only weakly singular. Therefore you can calculate this integral by considering only the element itself.

9. Castigliano's Theorem (the principle is a different statement) is not applicable in classical elasticity because, as the reader knows, the displacements beneath concentrated forces are infinite

$$U_{ij} = O(\ln r) \quad \text{(2-D)} \quad U_{ij} = O(r^{-1}) \quad \text{(3-D)}.$$

If you model a beam as an elastic plate and if you adaptively refine the FE-mesh near a point support in the hope to catch the true solution then the support reaction will slowly cut through the beam like through a piece of cake and it will drift up up and away.

But if you place a concentrated force on a Kirchoff plate then the deflection is bounded

$$g_0(y, x) = O(r^2 \ln r)$$

and so Castigliano's Theorem applies. Can you explain this different behavior? In each case, in plate bending as in plate stretching, the stresses must tend to infinity as $1/r$ (we are in 2-D) to balance the point load. But why are the in-plane displacements infinite and the lateral deflection finite? (Hint: The solution lies in the different order of the differential equations, see [2] p.177. In a Reissner type theory the lateral deflection too would be infinite). Castigliano's Theorem applies if and only if $m - i > n/2$ where m is half the order of the governing operator, i the order of the singularity (point forces $i = 0$) and n is the dimension of the continuum.

10. Using the program BE-PLATES calculate the stress field in the triangular plate in Fig. 4.48c. Compare your results with the exact solution: a biaxial uniform stress field $\sigma_{11} = \sigma_{22} = c., \sigma_{12} = 0$.

11. Using the program BE-PLATES calculate the stress field in the cross-section of the water dam in Fig. 4.48d.

5 Nonlinear problems

Influence functions, as the influence function for the longitudinal displacement $u(x)$ of a rod

$$u_1(x) = \int_0^l G_0(y,x)p_1(y)dy\,, \qquad (5.1)$$

are L_2-scalar products between the Green's function and the exterior load p_1 and because the scalar product is distributive

$$\int_0^l G_0(p_1 + p_2)\,dy = \int_0^l G_0 p_1\,dy + \int_0^l G_0 p_2\,dy\,,$$

the influence function for a nonlinear equation cannot be of the form (5.1). If Eq.(5.1) were the solution of the problem

$$D^{NL}u = p_1\,,$$

and

$$u_2(x) = \int_0^l G_0(y,x)p_2(y)\,dy$$

the solution of the problem
$$D^{NL}u = p_2\,,$$

then the function

$$u(x) = u_1(x) + u_2(x) = \int_0^l G_0(y,x)(p_1(y) + p_2(y))\,dy$$

would be the solution of the problem

$$D^{NL}(u_1 + u_2) = p_1 + p_2.$$

But this is a contradiction.

Equivalent to this is that Betti's principle no longer applies. In linear mechanics we obtain Betti's principle for self-adjoint operators by interchanging the places of u and $\hat{u}$ in Green's first identity and by subtracting the two identities, see [2],

$$B(\hat{u}, u) = G(\hat{u}, u) - G(u, \hat{u}) = 0$$

Because of the symmetry of the energy product, $E(u, \hat{u}) = E(\hat{u}, u)$, the reciprocal internal energies drop out and we are only left with the reciprocal external work. Nonlinear problems violate the symmetry condition

$$E(u, \hat{u}) \doteq E(\hat{u}, u)$$

and also the boundary operators ∂^i are no longer linear so that the same manoeuvre is no longer possible.

Some problems are only nonlinear on the boundary as the problem

$$\Delta u = 0 \qquad \text{on } \Omega$$

$$-\frac{\partial u}{\partial n} = u + \sin(u) - f \qquad \text{on } \Gamma$$

which leads to the nonlinear integral equation

$$c(\boldsymbol{x})u(\boldsymbol{x}) - \int_\Gamma \frac{\partial}{\partial \nu} g_0(\boldsymbol{y}, \boldsymbol{x}) u(\boldsymbol{y}) \, ds\boldsymbol{y}$$

$$+ \int_\Gamma g_0(\boldsymbol{y}, \boldsymbol{x})(u + \sin(u)) ds\boldsymbol{y} = \int_\Gamma g_0(\boldsymbol{y}, \boldsymbol{x}) f ds\boldsymbol{y}$$

This equation was reduced by a Galerkin method and piecewise constant trial functions to a discrete system of nonlinear equations and then solved by Newton's iteration, [64].

Other nonlinear problems can be reduced by a proper transformation to equivalent linear problems, [65], [66], but in general the boundary element method can only be applied to nonlinear problems if we also discretize the interior. The primary unknowns, the unknowns which determine the size of the system matrix, remain the boundary terms. These together with the secondary unknowns, the plastic deformations and stresses in the interior, must now be determined iteratively. Simply stated we put all the nonlinear terms on the right-hand side and correct the solution till the defect becomes zero.

The number of possible nonlinear models which describe the behavior of material is too large to be covered, even only approximately, in this chapter.

We, therefore, restrict ourselves to the numerical analysis. This is essentially always the same. For particular applications we refer the interested reader to [54] and [67].

5.1 The principle of virtual forces

In nonlinear elasticity we formulate the differential equations not in terms of u, ε, σ but in terms of the the the rates of change, $\dot{u}, \dot{\varepsilon}, \dot{\sigma}$, of the actual stress state. In terms of a one-dimensional model the differential equation we have to solve looks like

$$-EA\dot{u}'' = \dot{p} - \dot{\sigma}'_{pl}$$

Due to the peculiar nature of the right-hand side the domain integral in the influence function for $\dot{u}$ can be split into two integrals of which the second can be integrated by parts

$$\int\limits_0^l g_0(\dot{p} - \dot{\sigma}'_{pl})dy = \ldots + \int\limits_0^l g_0 \dot{p}dy + \int\limits_0^l g_0' \dot{\sigma}_{pl}dy$$

The new domain integral now obtained is the energy product between the fundamental solution and the plastic stresses and to extract from this energy product the information about the rate $\dot{u}$ of the displacement at a specific point we use Green's first identity, or, in terms of mechanics, we employ the principle of virtual forces.

In mechanics, if we calculate the displacement of a rod at a specific point, we form the L_2-scalar product between the normal force $N(y)$ of the distributed load and the normal force $\hat{N}(y, x)$ of the unit point-force $\hat{P} = 1$, see Fig. 5.1,

$$u(x) = \int\limits_0^l \frac{\hat{N}(y, x)N(y)}{EA} \, dy \tag{5.2}$$

We call this method an application of the principle of virtual forces.

In mathematical terms this is an application of Green's first identity

$$G(\hat{u}, u) = \int\limits_0^l EA\hat{u}''u \, dx + [\hat{N}u]_0^l - \int\limits_0^l \frac{\hat{N}N}{EA} \, dx = 0 \, ,$$

$$G(G_0, u) = \qquad 1 \times u(x) \qquad - \int\limits_0^l \frac{N_0 N}{EA} \, dy = 0$$

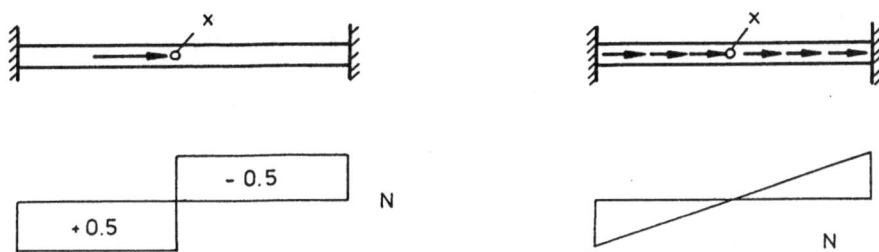

Figure 5.1 Application of the principle of virtual forces to calculate the displacement u

For $\hat{u}$ we substitute Green's function G_0. This function satisfies the boundary conditions and therefore the reciprocal work on the boundary is zero so that Eq.(5.2) is obtained.

Hence, to extend the boundary element method to nonlinear problems in elasticity we must derive an influence function for the displacement field of an elastic body in terms of the energy product between the auxiliary singular elastic state and the real elastic state, i.e. we must employ Green's first identity.

The first identity of an elastic body is the expression

$$G(\hat{u}, u) = \int_{\Omega} -L\hat{u} \cdot u \, d\Omega + \int_{\Gamma} \tau(\hat{u}) \cdot u \, ds - \int_{\Omega} \hat{S} \cdot E \, d\Omega = \delta W_e^c - \delta W_i^c = 0$$

The last integral is the energy product $E(\hat{u}, u)$ of the stress and strain tensors respectively of the two fields $\hat{u}$ and u, see [2] p. 37,

$$\hat{S} = C[E(\hat{u})], \qquad E = E(u) = \frac{1}{2}(\nabla u + \nabla u^T)$$

The dot denotes the scalar product

$$\int_{\Omega} \hat{S} \cdot E \, d\Omega = \int_{\Omega} \hat{\sigma}_{ij}\varepsilon_{ij} \, d\Omega$$

$$= \int_{\Omega} (\hat{\sigma}_{11}\varepsilon_{11} + \hat{\sigma}_{12}\varepsilon_{12} + \cdots + \hat{\sigma}_{33}\varepsilon_{33}) \, d\Omega$$

In the following we substitute for the field $\hat{u}$ one of the three fundamental solutions g_0^i and for the field u the field whose displacements we want to calculate.

To the fundamental solutions

$$g_0^i(y, x) = \{U_{ij}\}$$

belong the strain tensors

$$E^i = -\frac{1}{8\alpha\pi(1-\nu)\mu r^\alpha}\{\beta(\nabla r \cdot e_i)\nabla r \otimes \nabla r + (1-2\nu)(\nabla r \otimes e_i + e_i \otimes \nabla r)$$

$$-(\nabla r \cdot e_i)I\}$$

and stress tensors

$$S^i = -\frac{1}{4\alpha\pi(1-\nu)r^\alpha}\{\beta(\nabla r \cdot e_i)\nabla r \otimes \nabla r + (1-2\nu)(\nabla r \otimes e_i + e_i \otimes \nabla r)$$

$$-(\nabla r \cdot e_i)I\},$$

$$\alpha = 1,2, \quad \beta = 2,3, \quad (2-D, 3-D)$$

In the following we make use of a special property of the stress tensors S^i: The equilibrium condition requires that the integral of the surface stresses of any volume that does not contain the source point x must be zero. Therefore the integrals of the stress tensors over any ring-shaped domain as in Fig. 5.2,

$$N_{1,\varepsilon} = N_1(x) - N_\varepsilon(x), \quad 0 < \varepsilon < 1$$

must vanish

$$\int_{N_{1,\varepsilon}} S^i \, d\Omega_y = \int_\varepsilon^1 \frac{1}{r^\alpha} r^\alpha \, dr \int_{S_1} S^i \, dS_1 = o \tag{5.3}$$

To derive now the influence function for $u_i(x)$ we first consider the identity in the punctured domain $\Omega_\varepsilon(x)$

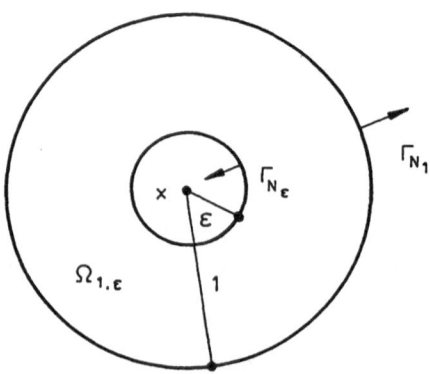

Figure 5.2 Ring shaped domain

$$G(g_0^i, \boldsymbol{u})_{\Omega_\epsilon} = \int\limits_{\Gamma_\epsilon} T_{ij} u_j \, ds_{\boldsymbol{y}} - \int\limits_{\Omega_\epsilon} \sigma_{jk}^i [\varepsilon_{jk}(\boldsymbol{y}) - \varepsilon_{jk}(\boldsymbol{x})] \, d\Omega_{\boldsymbol{y}}$$

$$- \int\limits_{\Omega_\epsilon} \sigma_{jk}^i \, d\Omega_{\boldsymbol{y}} \, \varepsilon_{jk}(\boldsymbol{x}) = 0$$

and we then let the radius ε tend to zero. Because of

$$\sigma_{jk}^i = O(r^{-\alpha}), \qquad d\Omega = r^\alpha \, dr \, dS_1, \qquad \varepsilon_{jk}(\boldsymbol{y}) - \varepsilon_{jk}(\boldsymbol{x}) = O(r),$$

and Eq.(5.3) the limit of the domain integral is bounded

$$\lim_{\varepsilon \to 0} \int\limits_{\Omega_\epsilon} \sigma_{jk}^i \varepsilon_{jk} \, d\Omega_{\boldsymbol{y}} = \int\limits_{\Omega} \sigma_{jk}^i \varepsilon_{jk} \, d\Omega_{\boldsymbol{y}}$$

The limit of the boundary integral is

$$\lim_{\varepsilon \to 0} \int\limits_{\Gamma_\epsilon} T_{ij}(\boldsymbol{y}, \boldsymbol{x}) u_j(\boldsymbol{y}) \, ds_{\boldsymbol{y}} = C_{ij}(\boldsymbol{x}) u_j(\boldsymbol{x}) + \int\limits_{\Gamma} T_{ij}(\boldsymbol{y}, \boldsymbol{x}) u_j(\boldsymbol{y}) \, ds_{\boldsymbol{y}},$$

so that we obtain the following result:

Influence function for the displacement field

p: $\boldsymbol{u} \in C^1(\bar{\Omega})$

q: $\lim\limits_{\varepsilon \to 0} G(g_0^i, \boldsymbol{u})_{\Omega_\epsilon} = C_{ij}(\boldsymbol{x}) u_j(\boldsymbol{x}) + \int\limits_{\Gamma} T_{ij}(\boldsymbol{y}, \boldsymbol{x}) u_j(\boldsymbol{y}) \, ds_{\boldsymbol{y}}$

$$- \int\limits_{\Omega} \sigma_{jk}^i \, \varepsilon_{jk} \, d\Omega_{\boldsymbol{y}} = 0 \tag{5.4}$$

This equation is the analog of Eq.(5.2), of the L_2-scalar product between the bending moments of the beam. Only that, here, additional boundary integrals appear because we used fundamental solutions instead of Green's functions.

Later we also want to calculate stresses and we, therefore, also need influence functions for the first derivatives of the displacement field, i.e. the components of the gradient, a 3×3 matrix. To this end we differentiate the fundamental solution g_0^i with respect to x_d (d is any of the three indices $1, 2, 3$) and we consider the identity

$$G(g^i_{0,x_d}, u(y) - u(x))_{\Omega_\epsilon} = \int\limits_{\Gamma_\epsilon} T_{ij,x_d}(u_j(y) - u_j(x))\, ds_y$$

$$-\int\limits_{\Omega_\epsilon} \sigma^i_{jk,x_d} \varepsilon_{jk}\, d\Omega_y = \int\limits_{\Gamma} T_{ij,x_d}(u_j(y) - u_j(x))\, ds_y$$

$$+\int\limits_{\Gamma_{N\epsilon}} T_{ij,x_d}(u_j(y) - u_j(x))\, ds_y - \int\limits_{\Omega_1} \sigma^i_{jk,x_d} \varepsilon_{jk}\, d\Omega_y$$

$$-\int\limits_{N_{1,\epsilon}} \sigma^i_{jk,x_d} \varepsilon_{jk}\, d\Omega_y = 0$$

in the punctured domain

$$\Omega_\epsilon = \Omega - N_\epsilon(x) = \Omega_1 \cup N_{1,\epsilon}\,, \qquad \epsilon < 1\,,$$

which splits into the domain $\Omega_1 = \Omega - N_1(x)$ and the circular domain

$$N_{1,\epsilon} = N_1(x) - N_\epsilon(x)$$

The latter is bounded by the two spheres $\Gamma_{N1}(x)$ und $\Gamma_{N\epsilon}(x)$, see Fig. 5.3. The strain tensor can be expanded into a Taylor series around x

$$\varepsilon_{jk}(y) = \varepsilon_{jk}(x) + \varepsilon^{(1)}_{jk}(y)\,,$$

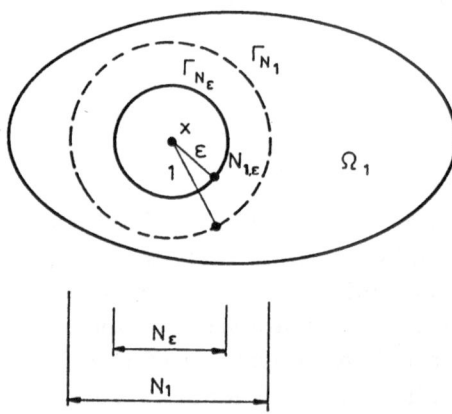

Figure 5.3 Punctured domain

where the remainder

$$\varepsilon_{jk}^{(1)}(\boldsymbol{y}) = \varepsilon_{jk,l}(\boldsymbol{x})(y_l - x_l) + \cdots = O(r), \qquad r = |\boldsymbol{y} - \boldsymbol{x}|,$$

denotes the linear and higher terms. We now have

$$\sigma_{jk,x_d}^i \, \varepsilon_{jk}^{(1)} \, d\Omega y = O(r^{-3}) O(r^1) O(r^2) = O(1)$$

and it therefore holds

$$\lim_{\epsilon \to 0} \int_{N_{1,\epsilon}} \sigma_{jk,x_d}^i \, \varepsilon_{jk}^{(1)} \, d\Omega y = \int_{N_1} \sigma_{jk,x_d}^i \, \varepsilon_{jk}^{(1)} \, d\Omega y$$

so that we are only concerned with the limit of the constant term

$$\lim_{\epsilon \to 0} \int_{N_{1,\epsilon}} \sigma_{jk,x_d}^i \, d\Omega y \, \varepsilon_{jk}(\boldsymbol{x})$$

We start with some remarks: because of

$$r_{,x_d} = -r_{,y_d} \; ,$$

$$\varepsilon_{ij}\sigma_{ij} = \frac{1}{2}(u_{i,j} + u_{j,i})\sigma_{ij} = u_{i,j}\, \sigma_{ij} \, , \quad (\text{if } \sigma_{ij} = \sigma_{ji}),$$

$$\varepsilon_{ij}\hat{\sigma}_{ij} = \sigma_{ij}\hat{\varepsilon}_{ij} \, ,$$

and the rules of integration by parts it follows

$$\int_{N_{1,\epsilon}} \sigma_{jk,x_d}^i \, \varepsilon_{jk} \, d\Omega y = \int_{N_{1,\epsilon}} -\varepsilon_{jk,d}^i \, \sigma_{jk} \, d\Omega y =$$

$$= - \int_{\Gamma_{N_1}} U_{ij,k} \, \sigma_{jk} \nu_d \, ds y + \int_{\Gamma_{N_\epsilon}} U_{ij,x_k} \, \sigma_{jk} \nu_d \, ds y$$

On $\Gamma_{N_\epsilon}(\boldsymbol{x})$ it holds that

$$\nu_d = r_{,x_d}$$

and, therefore, we have, as is easy to demonstrate

$$\int_{\Gamma_{N_\epsilon}} U_{ij,x_k} \, \sigma_{jk} \nu_d \, ds y = \int_{\Gamma_{N_\epsilon}} U_{ij,x_d} \, \sigma_{jk} \nu_k \, ds y = \int_{\Gamma_{N_\epsilon}} U_{ij,x_d} \, t_j \, ds y$$

and, see Eq.(4.13),

$$\lim_{\varepsilon \to 0} \int_{\Gamma_{N\varepsilon}} [U_{ij,x_d} t_j - T_{ij,x_d} (u_j(y) - u_j(x))] \, ds_y = -u_{i,d}(x) \qquad (5.5)$$

All this holds true for arbitrary (smooth) strains ε_{jk} and a fortiori for constant strains $\varepsilon_{jk}(x)$ too. Collecting our results we have:

Influence function for the first derivatives, based on the strain tensor

p: $u \in C^1(\bar{\Omega})$, x an interior point, N_1 is the unit sphere around x,

q: $\lim_{\varepsilon \to 0} G(g_{0,x_d}^i, u(y) - u(x))_{\Omega_\varepsilon} =$

$$u_{i,d}(x) + \int_\Gamma T_{ij,x_d} (u_j(y) - u_j(x)) \, ds_y - \int_{\Omega_1} \sigma_{jk,x_d}^i \varepsilon_{jk} \, d\Omega_y$$

$$- \int_{N_1} \sigma_{jk,x_d}^i \varepsilon_{jk}^{(1)} \, d\Omega_y + \int_{\Gamma_{N_1}} \sigma_{jk}^i \nu_d \, ds_y \varepsilon_{jk}(x) = 0 \qquad (5.6)$$

The last term is the scalar product

$$\int_{\Gamma_{N_1}} \sigma_{jk}^i \nu_d \, ds_y \varepsilon_{jk}(x) = C^{id} \cdot E(x)$$

between the strain tensor $E(x)$ at the point x and the matrix

$$C^{id} = -\frac{1}{15(1-\nu)}[(8 - 10\nu)e_i \otimes e_d - (1 - 5\nu)\delta_{id} I]$$

$$= \int_{S_1} S^i \nu_d \, ds_y,$$

which is the integral of the stress tensor S^i and the component ν_d of the normal vector over the unit sphere S_1.

If we interchange σ_{jk}^i with ε_{jk}^i and correspondingly ε_{jk} with σ_{jk}, then we obtain the following result:

Influence function for the first derivatives, based on the stress tensor

p: $u \in C^1(\bar{\Omega})$, x an interior point, N_1 is the unit sphere around x,

q: $\lim_{\varepsilon \to 0} G(g_{0,x_d}^i, u(y) - u(x))_{\Omega_\varepsilon} =$

$$u_{i,d}\left(\boldsymbol{x}\right) + \int_{\Gamma} T_{ij,x_d}\left(u_j(\boldsymbol{y}) - u_j(\boldsymbol{x})\right) ds_y - \int_{\Omega_1} \varepsilon^i_{jk,x_d}\, \sigma_{jk}\, d\Omega_y$$

$$- \int_{N_1} \varepsilon^i_{jk}\sigma^{(1)}_{jk}\, d\Omega_y + \int_{\Gamma_{N_1}} \varepsilon^i_{jk,x_d}\, \nu_d\, ds_y \sigma_{jk}(\boldsymbol{x}) = 0, \qquad (5.7)$$

The last term in Eq.(5.7)

$$\int_{\Gamma_{N_1}} \varepsilon^i_{jk}\nu_d\, ds_y \sigma_{jk}(\boldsymbol{x}) = \boldsymbol{D}^{id} \cdot \boldsymbol{S}(\boldsymbol{x})$$

is the scalar product between the stress tensor $\boldsymbol{S}(\boldsymbol{x})$ at $\boldsymbol{x}$ and the matrix

$$\boldsymbol{D}^{id} = -\frac{1}{30\mu(1-\nu)}[(8-10\nu)\boldsymbol{e}_i \otimes \boldsymbol{e}_d + \delta_{id}\, \boldsymbol{I}] = \int_{S_1} \boldsymbol{E}^i \nu_d\, ds_y$$

The equations of the stresses at an internal point are therefore

$$\sigma_{ij}(\boldsymbol{x}) = \int_{\Gamma} S_{ijk}u_k\, ds_y + \int_{\Omega_1} \sigma_{ijkl}\varepsilon_{kl}\, d\Omega_y + \int_{N_1} \sigma_{ijkl}\varepsilon^{(1)}_{kl}\, d\Omega_y + f_{ij}(\boldsymbol{x}),$$

where, in 3-D,

$$f_{ij}(\boldsymbol{x}) = -\frac{2\mu}{15(1-\nu)}[(7-5\nu)\varepsilon_{ij}(\boldsymbol{x}) + (1+5\nu)\varepsilon_{kk}(\boldsymbol{x})\delta_{ij}]$$

Because of the symmetry

$$\int_{\Omega} \boldsymbol{S}^i \cdot \boldsymbol{E}(\hat{\boldsymbol{u}})\, d\Omega_y = \int_{\Omega} \boldsymbol{E}^i(\boldsymbol{u}) \cdot \hat{\boldsymbol{S}}\, d\Omega_y$$

this is equivalent to

$$\sigma_{ij}(\boldsymbol{x}) = \int_{\Gamma} S_{ijk}u_k\, ds_y + \int_{\Omega_1} \varepsilon_{ijkl}\sigma_{kl}\, d\Omega_y + \int_{N_1} \varepsilon_{ijkl}\sigma^{(1)}_{kl}\, d\Omega_y + g_{ij}(\boldsymbol{x}),$$

where, in 3-D,

$$g_{ij}(\boldsymbol{x}) = -\frac{1}{15(1-\nu)}[(7-5\nu)\sigma_{ij}(\boldsymbol{x}) + (1-5\nu)\sigma_{kk}(\boldsymbol{x})\delta_{ij}]$$

The equations of the kernels are

$$\sigma_{ijkl} = \frac{\mu}{2\alpha\pi(1-\nu)r^\beta}\{\beta(1-2\nu)(\delta_{ij}r_{,k}\,r_{,l} + \delta_{kl}r_{,i}\,r_{,j})$$

$$+ \beta\nu(\delta_{li}r_{,j}\,r_{,k} + \delta_{jk}r_{,l}\,r_{,i} + \delta_{ik}r_{,l}\,r_{,j} + \delta_{jl}r_{,i}\,r_{,k}) - \beta\gamma r_{,i}\,r_{,j}\,r_{,k}\,r_{,l}$$

$$+ (1-2\nu)(\delta_{ik}\delta_{lj} + \delta_{jk}\delta_{li}) - (1-4\nu)\delta_{ij}\delta_{kl}\},$$

$$\varepsilon_{ijkl} = \frac{1}{4\alpha\pi(1-\nu)r^\beta}\{(1-2\nu)(\delta_{ik}\delta_{lj} + \delta_{jk}\delta_{li} - \delta_{ij}\delta_{kl} + \beta\delta_{ij}r_{,k}\,r_{,l})$$

$$+ \beta\nu(\delta_{li}r_{,j}\,r_{,k} + \delta_{jk}r_{,l}\,r_{,i} + \delta_{ik}r_{,l}\,r_{,j} + \delta_{jl}r_{,i}\,r_{,k})$$

$$+ \beta\delta_{kl}r_{,i}\,r_{,j} - \beta\gamma r_{,i}\,r_{,j}\,r_{,k}\,r_{,l}\},$$

where $\alpha = 1, 2$, $\beta = 2, 3$, $\gamma = 4, 5$ in 2-D and 3-D.

5.2 The calculation of the singular integrals

At the beginning of our derivation of the influence functions for the components $u_{i,d}$ of the gradient we have split the domain Ω into the unit ball $N_1(\boldsymbol{x})$ around $\boldsymbol{x}$ and a remainder, the punctured domain

$$\Omega_1 = \Omega - N_1(\boldsymbol{x}),$$

We, next, applied integration by parts to the strongly singular integral of the energy product over the unit ball and, thus, obtained after a limiting process the influence function (5.6).

To interpolate the strains in the interior we shall, in general, subdivide the interior into finite elements, i.e. cubes but not balls. This causes no difficulties because, as we shall show in the following, we can replace the punctured domain Ω_1 by the remainder, Ω_{FE}, of an arbitrary FE-domain.

In the following we denote by N_{FE} the elements which enclose the source point $\boldsymbol{x}$ and the domain with this cavity by

$$\Omega_{FE} = \Omega - N_{FE},$$

The surface of the cavity we denote by Γ_{FE}.

Because it makes no difference for the derivation of the equations how we split up the original domain Ω,

$$\Omega = \Omega_1 \cup N_1 \qquad \text{or} \qquad \Omega = \Omega_{FE} \cup N_{FE},$$

we have

$$u_{i,d}(x) + \int_{\Gamma} T_{ij,x_d}(u_j(y) - u_j(x))\,ds_y = \int_{\Omega_1} \sigma^i_{jk,x_d}\,\varepsilon_{jk}\,d\Omega_y$$

$$+ \int_{N_1} \sigma^i_{jk,x_d}\,\varepsilon^{(1)}_{jk}\,d\Omega_y - \int_{\Gamma_{N_1}} \sigma^i_{jk}\nu_d\,ds_y\varepsilon_{jk}(x)\,,$$

$$u_{i,d}(x) + \int_{\Gamma} T_{ij,x_d}(u_j(y) - u_j(x))\,ds_y = \int_{\Omega_{\rm FE}} \sigma^i_{jk,x_d}\,\varepsilon_{jk}\,d\Omega_y$$

$$+ \int_{N_{\rm FE}} \sigma^i_{jk,x_d}\,\varepsilon^{(1)}_{jk}\,d\Omega_y - \int_{\Gamma_{\rm FE}} \sigma^i_{jk}\nu_d\,ds_y\varepsilon_{jk}(x)\,,$$

and therefore as well

$$\int_{\Omega_1} \sigma^i_{jk,x_d}\,\varepsilon_{jk}\,d\Omega_y + \int_{N_1} \sigma^i_{jk,x_d}\,\varepsilon^{(1)}_{jk}\,d\Omega_y - \int_{\Gamma_{N_1}} \sigma^i_{jk}\nu_d\,ds_y\,\varepsilon_{jk}(x)$$

$$= \int_{\Omega_{\rm FE}} \sigma^i_{jk,x_d}\,\varepsilon_{jk}\,d\Omega_y + \int_{N_{\rm FE}} \sigma^i_{jk,x_d}\,\varepsilon^{(1)}_{jk}\,d\Omega_y$$

$$- \int_{\Gamma_{\rm FE}} \sigma^i_{jk}\nu_d\,ds_y\,\varepsilon_{jk}(x) \tag{5.8}$$

The crucial point is now that the integral over the surface of the two cavities are equal,

$$\int_{\Gamma_{N_1}} \sigma^i_{jk}\nu_d\,ds_y\,\varepsilon_{jk}(x) = \int_{\Gamma_{\rm FE}} \sigma^i_{jk}\nu_d\,ds_y\,\varepsilon_{jk}(x)\,, \tag{5.9}$$

and for that reason we may replace Ω_1 by $\Omega_{\rm FE}$ and N_1 by $N_{\rm FE}$.

To prove Eq.(5.9) we may assume, without any loss of generality, that the strains $\varepsilon^{(1)}_{jk}$ are zero, that the FE-cavity is larger than the unit ball and that we must, therefore, add to the FE-domain a domain $\Omega_{\rm FE,1}$, see Fig. 5.4, to obtain Ω_1

$$\Omega_1 = \Omega_{\rm FE} \cup \Omega_{\rm FE,1}$$

Hence, it holds that

$$\int_{\Omega_1} \sigma^i_{jk,x_d}\,\varepsilon_{jk}\,d\Omega_y = \int_{\Omega_{\rm FE}} \sigma^i_{jk,x_d}\,\varepsilon_{jk}\,d\Omega_y + \int_{\Omega_{\rm FE,1}} \sigma^i_{jk,x_d}\,\varepsilon_{jk}\,d\Omega_y \tag{5.10}$$

236

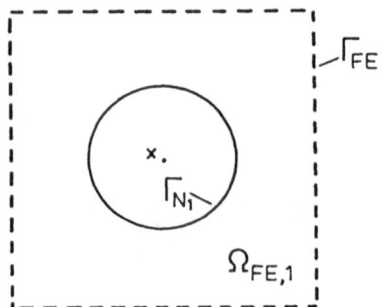

Figure 5.4 The domain between the edge of the FE-domain and the edge of the punctured domain Ω_1

The domain $\Omega_{\mathrm{FE},1}$ is bounded by the two surfaces Γ_{N_1} (interior boundary) and Γ_{FE} (exterior boundary) and, therefore, the integral over $\Omega_{\mathrm{FE},1}$ becomes

$$\int_{\Omega_{\mathrm{FE},1}} \sigma^i_{jk,x_d}\,\varepsilon_{jk}\,d\Omega y = -\int_{\Gamma_{N_1}} \sigma^i_{jk}\nu_d\varepsilon_{jk}\,ds y - \int_{\Gamma_{\mathrm{FE}}} \sigma^i_{jk}(-\nu_d)\varepsilon_{jk}\,ds y \qquad (5.11)$$

The minus sign in $(-\nu_d)$ reflects the fact that the normal vector points now away from the source point $\boldsymbol{x}$ because Γ_{FE} is the external boundary while in Eq.(5.8) it pointed to the source point because Γ_{FE} was then the internal boundary.

If we substitute (5.11) into Eq.(5.10) and Eq.(5.10) into Eq.(5.8), then the statement follows immediately:

$$\int_{\Omega_{\mathrm{FE}}} \sigma^i_{jk,x_d}\,\varepsilon_{jk}\,d\Omega y - \int_{\Gamma_{N_1}} \sigma^i_{jk}\nu_d\varepsilon_{jk}\,ds y - \int_{\Gamma_{\mathrm{FE}}} \sigma^i_{jk}(-\nu_d)\varepsilon_{jk}\,ds y$$

$$-\int_{\Gamma_{N_1}} \sigma^i_{jk}\nu_d\,ds y\,\varepsilon_{jk}(\boldsymbol{x}) = \int_{\Omega_{\mathrm{FE}}} \sigma^i_{jk,x_d}\,\varepsilon_{jk}\,d\Omega y - \int_{\Gamma_{\mathrm{FE}}} \sigma^i_{jk}\nu_d\,\varepsilon_{jk}(\boldsymbol{x})\,ds y\,,$$

or

$$2\int_{\Gamma_{N_1}} \sigma^i_{jk}\nu_d\,ds y\,\varepsilon_{jk}(\boldsymbol{x}) = 2\int_{\Gamma_{\mathrm{FE}}} \sigma^i_{jk}\nu_d\,ds y\,\varepsilon_{jk}(\boldsymbol{x}).$$

Hence, instead of Eq.(5.6), we can use the equation

$$u_{i,d}\,(\boldsymbol{x}) + \int_{\Gamma} T_{ij,x_d}\,(u_j(\boldsymbol{y}) - u_j(\boldsymbol{x}))\,ds y - \int_{\Omega_{\mathrm{FE}}} \sigma^i_{jk,x_d}\,\varepsilon_{jk}\,d\Omega y$$

$$- \int_{N_{\mathrm{FE}}} \sigma^i_{jk,x_d} \, \varepsilon^{(1)}_{jk} \, d\Omega_y + \int_{\Gamma_{N_1}} \sigma^i_{jk} \nu_d \, ds_y \, \varepsilon_{jk}(\boldsymbol{x}) = 0 \qquad (5.12)$$

The same modification applies to Eq.(5.7).

The only problematic term left in Eq.(5.12) is the integral over the domain N_{FE}. But if we switch to spherical coordinates then the integrand becomes regular.

5.3 The system of differential equations

In a material that plastifies or creeps the state variables $\boldsymbol{u}$, $\boldsymbol{E}$, $\boldsymbol{S}$ are functions of time, t, or a load parameter λ which we can treat like a pseudo time variable. The state of the material at the time step $t + \Delta t$ is approximately

$$\boldsymbol{u}(t + \Delta t) = \dot{\boldsymbol{u}}(t)\Delta t, \qquad \boldsymbol{E}(t + \Delta t) = \dot{\boldsymbol{E}}(t)\Delta t, \quad \text{etc.} \qquad (5.13)$$

The rates of the state variables satisfy the equations

$$\frac{1}{2}(\dot{u}_{i,j} + \dot{u}_{j,i}) - \dot{\varepsilon}_{ij} = \dot{\varepsilon}^p_{ij}, \qquad (5.14\mathrm{a})$$

$$C_{ijkl}\dot{\varepsilon}_{kl} - \dot{\sigma}_{ij} = \dot{\sigma}^p_{ij}, \qquad (5.14\mathrm{b})$$

$$-\dot{\sigma}_{ij,j} = \dot{p}_i, \qquad (5.14\mathrm{c})$$

and the boundary conditions

$$\dot{u}_i = \dot{\bar{u}}_i \quad \text{on } \Gamma_1 \qquad \dot{\sigma}_{ij} n_j = \dot{\bar{t}}_i \quad \text{on } \Gamma_2$$

The terms ε^p_{ij} and σ^p_{ij} are the plastic strains and stresses, which depend on the actual state of the material but also on the rates $\dot{\varepsilon}_{ij}$, $\dot{\sigma}_{ij}$, the quantities on the left side. This is why the problem is nonlinear. The left-hand side of Eq.(5.14) is the system of equations of the linear theory, and because in the linear theory terms as ε^p_{ij} and σ^p_{ij} have the meaning of initial strains and initial stresses we speak of an *initial strain* or *initial stress* algorithm.

The third equation, the equilibrium condition (5.14c), states that the increment of the divergence of the difference between the elastic stresses and the plastic stresses

$$\dot{\sigma}_{ij} = C_{ijkl}\dot{\varepsilon}_{kl} - \dot{\sigma}^p_{ij} = \dot{\sigma}^{el}_{ij} - \dot{\sigma}^p_{ij}$$

is equal to the exterior load.

In the *initial stress* algorithm the terms ε^p_{ij} are zero and we obtain, if we substitute the equations into each other, the following system of differential equations for the rates of the displacement field alone

$$-L_{ij}\dot{u}_j = \dot{p}_i - \dot{\sigma}^p_{ij,j}$$

Hence, the integral representation of the field $\dot{u}$ is

$$C_{ij}(\boldsymbol{x})\dot{u}_j(\boldsymbol{x}) = \int_\Gamma [U_{ij}\dot{t}^{el}_j - T_{ij}\dot{u}_j]\, ds\boldsymbol{y} + \int_\Omega U_{ij}(\dot{p}_j - \dot{\sigma}^p_{jk,k})\, d\Omega\boldsymbol{y}$$

Because of

$$\int_\Omega U_{ij}(-\dot{\sigma}^p_{jk,k})\, d\Omega\boldsymbol{y} = \int_\Gamma U_{ij}(-\dot{\sigma}^p_{jk})\nu_k\, ds\boldsymbol{y} + \int_\Omega U_{ij,k}\,\dot{\sigma}^p_{jk}\, d\Omega\boldsymbol{y}$$

$$= \int_\Gamma U_{ij}(-\dot{\sigma}^p_{jk})\nu_k\, ds\boldsymbol{y} + \int_\Omega \frac{1}{2}(U_{ij,k}+U_{ik,j})\dot{\sigma}^p_{jk}\, d\Omega\boldsymbol{y}$$

$$= \int_\Gamma U_{ij}(-\dot{\sigma}^p_{ik})\nu_k\, ds\boldsymbol{y} + \int_\Omega \varepsilon^i_{jk}\dot{\sigma}^p_{jk}\, d\Omega\boldsymbol{y}$$

$$= -\int_\Gamma U_{ij}\dot{t}^p_j\, ds\boldsymbol{y} + \int_\Omega \varepsilon^i_{jk}\dot{\sigma}^p_{jk}\, d\Omega\boldsymbol{y}$$

and

$$\dot{t}_j = \dot{t}^{el}_j - \dot{t}^p_j$$

this is equivalent to

$$C_{ij}(\boldsymbol{x})\dot{u}_j(\boldsymbol{x}) = \int_\Gamma [U_{ij}\dot{t}_j - T_{ij}\dot{u}_j]\, ds\boldsymbol{y} + \int_\Omega U_{ij}\dot{p}_j\, d\Omega\boldsymbol{y}$$

$$+ \int_\Omega \varepsilon^i_{jk}\dot{\sigma}^p_{jk}\, d\Omega\boldsymbol{y} \tag{5.15}$$

Consequently the influence functions for the rates are, see Eq.(5.7),

$$\dot{u}_{i,d}(\boldsymbol{x}) = \int_\Gamma [U_{ij,x_d}\,\dot{t}_j - T_{ij,x_d}\,\dot{u}_j]\, ds\boldsymbol{y} + \int_\Omega U_{ij,x_d}\,\dot{p}_j\, d\Omega\boldsymbol{y}$$

$$+ \int_{\Omega_1} \varepsilon^i_{jk,x_d}\,\dot{\sigma}^p_{jk}\, d\Omega\boldsymbol{y} - \int_{N_1} \varepsilon^i_{jk,x_d}\,\dot{\sigma}^{p(1)}_{jk}\, d\Omega\boldsymbol{y}$$

$$- D^{id}\cdot\dot{S}^p(\boldsymbol{x}) \tag{5.16}$$

In a *initial-strain* formulation the terms σ_{ij}^p are zero and we obtain the system of differential equations

$$-L_{ij}\dot{u}_j = \dot{p}_i + C_{ijkl}\dot{\varepsilon}_{kl,j}^p$$

Correspondingly we have to replace the inelastic terms in Eqs.(5.15) and (5.16) by

$$\int\limits_{\Omega} \sigma_{jk}^i \dot{\varepsilon}_{jk}^p \, d\Omega_y$$

and

$$\int\limits_{\Omega_1} \sigma_{jk,x_d}^i \dot{\varepsilon}_{jk}^p \, d\Omega_y - \int\limits_{N_1} \sigma_{jk,x_d}^i \dot{\varepsilon}_{jk}^{p(1)} \, d\Omega_y - C^{id} \cdot \dot{E}^p(x)$$

5.4 Numerical treatment

by H. Sippel

The domain integrals which represent the influence of the plastic stresses or strains respectively, cannot be transformed into equivalent boundary integrals, so that we must interpolate the plastic terms in the plastic zone as well. To do this we subdivide the plastic zone into cells and we approximate the strain rates $\dot{\varepsilon}_{ij}^p$ or stress rates $\dot{\sigma}_{ij}^p$ within these cells by piecewise polynomials so that the discretized coupling condition on the boundary becomes

$$H\dot{u} = G\dot{t} + \begin{cases} K\dot{\varepsilon}^p & \text{(initial strain alg.)} \\ Q\dot{\sigma}^p & \text{(initial stress alg.)} \end{cases} \tag{5.17}$$

where the vectors $\dot{\varepsilon}^p$ and $\dot{\sigma}^p$ denote the rates of the plastic strains and stresses at the interior points and the boundary points. The vectors $\dot{u}$ and $\dot{t}$ are the rates of the boundary displacements and tractions.

Similarily we obtain for the stresses at interior points the equation

$$\dot{\sigma} = H'\dot{u} + G'\dot{t} + \begin{cases} (D+F)\dot{\varepsilon}^p \\ (C+G)\dot{\sigma}^p \end{cases} \tag{5.18}$$

where the matrices D and C represent the contributions of the domain integrals and the diagonal matrics F and G the integralfree terms. If we arrange Eq.(5.17) according to unknowns

$$\tilde{A}\,\dot{x} = \dot{r} + \begin{cases} K\dot{\varepsilon}^p \\ Q\dot{\sigma}^p \end{cases} \tag{5.19a}$$

and solve for the unknown boundary values $\dot{x}$ then follows

$$\dot{x} = \dot{x}^e + \begin{cases} \tilde{K}\dot{\varepsilon}^p \\ \tilde{Q}\dot{\sigma}^p \end{cases} \tag{5.19b}$$

where $\dot{x}^e = \tilde{A}^{-1}\dot{r}$ is the elastic solution and where the matrices $\tilde{K} = \tilde{A}^{-1}K$ and $\tilde{Q} = \tilde{A}^{-1}Q$ describe the influence of the plastic terms on the solution. If we arrange Eq.(5.18) accordingly then we obtain

$$\dot{\sigma} = \tilde{A}'\,\dot{x} + \dot{r}' + \begin{cases} (D+F)\dot{\varepsilon}^p \\ (C+G)\dot{\sigma}^p \end{cases} \tag{5.20}$$

or if we employ Eq.(5.19b)

$$\dot{\sigma} = \tilde{A}'\,\dot{x}^e + \dot{r}' + \begin{cases} (D+F+\tilde{A}'\tilde{K})\dot{\varepsilon}^p \\ (C+G+\tilde{A}'\tilde{Q})\dot{\sigma}^p \end{cases} = \dot{\sigma}^e + \begin{cases} S\dot{\varepsilon}^p \\ U\dot{\sigma}^p \end{cases} \tag{5.21}$$

If we use a constant interpolation for the platic terms $\dot{\varepsilon}^p$ or $\dot{\sigma}^p$ then the matrix equation for the solution algorithm contains only the stresses at interior points. If we use a higher approximation instead then the boundary stresses must be considered as well.

It remains to transform the foregoing rate formulations into incremental formulations. This is simply done by multiplying the equations with the time step Δt.

Before we discuss this incremental formulation let us make two concluding remarks:

(i) In the BE-approach only the plastified region must be subdivided into cells so that we may expect an advantage if the plastic zone is of a local character as for example in stress concentration problems.

(ii) We must only perform once a full elastic calculation with an inversion of the matrix $\tilde{A}$. Note also that the elements of the matrices (5.17) and (5.20) remain constant.

5.4.1 Initial strain algorithm

In the following we shall formulate first the *method of accumulating plastic deformations* and then, as an alternative, the initial stress algorithm. The method of accumulating plastic deformations is based on the initial strain formulation and the Prandtl-Reuss equations (von Mises yield criterion.)

Before the elasto-plastic calculation begins the full load is applied and a pure elastic solution is obtained. In the next step the elastic solution is so scaled that the most highly stressed node reaches the uniaxial yield limit Y. This is achieved by multiplying the elastic solution with the load factor

$$L_0 = \frac{Y}{\sigma_{V_{\max}}}$$

where $\sigma_{V_{\max}}$ is the equivalent uniaxial stress at the specific node. From this point onward the remaining load is applied incrementally in N loadsteps. At the $i-$th loadstep L the loadfactor is

$$L_i = L_{i-1} + \frac{1}{N}(L - L_0)$$

and we have

$$\sigma_i = L_i\sigma^e + \tilde{S}(\varepsilon^p_{i-1} + \Delta\varepsilon^p_i) \qquad (5.22)$$

where ε^p_{i-1} are the *accumulated* plastic strain increments of the previous loadsteps and $\Delta\varepsilon^p_i$ are the plastic strain increments at the *current* loadstep i. In each load increment at all nodes the elastic stress surplus $\Delta\sigma^e$ which exceeds the yield limit must be transformed into plastic strain increments $\Delta\varepsilon^p_i$. To calculate these increments $\Delta\varepsilon^p_i$ we make an initial guess and to check this guess we calculate with (5.22) the stresses σ_i substitute these stresses into the suitably modified Prandtl-Reuss equations and we so calculate the improved plastic deformation increments $\Delta\varepsilon^p_i$.

The convergence is checked by comparing the uni-axial strain $\Delta\varepsilon^p_v$ with the value of the previous iteration. If at one node the difference exceeds a specified margin then the iteration is repeated with the improved value $\Delta\varepsilon^p_i$.

If convergence has been achieved the plastic deformations are accumulated and the next load increment applied. This process is repeated until the maximum load level is reached. The unknown boundary values then follow from Eq.(5.19).

5.4.2 Initial stress algorithm

The initial stress algorithm allows to implement a variety of yield criteria: the von Mises yield criterion, the Tresca criterion, the Mohr-Coulomb criterion or the Drucker-Prager criterion. The algorithm is nearly identical with that realized in FE-codes.

We start as before: the stresses caused by the full load are so scaled that the most highly stressed node reaches the yield stress Y. Beginning with this load level the iterative solution procedure sets in. Each load step (i) consists of $j = 1, \ldots$ iterations. At the first of these iterations we calculate the total fictitious elastic stresses

$$\Delta \tilde{\sigma}^e = \Delta \sigma^e + (V + I)\Delta \sigma^p$$

where $\Delta \sigma^e$ is the elastic stress increment and $\Delta \sigma^p$ the residual plastic stresses of the previous $(i - 1)$ iterations (the matrix I is the unit matrix). With the increment of the true stresses $\Delta \sigma$, which follow from the constitutive equations, the plastic stress increments can be calculated

$$\Delta \tilde{\sigma}^p = \Delta \tilde{\sigma}^e - \Delta \sigma$$

If the norm of these increments is at any node less than a predefined margin the next load step can begin.

If this is not the case an iteration starts in which the plastic stresses are repeatedly transformed into fictitious elastic stresses

$$\Delta \tilde{\sigma}^e = (V + I)\Delta \sigma^p$$

which in turn are substituted into the constitutive equations so that the true stresses can be updated till convergence is achieved.

After each iteration the true stresses are accumulated, so that at the end of the last load step the actual stresses are calculated.

5.4.3 Examples

The following examples, a notched beam and a plate with an elliptical hole, were calculated with the computer program BLASI. This program serves as a post-processor program to the BE-program BETSY-2D. BLASI consists of two stand-alone separate modules:

BLASIAV which is an implementation of the initial strain algorithm and the Prandtl-Reuss equations.

BLASIAS which is an implementation of the initial stress algorithm and allows different constitutive equations.

a) Notched beam

Because of the symmetry of the specimen and the load, see Fig. 5.5a, it sufficed to discretize only one half of the beam, see Fig. 5.5b. The material

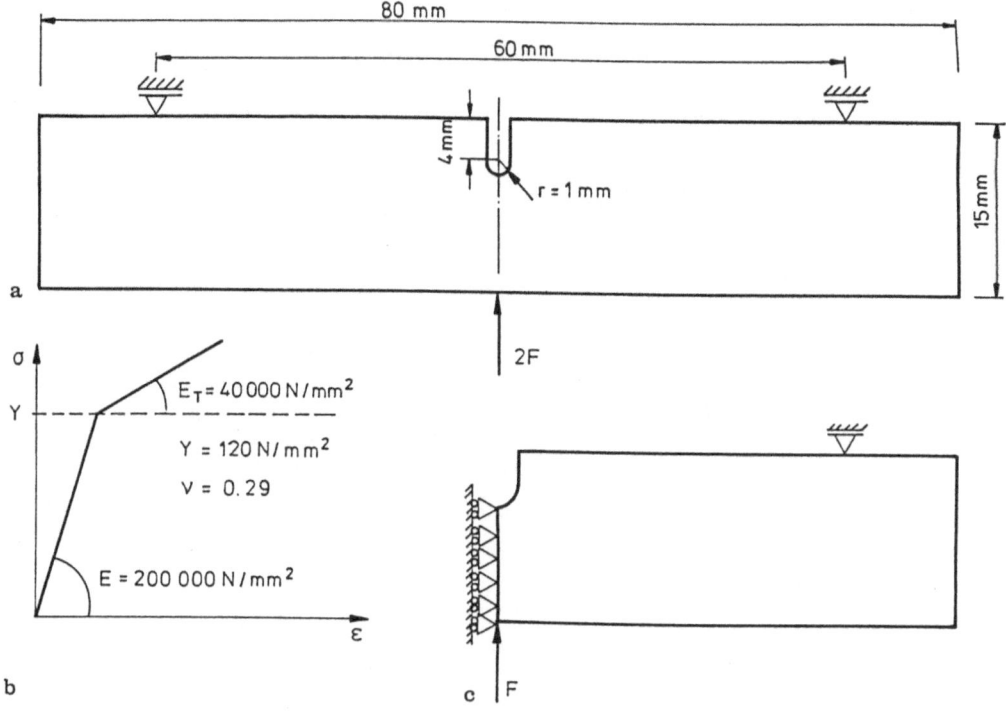

Figure 5.5 Notched beam

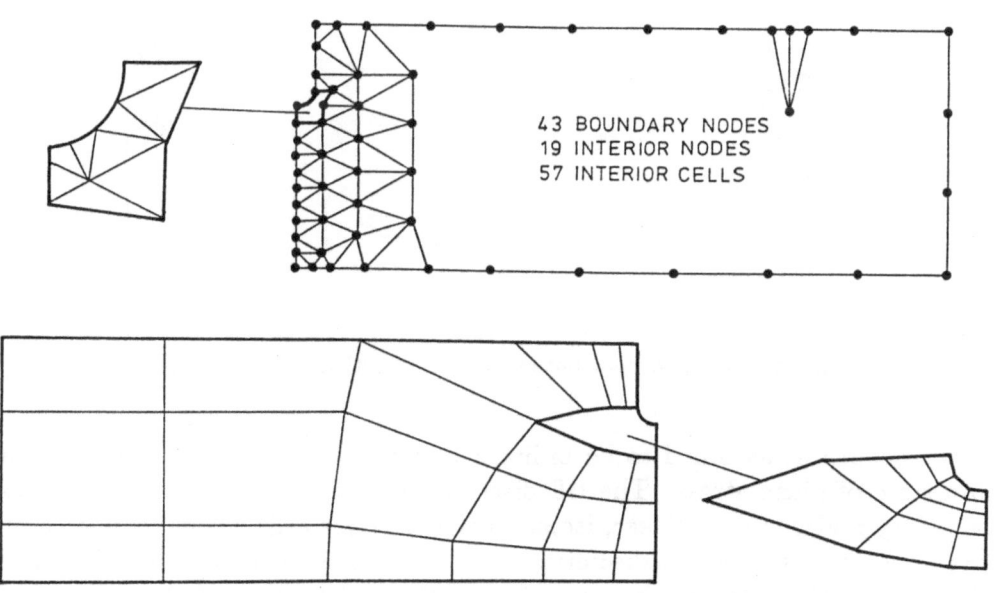

43 BOUNDARY NODES
19 INTERIOR NODES
57 INTERIOR CELLS

4-NODES: 34 ELEMENT 48 NODES
8-NODES: 34 ELEMENT 129 NODES

Figure 5.6 BE-discretization and FE-discretization

244

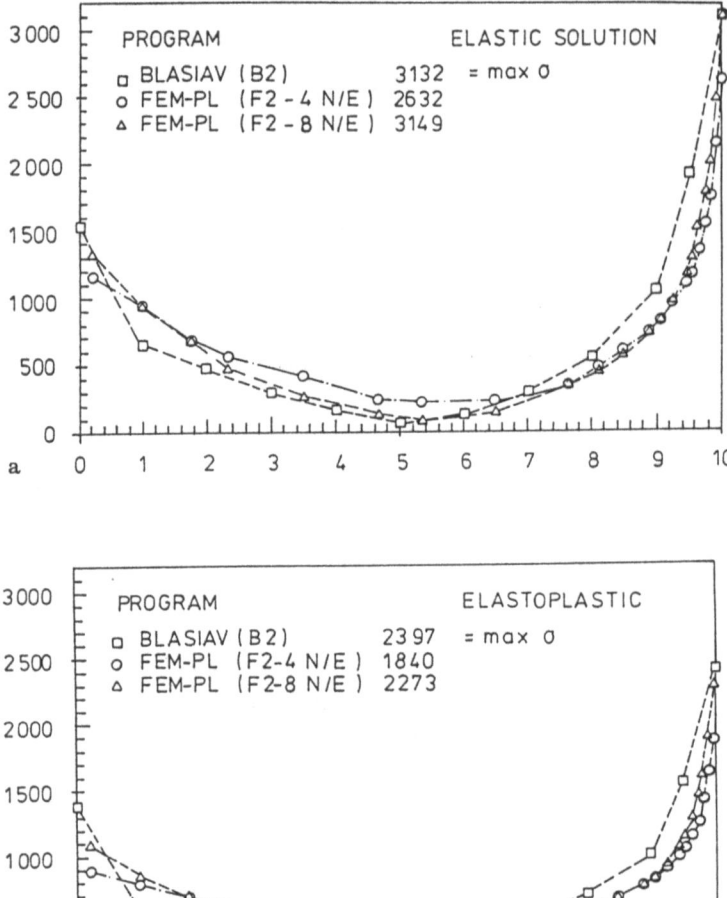

Figure 5.7 Stress distribution along the vertical axis of symmetry

of the beam was assumed to be bi-linear and the beam was considered to be in a state of plane stress. The BE-discretization of the beam consisted of 43 boundary nodes and 57 linear, isoparametric interior cells which connected at 19 interior nodes. The FE-discretization of the same beam is seen in Fig. 5.6. To compare the two solutions we studied the distribution of the von Mises stresses

$$\sigma_V^2 = \sigma_{xx}^2 + \sigma_{yy}^2 + 3\sigma_{xy}^2 - \sigma_{xx}\sigma_{yy}$$

along the axis of symmetry, see Fig. 5.7. The first plot, Fig. 5.7a, shows the

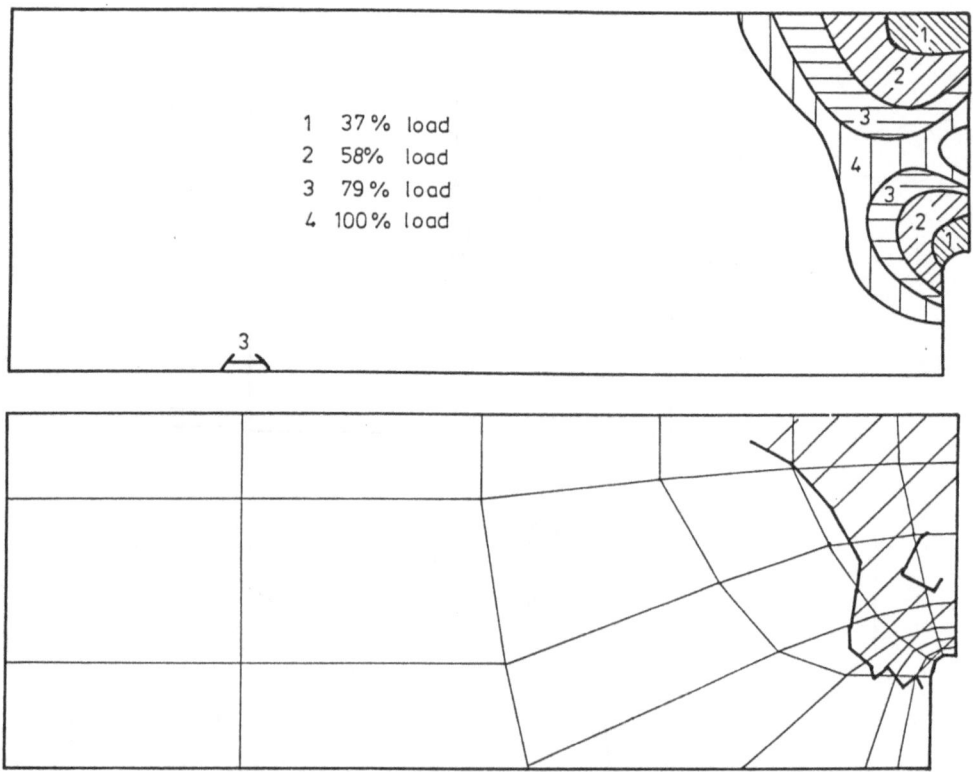

1   37 % load
2   58%   load
3   79 % load
4   100% load

Figure 5.8 a-b Spread of the plastic zone: **a** BE-model, **b** FE-model

distribution of the elastic stresses and the second plot, Fig. 5.7b, shows the elasto-plastic solution. The spread of the plastic zone in the four consecutive loadsteps is seen in Fig. 5.8.

The fact that in the BE-method the boundary stresses are directly evaluated on the boundary obviously leads to a better approximation of the stress distribution, in particular near the notch. Note that the difference between the two solutions, the gap between the two plots, outside of the plastic zone is the same as in the case of the purely elastic solutions. This is so because the elastic solution is the basis of the plastic solution. Note also that the rather crude 4-nodes-per-element FE-solution cannot handle all the fine details.

a) Plates with elliptic holes

We investigated a series of plates with elliptic holes under uniform tension $\sigma^\infty = 70N/mm^2$, see Fig 5.9. We show the results for a hole with a ratio 4:1 of the axis and we compare the results between BLASIAS (initial stress algorithm) and BLASIAV (initial strain algorithm). The relevant quantity we study is the normalized boundary traction $t_n/\sigma^\infty = \alpha_k$.

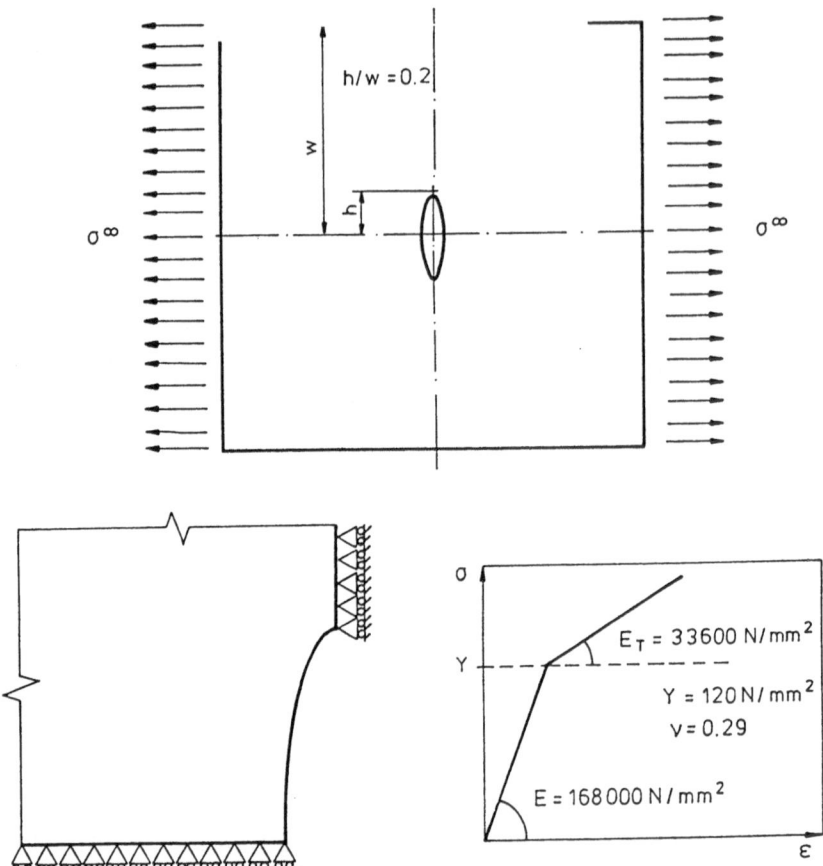

Figure 5.9 Plate with an elliptic hole

Beginning with the same elastic solution ($\alpha_k = 9.3$) two slightly different discretizations, b_1 and b_2 were investigated. These discretizations only differed by the distance of the interior node closest to the notch, see Fig. 5.10.

While the results of BLASIAS and BLASIAV for version b_1 differ significantly the results for version b_2 show a good agreement, see Fig. 5.11. The reason for this different behavior is that the initial stress algorithm is only applicable to the stresses which are computed with a quasi-Hooke's law. For to apply it also in the program BLASIAS the boundary stresses were calculated from (5.17) after the total load was applied. If the interior nodes have a larger distance to the boundary (version b_1 as compared with version b_2) then the plastic influence is not that strongly felt and the computed boundary tractions become too large. If interior nodes are put close to the boundary this effect is no longer observed.

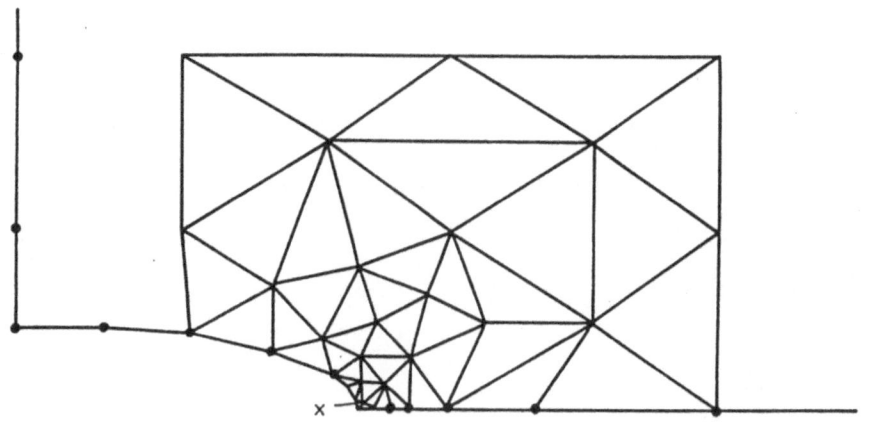

32 BOUNDARY NODES   VERSION B1 : X (1.005 ; 0.005)

20 INTERIOR NODES   VERSION B2 : X (1.001 ; 0.001)

45 INTERIOR CELLS

Figure 5.10 Details of the BE-discretization

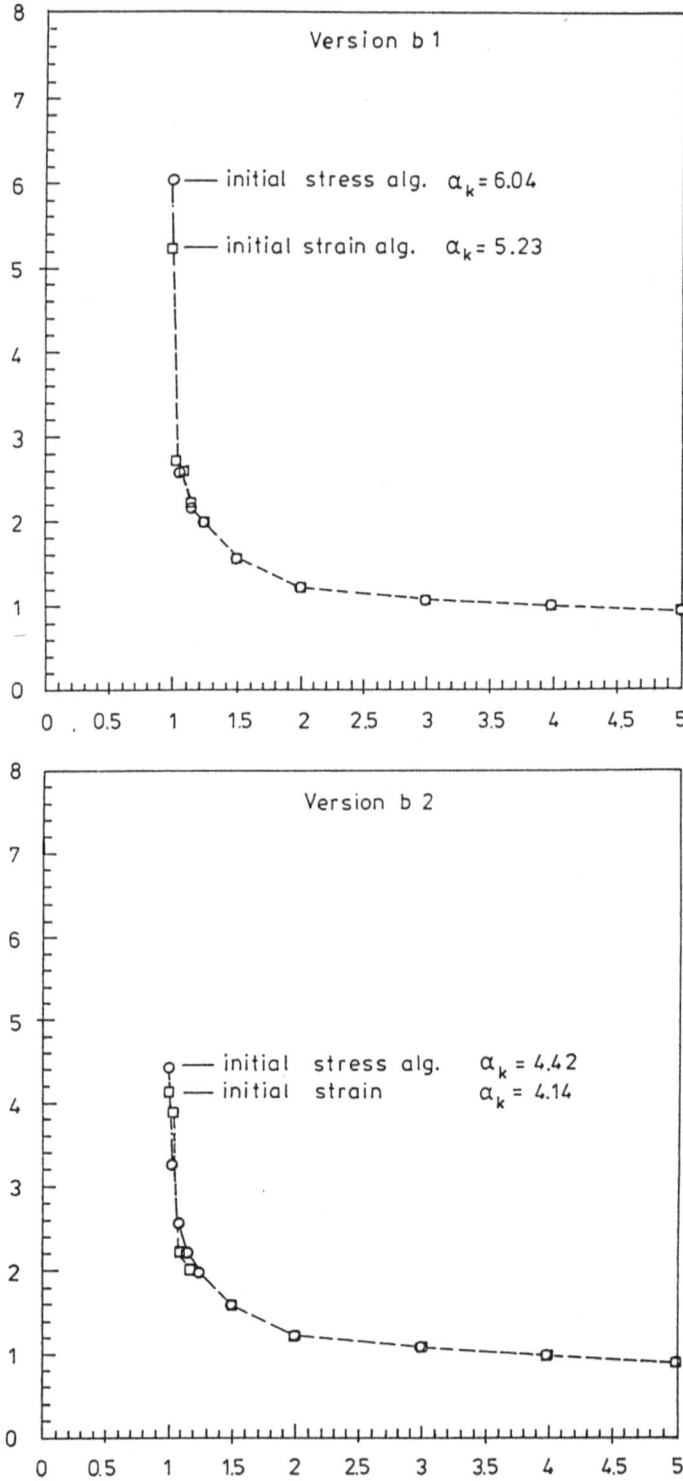

Figure 5.11 Comparison of the maximum stresses

6 Plates

In this chapter we apply the boundary element method to Kirchhoff plates. These plates are governed by a fourth-order equation, the bi-harmonic equation.

6.1 Introduction

In structural mechanics we calculate the deflection $\delta = w(x)$ of a beam as in Fig. 6.1, by applying a concentrated force $\hat{P} = 1$ at the point x and by formulating the L_2-scalar product between the bending moment of the real load and the concentrated force ("the dummy-load")

$$1 \times \delta = \int\limits_0^l \frac{\hat{M}(y,x)M(y)}{EI}\, dy$$

However, the deflection can also be calculated with Betti's principle

$$W_{1,2} = W_{2,1}$$

because according to this principle the work done by the concentrated force $\hat{P} = 1$ on acting through δ is equal to the work done by the distributed forces p on acting through $\hat{w}$, the deflection of the beam under the attack of the concentrated force $\hat{P} = 1$,

$$1 \times \delta = \int\limits_0^l \hat{w}\, p\, dy$$

If we try to extend this approach to a hinged rectangular plate, see Fig. 6.2,

$$1 \times \delta = \int\limits_\Omega \hat{w}\, p\, d\Omega$$

then this must fail because we have no analytical expression for the deflection

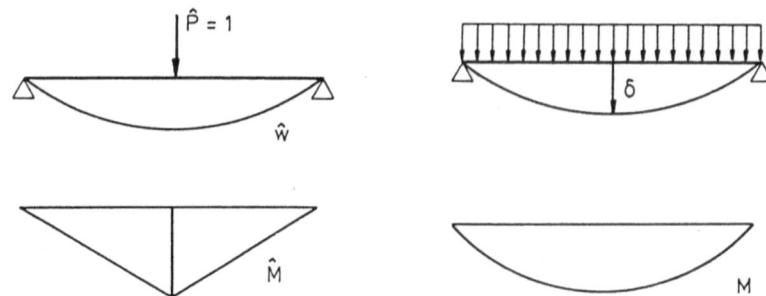

Figure 6.1 Calculation of the deflection δ with the unit-dummy-load method

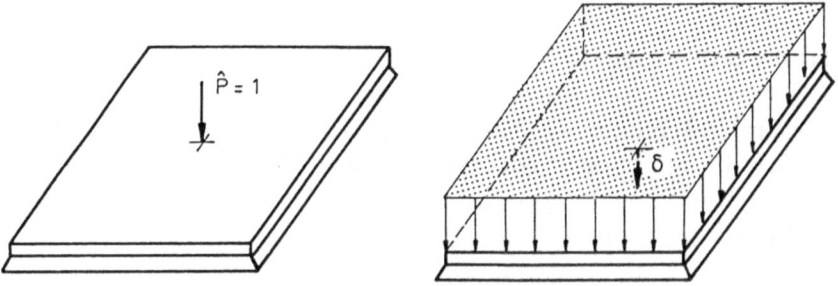

Figure 6.2 Application of Betti's principle to a hinged plate

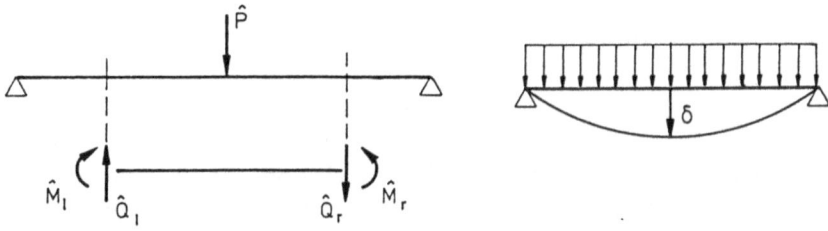

Figure 6.3 The dummy-load $\hat{P} = 1$ applied to a beam with a greater span

of a rectangular plate under the action of a concentrated force $\hat{P}$. The function $\hat{w}$ is unknown. How may we proceed? Let us return to the beam and let us imagine that the concentrated force $\hat{P} = 1$ is not applied to the original beam but to a beam with a greater span, see Fig. 6.3. Let us assume that for some unknown reason we cannot calculate the deflection $\hat{w}$ of the original beam under the attack of this force but we know what the deflection $\hat{w}$ of the second, longer beam looks like. In this situation Betti's principle still applies. It is only that first we must cut off the two projecting ends of the longer beam. By this manoeuvre the internal actions at the cuts become external forces and because now these too contribute work the statement of Betti's principle reads

$$W_{1,2} = 1 \times \delta - \hat{M}_r w'_r + \hat{M}_l w'_l = \int_0^l \hat{w}\, p\, dy - Q_l \hat{w}_l + Q_r \hat{w}_r = W_{2,1}$$

If we solve this expression for δ, then it follows that

$$1 \times \delta = \int_0^l \hat{w}\, p\, dy - Q_l \hat{w}_l + Q_r \hat{w}_r + \hat{M}_r w'_r - \hat{M}_l w'_l,$$

or, in a symbolic notation,

$$1 \times \delta = \int_0^l \hat{w}\, p\, dy + \text{work on the boundary } (W_{2,1})$$

$$- \text{work on the boundary } (W_{1,2})$$

Hence, the deflection can also be determined if we let the concentrated force $\hat{P} = 1$ act on an auxiliary beam that has a wider span. This insight is the key to the solution of the plate problem.

We do not know the shape of the deflection surface $\hat{w}$ of the rectangular plate but we know how a circular plate deflects

$$\hat{w}(r) = \frac{1}{16\pi K} \left\{ \frac{3+\nu}{1+\nu} R^2 (1 - \frac{r^2}{R^2}) - 2r^2 \ln R \right\} + \frac{1}{8\pi K} r^2 \ln r \qquad (6.1)$$

under the action of a concentrated force. But this information suffices: we let the concentrated force $\hat{P} = 1$ act at the centre of a circular plate whose radius R we choose to be so large that the real plate fits into the interior, see Fig. 6.4. We then separate that subregion of the circular plate which coincides with the real plate by a cut from the circular plate so that the prior internal actions become exterior forces on the edge of the cut. Analogously we separate the real plate from its supports. Both plates are in equilibrium and we therefore must have

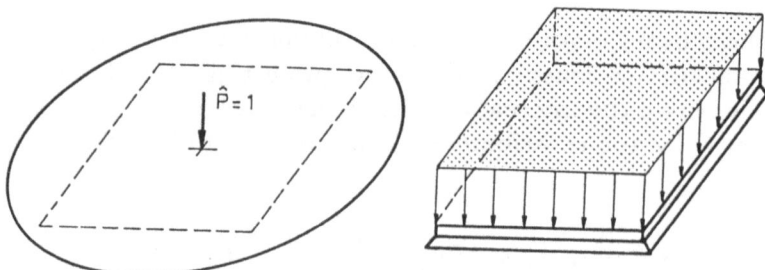

Figure 6.4 We choose the radius of the circular plate so large that the real plate fits into the interior

$$W_{1,2} = 1 \times \delta + \int_{\Gamma} (\hat{V}_n w - \hat{M}_n \frac{\partial w}{\partial n})\, ds + \sum_{e=1}^{4} \hat{F}_e w_e =$$

$$= \int_{\Omega} \hat{w} p\, d\Omega + \int_{\Gamma} (V_n \hat{w} - M_n \frac{\partial \hat{w}}{\partial n})\, ds + \sum_{e=1}^{4} F_e \hat{w}_e = W_{2,1}\,,$$

or, if we solve this equation for δ,

$$1 \times \delta = \int_{\Omega} \hat{w}\, p\, d\Omega - \int_{\Gamma} (\hat{V}_n w - \hat{M}_n \frac{\partial w}{\partial n} + \frac{\partial \hat{w}}{\partial n} M_n - \hat{w} V_n)\, ds$$

$$+ \sum_{e=1}^{4} (F_e \hat{w}_e - w_e \hat{F}_e) \tag{6.2}$$

The terms F_e are the corner forces and the terms $w_e = w(x^e)$ the corner deflections. To be complete we mentioned also those terms which are zero on the boundary as the deflection w and the bending moment M_n.

Note that Eq.(6.2) is of the same type as the expression for the beam

$$1 \times \delta = \int_{\Omega} \hat{w}\, p\, d\Omega + \text{work on the boundary } (W_{2,1})$$

$$- \text{work on the boundary } (W_{1,2})$$

Before proceeding we place the edge of the circular plate at infinity. Because we are only interested in that part of the circular plate that coincides with the real plate we do not care how large the radius R is. This means that we can cancel in Eq.(6.1) the first term and work only with the remainder, the function

$$\hat{w}(r) = \frac{1}{8\pi K} r^2 \ln r, \qquad (6.3)$$

This function represents the deflection of an infinite plate under the action of a concentrated force $\hat{P} = 1$.

To apply Eq.(6.2) we need to know the displacement and force terms on the boundary of the real plate. But this poses a problem because the slope $\partial w/\partial n$ and the Kirchhoff shear V_n are unknown. How can we determine these quantities? First, we simplify the problem: we interpolate the unknown boundary functions at n nodes by piecewise linear functions so that the problem reduces to the task to determine the $2n$ unknown nodal values of the two functions and this we do as follows: the unknown functions $\partial w/\partial n$ and V_n must comply with the requirement that the elastic state of the plate is an equilibrium state, i.e., that the plate satisfies Betti's principle. Let us assume we know a second equilibrium state of the plate, we call this auxiliary state the state 2 and the real state the state 1, then we must have $W_{1,2} = W_{2,1}$. Because this test can be done with other auxiliary states as well we can determine the $2n$ nodal values by repeating this test with $2n$ different equilibrium states 2. For the states 2 we choose consecutively the elastic state of an infinite plate if a concentrated force $\hat{P} = 1$ or a concentrated couple $\hat{M}_n = 1$, respectively, acts at one of the n nodes of the dashed line. Figure 6.5 shows a couple of boundary functions which are pairwise conjugated and which contribute to the external work.

We can solve these $2n$ equations $W_{1,2} = W_{2,1}$ for the $2n$ unknown nodal values of the slope $\partial w/\partial n$ and the Kirchhoff shear V_n. From Eq.(6.2) we then obtain the deflection at any interior point.

If we solve the plate problem with finite elements then we subdivide the plate into, say, triangular conforming elements as indicated in Fig. 6.6 and we construct a set of basis functions by patching together the local basis functions at the nodes.

These basis functions represent unit deflections $w = 1$ and unit rotations $w_{,1} = 1$ and $w_{,2} = 1$. They are non-zero only on those elements, which have the particular node, that is deflected or rotated, as one of their vertices. The main part of the plate remains flat. On the mesh lines which connect the neighboring nodes with the active node and on the lines which connect the neighboring nodes among themselves, the second and third derivatives of the basis functions are discontinuous and so too are the internal actions. From a mechanical point of view this means that exterior moments M_Δ and forces V_Δ are acting along these lines while the enclosed elments bear distributed forces p_h^e. These forces are simply the right-hand sides of the pertinent local basis functions

$$p_h^e = \sum_i K \, \Delta\Delta \, \varphi_i^e$$

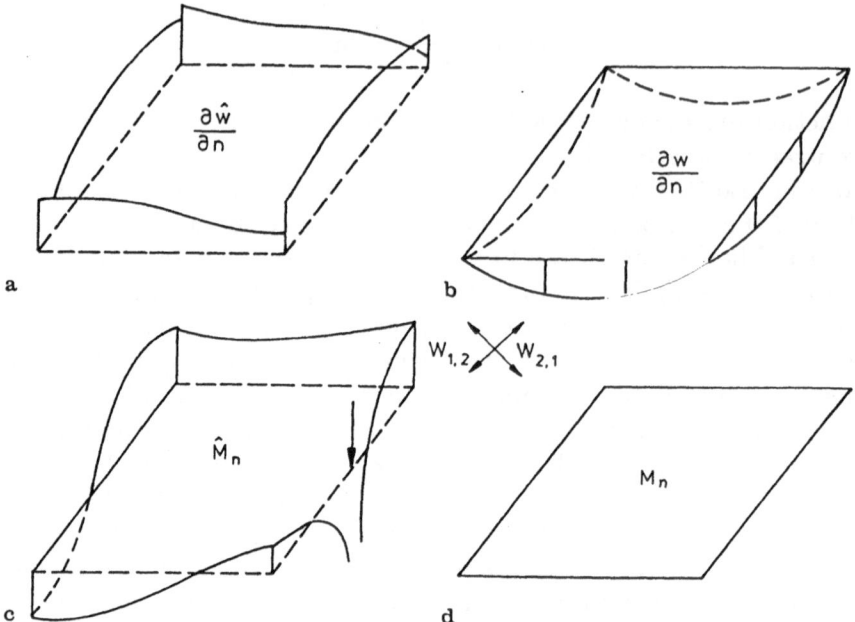

Figure 6.5 a-d. The subdomain and the real plate: **a-b** the unknown boundary terms of the real plate; **c-d** the conjugated boundary terms on the edge of the subdomain of the infinite plate

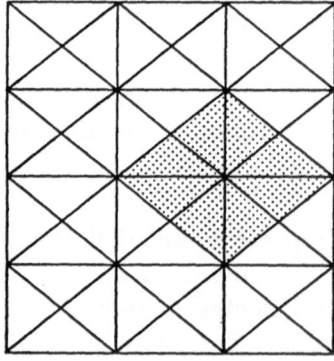

Figure 6.6 The patch of the FE-mesh which moves if the node at its middle is pushed down (conforming elements)

If the local basis functions φ_i^e are polynomials of degree three then, naturally, the element loads p_h^e are zero.

In the Kirchhoff-plate theory the *energy product* (= virtual strain energy) of two functions w and $\hat{w}$ is the expression

$$E(w, \hat{w}) = \int_{\Omega} [w,_{11} (\hat{w},_{11} + \nu\hat{w},_{22}) + 2(1 - \nu)w,_{12} \, \hat{w},_{12}$$

$$+ w,_{22} (\hat{w},_{22} + \nu\hat{w},_{11})] \, d\Omega$$

Because it is a bilinear form and the FE-approximation

$$w_h = u_i \varphi_i(\boldsymbol{x})$$

a sum of n basis functions the internal energy of the FE-approximation can be expanded with respect to the energy products of its basis functions

$$\frac{1}{2} E(w_h, w_h) = \frac{1}{2} E(u_i \varphi_i, u_j \varphi_j) = \frac{1}{2} u_i E(\varphi_i, \varphi_j) u_j = \frac{1}{2} \boldsymbol{u}^T \boldsymbol{K} \boldsymbol{u}$$

The vector $\boldsymbol{u}$ is the vector of the nodal degrees of freedom and $\boldsymbol{K}$ is the stiffness matrix of the plate,

$$K_{ij} = E(\varphi_i, \varphi_j)$$

The external work is a linear form of the nodal values u_i

$$\int_{\Omega} p \, w_h \, d\Omega = \int_{\Omega} p \, u_i \varphi_i \, d\Omega = \boldsymbol{f}^T \boldsymbol{u}$$

where

$$f_i = \int_{\Omega} p \, \varphi_i \, d\Omega$$

so that the potential energy of the FE-solution is the expression

$$\Pi_1(w_h) = \frac{1}{2} \boldsymbol{u}^T \boldsymbol{K} \boldsymbol{u} - \boldsymbol{f}^T \boldsymbol{u}$$

and the requirement that the potential energy becomes a minimum leads to the system of equations

$$K_{ij} u_j = f_i, \qquad i = 1, 2 \dots, n,$$

and we conclude as before: *The FE-solution is so tuned that the work of its exterior forces is equal to the work of the distributed load p with respect to all virtual deflections φ_i.*

Note that the exterior forces which are associated with the FE-solution are *not* the equivalent nodal forces (and couples) f_i, but

a) the element forces p_h^e,

b) the moments M_Δ and vertical forces V_Δ acting on the element boundaries.

c) the concentrated forces at the nodes. These forces are the sums of the

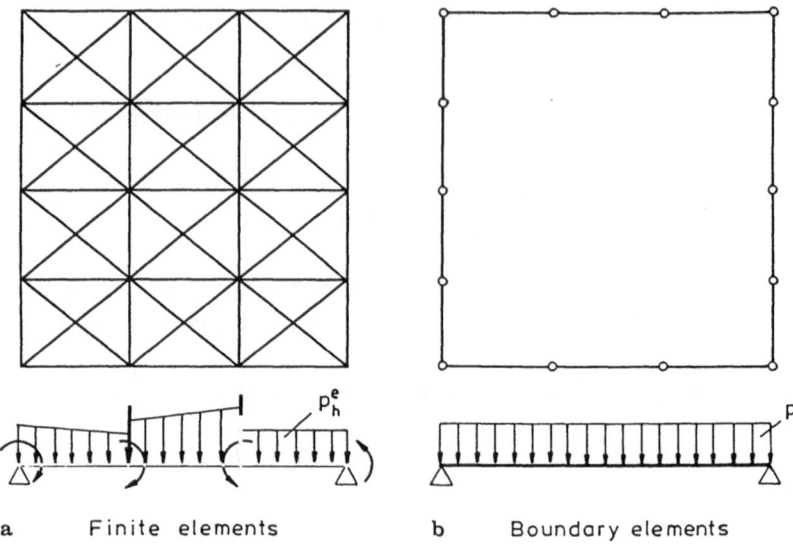

Figure 6.7 Comparison between finite elements and boundary elements

corner forces of the single elements which connect with the particular node. If the derivative $w_{,xy}$ of the FE-solution is continuous at the nodes then these forces are zero.

If the elements were nonconforming then the slope would be discontinuous across the element boundaries and this would give the deflection surface an angular character.

Summary: To approximate the deflection of a plate with finite elements means to replace the original distribution of forces by element loads p_h^e (whose magnitude varies between the elements) and by forces V_Δ and moments M_Δ which act along the element boundaries. The geometric support condition $w = 0$ is satisfied exactly, but the statical condition $M_n = 0$ only approximately, see Fig. 6.7.

In contrast to the FE-solution the BE-solution is smooth and the internal actions are also smooth. The BE-load is identical with the real load because the BE-solution satisfies the plate equation $K\Delta\Delta w = p$. But on the edge errors creep in. The BE-solution neither satisfies the geometrical boundary conditions nor the statical boundary conditions exactly (it satisfies the geometric conditions only at the collocation points). These deviations are in general less pronounced than in the case of a FE-solution but they are definitely there. Otherwise we would have the exact solution.

After these introductory remarks the method is explained more systematically and in more detail.

6.2 Fundamentals

The deflection w, the curvatures κ_{ij} and the bending moments M_{ij} of a plate satisfy the equations

$$\kappa_{ij} - w_{,ij} = 0, \qquad \text{(3 Eqs.)},$$

$$K\{(1-\nu)\kappa_{ij} + \nu\kappa_{kk}\delta_{ij}\} + M_{ij} = 0, \qquad \text{(3 Eqs.)},$$

$$-M_{ij,ji} = p, \qquad \text{(1 Equ.)}.$$

These seven equations are equivalent to a fourth-order differential equation for the deflection w

$$K(w_{,1111} + 2w_{,1122} + w_{,2222}) = K\Delta\Delta w = p$$

alone. The displacements on the boundary are the deflection and the slope

$$w, \qquad \frac{\partial w}{\partial n} = w_{,1}n_1 + w_{,2}n_2,$$

and the forces are the bending moment and the Kirchhoff shear

$$M_n(w) = M_{ij}n_in_j, \qquad V_n(w) = \frac{d}{ds}M_{nt} + Q_n$$

The Kirchhoff shear is the sum of the tangential derivative of the twisting moment M_{nt} and the shear force Q_n,

$$M_{nt} = M_{ij}n_it_j, \qquad Q_n = Q_1n_1 + Q_2n_2,$$

where

$$M_{11} = -K(w_{,11} + \nu w_{,22}), \qquad M_{22} = -K(w_{,22} + \nu w_{,11}),$$

$$M_{12} = -(1-\nu)Kw_{,12},$$

$$Q_1 = -K(w_{,111} + w_{,221}), \qquad Q_2 = -K(w_{,112} + w_{,222})$$

The basis of the boundary element method is Betti's principle or Green's second identity. This identity is obtained if we shift in the integral

$$\int_\Omega K\Delta\Delta\hat{w}\, w\, d\Omega = \text{force} \times \text{displacement} \qquad (6.4)$$

the operator $K\Delta\Delta$ by integration by parts onto the function w.

p: $\hat{w}, w \in C^4(\bar{\Omega})$,

$$q: B(\hat{w}, w) = \int\limits_{\Omega} K\Delta\Delta\hat{w}\, w\, d\Omega + \int\limits_{\Gamma} \left(\hat{V}_n w - \hat{M}_n \frac{\partial w}{\partial n} + \frac{\partial \hat{w}}{\partial n} M_n - \hat{w} V_n\right) ds$$

$$+ \sum_c [F(\hat{w})(\boldsymbol{x}^c) w(\boldsymbol{x}^c) - \hat{w}(\boldsymbol{x}^c) F(w)(\boldsymbol{x}^c)]$$

$$- \int\limits_{\Omega} \hat{w} K\Delta\Delta w\, d\Omega = 0 \tag{6.5}$$

The term

$$F(\hat{w})(\boldsymbol{x}^c) = M_{nt}(\hat{w})(\boldsymbol{x}^c_+) - M_{nt}(\hat{w})(\boldsymbol{x}^c_-)$$

is the corner force at a corner point $\boldsymbol{x}^c$. The notation $F(\hat{w})$ (as also $M_{nt}(\hat{w})$ etc.) is to denote that F is associated with the deflection $\hat{w}$. The sum $\sum$ in Eq.(6.5) is to be taken over all corner points $\boldsymbol{x}^c$ of the plate.

Note that it is the Kirchhoff shear which is multiplied with the deflection, so the Kirchhoff shear is conjugated to the deflection and not the shear force Q_n. This means that the equilibrium condition of a plate is to be formulated in terms of the Kirchhoff shear and not the shear force Q_n.

6.3 Influence functions for w and $\partial w/\partial n$

Of the four Betti-data of a plate two are prescribed and two are unknown so that we now need two integral equations. One renders the influence function for the deflection w and the second the influence function for the normal derivative $\partial w/\partial n$, the slope on the edge of the plate. The derivation of these two influence functions is done as follows:

a) We load an infinite plate at the point $\boldsymbol{x}$ with a concentrated force $\hat{P} = 1$ or a concentrated couple $\hat{M}_n = 1$, respectively. The vector $\boldsymbol{n}$ denotes the direction into which the "wheel" M is turning, see Fig. 6.8. Originally it connected the foot points of the two opposite forces which represented the couple.

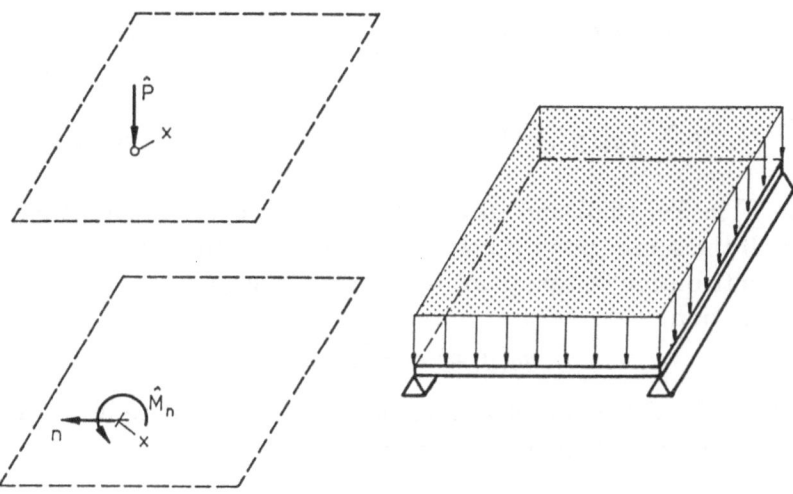

Figure 6.8 Application of Betti's principle

b) The subregion of the infinite plate that coincides with the real plate is separated from the rest of the plate and the internal actions along the edge become external forces.

c) Analogously we separate the real plate from its supports and let the support reactions become exterior boundary forces.

d) We formulate with these two solutions Betti's principle.

After the usual limiting processes, see [69], we thus obtain an influence function for the deflection $w(x)$

$$c(x)w(x) = \int_{\Gamma} [\, g_0(y,x)V_\nu(w)(y) - \frac{\partial}{\partial\nu}g_0(y,x)M_\nu(w)(y) - V_\nu(g_0(y,x))w(y)$$

$$+ M_\nu(g_0(y,x))\frac{\partial w}{\partial\nu}(y)]\,ds_y + \int_{\Omega} g_0(y,x)p(y)\,d\Omega_y$$

$$+ \sum_c [\, g_0(y^c,x)F(w)(y^c) - w(y^c)F(g_0)(y^c,x)] \tag{6.6}$$

and the normal derivative $\partial w(x)/\partial n$, respectively,

$$c_1(x)w_{,1}(x) + c_2(x)w_{,2}(x) = \int_{\Gamma} [\, g_1(y,x)V_\nu(w)(y)$$

$$- \frac{\partial}{\partial\nu}g_1(y,x)M_\nu(w)(y) - V_\nu(g_1(y,x))[w(y) - w(x)]$$

$$+ M_\nu(g_1(\boldsymbol{y}, \boldsymbol{x}))\frac{\partial w}{\partial \nu}(\boldsymbol{y})]\, ds_{\boldsymbol{y}} + \int_\Omega g_1(\boldsymbol{y}, \boldsymbol{x})p(\boldsymbol{y})\, d\Omega_{\boldsymbol{y}}$$

$$+ \sum_c [g_1(\boldsymbol{y}^c, \boldsymbol{x})F(w)(\boldsymbol{y}^c) - [w(\boldsymbol{y}^c) - w(\boldsymbol{x})]F(g_1)(\boldsymbol{y}^c, \boldsymbol{x})] \qquad (6.7)$$

The sum $\sum$ in Eqs.(6.6) and (6.7) is to be taken over all corner points $\boldsymbol{y}^c$ of the plate. If the source point $\boldsymbol{x}$ coincides with one of the corner points, $\boldsymbol{x} = \boldsymbol{y}^c$, then the contribution of this point to the sum is neglected.

The function

$$g_0(\boldsymbol{y}, \boldsymbol{x}) = \frac{1}{8\pi K}r^2 \ln r$$

in Eq.(6.6) is the deflection at a point $\boldsymbol{y}$ of an infinite plate if a concentrated force $\overset{\frown}{P} = 1$ is acting at the point $\boldsymbol{x}$ and the functions

$$\frac{\partial}{\partial \nu}g_0(\boldsymbol{y}, \boldsymbol{x}) = \frac{1}{8\pi K}rr_\nu(1 + 2\ln r),$$

$$M_\nu(g_0(\boldsymbol{y}, \boldsymbol{x})) = -\frac{1}{8\pi}[2(1 + \nu)\ln r + (3 + \nu)r_\nu^2 + (1 + 3\nu)r_\tau^2],$$

$$V_\nu(g_0(\boldsymbol{y}, \boldsymbol{x})) = -\frac{2}{8\pi r}[2r_\nu + (1 - \nu)(r_\nu - \kappa r)(r_\nu^2 - r_\tau^2)],$$

$$M_{\nu\tau}(g_0(\boldsymbol{y}, \boldsymbol{x})) = -\frac{1}{4\pi}(1 - \nu)r_\nu r_\tau$$

are the corresponding displacement and force terms at a boundary point $\boldsymbol{y}$ with the normal vector ν and tangent vector τ, see Fig. 6.9. The term $\kappa = 1/R$ is the curvature of the boundary at the point $\boldsymbol{y}$.

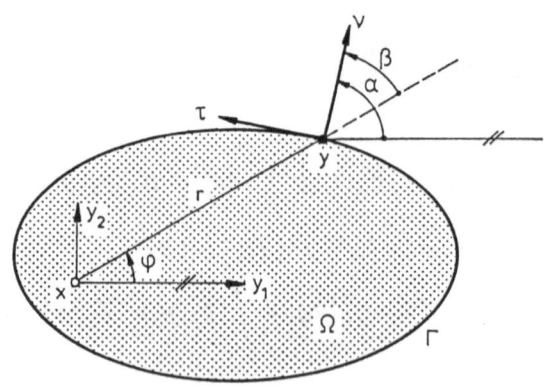

Figure 6.9 Source point $\boldsymbol{x}$ and integration point $\boldsymbol{y}$

The fundamental solution

$$g_1(y,x) = \frac{\partial}{\partial n_x} g_0(y,x) = \frac{1}{8\pi K} r(1 + 2\ln r)r_n$$

in the second equation (6.7) is the deflection at the point y if at some distant point x a concentrated couple of magnitude 1 is bending the plate into the direction n. The corresponding displacements and forces on the boundary are

$$\frac{\partial}{\partial \nu} g_1(y,x) = \frac{1}{8\pi K}[2(r_\nu r_n + r_t r_\tau)\ln r + 3r_\nu r_n + r_t r_\tau],$$

$$M_\nu(g_1(y,x)) = -\frac{1}{4\pi r}[(1+\nu)r_n + 2(1-\nu)r_\nu r_\tau r_t],$$

$$V_\nu(g_1(y,x)) = -\frac{1}{4\pi r^2}[\{3 - \nu - 2(1-\nu)r_\tau^2\}(r_\tau r_t - r_\nu r_n)$$

$$+ 4(1-\nu)(r_\nu - \kappa r)r_\nu r_\tau r_t],$$

$$M_{\nu\tau}(g_1(y,x)) = -\frac{(1-\nu)}{4\pi r}(r_\tau^2 - r_\nu^2)r_t$$

The characteristic functions in Eqs.(6.6) and (6.7) are

$$c(x) = \begin{cases} 1, & x \in \Omega, \\ \Delta\varphi/2\pi, & x \in \Gamma, \\ 0, & x \in \Omega^c, \end{cases} \quad \text{(complement)},$$

and

$$c_1(x) = \begin{cases} n_1, \\ \dot{c}_1(x), \\ 0, \end{cases} \quad c_2(x) = \begin{cases} n_2, & x \in \Omega, \\ \dot{c}_2(x), & x \in \Gamma, \\ 0, & x \in \Omega^c \end{cases}$$

The boundary values of the latter are

$$\dot{c}_1(x) = \frac{\Delta\varphi}{2\pi} n_1 + \frac{\nu}{2\pi}[\frac{1}{2}\sin 2\varphi\, n_1 + \sin^2\varphi\, n_2]_{\varphi_2}^{\varphi_1}, \qquad (6.8)$$

$$\dot{c}_2(x) = \frac{\Delta\varphi}{2\pi} n_2 + \frac{\nu}{2\pi}[\sin^2\varphi\, n_1 - \frac{1}{2}\sin 2\varphi\, n_2]_{\varphi_2}^{\varphi_1} \qquad (6.9)$$

The angles φ_1 and φ_2 are the angles which the two tangents at the boundary point form with the x_1-axis, see Fig. 3.6. At smooth points the difference between the angles $\Delta\varphi = \varphi_1 - \varphi_2$ is $180° = \pi$. In such a case the brackets in Eqs.(6.8) and (6.9) vanish,

$$[\sin 2\varphi]_{\varphi_2}^{\varphi_1} = \sin 2\varphi_1 - \sin 2\varphi_2 = 0 \,,$$

$$[\sin^2 \varphi]_{\varphi_2}^{\varphi_1} = \sin^2 \varphi_1 - \sin^2 \varphi_2 = 0$$

This means that at smooth points the c-functions are simply

$$\dot{c}_i(\boldsymbol{x}) = \frac{1}{2}n_i(\boldsymbol{x}) \,,$$

and, therefore, the right side of the influence function (6.7) is at such points half the slope

$$\dot{c}_1(\boldsymbol{x})w_{,1}(\boldsymbol{x}) + \dot{c}_2(\boldsymbol{x})w_{,2}(\boldsymbol{x}) = \frac{1}{2}\frac{\partial w}{\partial n}(\boldsymbol{x})$$

At corner points this is no longer true,

$$\dot{c}_1(\boldsymbol{x})w_{,1}(\boldsymbol{x}) + \dot{c}_2(\boldsymbol{x})w_{,2}(\boldsymbol{x}) \neq \frac{1}{2}\frac{\partial w}{\partial n}(\boldsymbol{x})$$

To calculate the slope at the two sides of the corner point we must, therefore, formulate Eq.(6.7) twice. The first time we substitute for the vector $\boldsymbol{n}$ in the fundamental solution g_1 the normal vector $\boldsymbol{n} = \boldsymbol{n}^l$ on the left-hand side and the second time the normal vector on the right-hand side $\boldsymbol{n} = \boldsymbol{n}^r$ of the corner point and we then solve the two equations

$$\dot{c}_1(\boldsymbol{x})w_{,1}(\boldsymbol{x}) + \dot{c}_2(\boldsymbol{x})w_{,2}(\boldsymbol{x}) = r^l \,,$$

$$\dot{c}_1(\boldsymbol{x})w_{,1}(\boldsymbol{x}) + \dot{c}_2(\boldsymbol{x})w_{,2}(\boldsymbol{x}) = r^r \,,$$

(r^l and r^r are the different right-hand sides) for the unknowns $w_{,1}(\boldsymbol{x})$ and $w_{,2}(\boldsymbol{x})$. These in turn uniquely determine the slope of the plate in any given direction.

6.4 Coupling on the boundary

If we place in Eqs.(6.6) and (6.7) the point $\boldsymbol{x}$ on the boundary then the boundary functions on the left side are the same boundary functions as on the right side so that the two equations formulate a coupling condition between the boundary values of a plate. If we arrange these equations according to displacement terms and force terms then these integral equations can be written in a somewhat symbolic notation as

$$\begin{bmatrix} H \end{bmatrix} \begin{bmatrix} w \\ \frac{\partial w}{\partial \nu} \end{bmatrix} = \begin{bmatrix} G \end{bmatrix} \begin{bmatrix} M_\nu \\ V_\nu \end{bmatrix} + \begin{bmatrix} d \end{bmatrix} \,,$$

and it now holds that if four boundary functions w, $\partial w/\partial \nu$, M_ν and V_ν (and the corner forces F_e) satisfy these two integral equations then the influence function (6.6) based on these functions have these functions as boundary values.

Because two of the four functions are determined by support and loading conditions the two integral equations can be solved for the two unknown functions and with the influence function (6.6) we then obtain the deflection at any interior point.

6.5 Discretization

For the boundary integrals to exist, the piecewise polynomial basis functions which interpolate the force and displacement terms must be sufficiently smooth at the collocation points. This means that the approximation of the deflection must be C^1 at the node (the tangential derivative must be continuous) and the approximation of the slope must be C^0, i.e. continuous. However the approximation of the bending moment may be discontinuous at the collocation point and the Kirchhoff shear, theoretically at least, may consist of Dirac-functions, i.e. point loads, only.

To comply with these conditions we approximate the deflection w along the boundary by Hermite-polynomials,

$$w(\boldsymbol{x}) = w_i\psi_i(\boldsymbol{x}) + w_i'\chi_i(\boldsymbol{x}), \qquad w' = \frac{dw}{ds},$$

that is, C^1-functions with the interpolation properties

$$\psi_i(\boldsymbol{x}^j) = \delta_{ij}, \qquad \psi_i'(\boldsymbol{x}^j) = 0, \qquad (\ )' = \frac{d}{ds},$$

$$\chi_i(\boldsymbol{x}^j) = 0, \qquad \chi_i'(\boldsymbol{x}^j) = \delta_{ij},$$

and the other boundary functions by piecewise linear polynomials see Fig. 6.10c.

With respect to the coordinate ξ of the master element $0 \le \xi \le 1$ the functions ψ_i and χ_i are given as

$$\psi_i(\boldsymbol{x}) = \begin{cases} \xi^2(3 - 2\xi), & \boldsymbol{x} \in \Gamma_{i-1}, \\ (\xi - 1)^2(1 + 2\xi), & \boldsymbol{x} \in \Gamma_i, \\ 0, & \text{otherwise}, \end{cases}$$

$$\chi_i(\boldsymbol{x}) = \begin{cases} \xi^2(\xi - 1)l_{i-1}, & \boldsymbol{x} \in \Gamma_{i-1}, \\ \xi(\xi - 1)^2 l_i, & \boldsymbol{x} \in \Gamma_i, \\ 0, & \text{otherwise}, \end{cases}$$

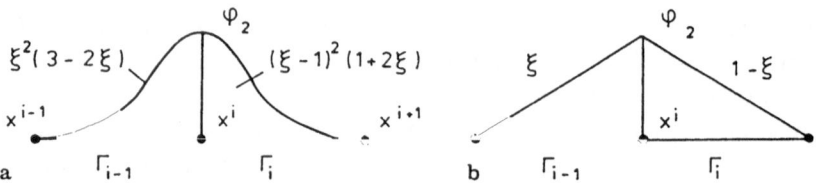

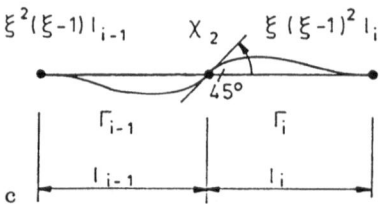

Figure 6.10 a-c. The basis functions: **a-b** Hermite polynomials interpolate the deflection w; **c** linear functions interpolate the other functions

and the linear functions $\varphi_i(\boldsymbol{x})$ are given as

$$\varphi_i(\boldsymbol{x}) = \begin{cases} \xi, & \boldsymbol{x} \in \Gamma_{i-1}, \\ 1 - \xi, & \boldsymbol{x} \in \Gamma_i, \\ 0, & \text{otherwise.} \end{cases}$$

All boundary functions, with the exception of the deflection w, are discontinuous at corner points. To be able to handle such discontinuities we assign to each function at each node two nodal values so that the approximation of, say, the Kirchhoff shear looks like

$$V_n(\boldsymbol{x}) = \varphi_1^i(\boldsymbol{x})V_1^i + \varphi_2^i(\boldsymbol{x})V_2^i ,$$

where φ_1^i and φ_2^i are the restrictions of the hat-function φ_i, see Fig. 6.10c, onto the element to left of the node, Γ_{i-1} and to the right of the node, Γ_i. A lower index 1 or 2 in the following denotes always a value to the *left* or to the *right* of the node.

6.6 Singular integrals

The integrals over the two elements Γ_{i-1} and Γ_i which enclose the collocation point $\boldsymbol{x}^i$, see Fig. 6.11, are integrated analytically, see [70].

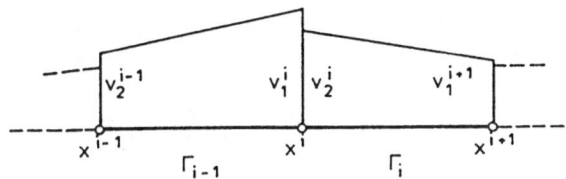

Figure 6.11 Piecewise linear modelling of the Kirchhoff shear

Employing the notations

$$l_i = \text{length of element } i, \qquad \lambda_i = \ln l_i$$

for the integrals over $\Gamma_{i-1} \cup \Gamma_i$ we obtain

$$\int g_0(\boldsymbol{y}, \boldsymbol{x}^i) V_\nu(w)(\boldsymbol{y})\, ds_{\boldsymbol{y}} = \{[\frac{1}{4}(\lambda_{i-1} - \frac{1}{4})V_2^{i-1}$$

$$+ \frac{1}{12}(\lambda_{i-1} - \frac{7}{12})V_1^i]l_{i-1}^3 + [\frac{1}{12}(\lambda_i - \frac{7}{12})V_2^i + \frac{1}{4}(\lambda_i - \frac{1}{4})V_1^{i+1}]l_i^3\}\frac{1}{8\pi K},$$

$$\int \frac{\partial}{\partial\nu} g_0(\boldsymbol{y}, \boldsymbol{x}^i) M_\nu(w)(\boldsymbol{y})\, ds_{\boldsymbol{y}} = \int V_\nu(g_0(\boldsymbol{y}, \boldsymbol{x}^i)) w(\boldsymbol{y})\, ds_{\boldsymbol{y}} = 0,$$

$$\int M_\nu(g_0(\boldsymbol{y}, \boldsymbol{x}^i)) \frac{\partial w}{\partial\nu}(\boldsymbol{y})\, ds_{\boldsymbol{y}} = -\{[(\lambda_{i-1}(1+\nu)+\nu)w_{\nu_2}^{i-1}$$

$$+ (\lambda_{i-1}(1+\nu)-1)w_{\nu_1}^i]l_{i-1} + [(\lambda_i(1+\nu)-1)w_{\nu_2}^i + (\lambda_i(1+\nu)+\nu)w_{\nu_1}^{i+1}]l_i\}\frac{1}{8\pi}$$

and

$$\int g_1(\boldsymbol{y}, \boldsymbol{x}^i) V_\nu(w)(\boldsymbol{y})\, ds_{\boldsymbol{y}} = \frac{1}{3}\{\boldsymbol{\tau}^{i-1} \cdot \boldsymbol{n}\, l_{i-1}^2[(2\lambda_{i-1} + \frac{1}{3})V_2^{i-1}$$

$$+ (\lambda_{i-1} - \frac{1}{3})V_1^i] - \boldsymbol{\tau}^i \cdot \boldsymbol{n}\, l_i^2[(\lambda_i - \frac{1}{3})V_2^i + (2\lambda_i + \frac{1}{3})V_1^{i+1}]\}\frac{1}{8\pi K}$$

$$\int \frac{\partial}{\partial\nu} g_1(\boldsymbol{y}, \boldsymbol{x}^i) M_\nu(w)(\boldsymbol{y})\, ds_{\boldsymbol{y}} = \{\boldsymbol{\tau}^{i-1} \cdot \boldsymbol{t}\, l_{i-1}[-\lambda_{i-1}M_2^{i-1}$$

$$+ (1 - \lambda_{i-1})M_1^i] + \boldsymbol{\tau}^i \cdot \boldsymbol{t}\, l_i[(1 - \lambda_i)M_2^i - \lambda_i M_1^{i+1}]\}\frac{1}{8\pi K},$$

$$\int V_\nu(g_1(\boldsymbol{y}, \boldsymbol{x}^i))[w(\boldsymbol{y}) - w(\boldsymbol{x}^i)]\, ds_{\boldsymbol{y}} - \int M_\nu(g_1(\boldsymbol{y}, \boldsymbol{x}^i))\frac{\partial w}{\partial\nu}\, ds_{\boldsymbol{y}}$$

$$= 2(1+\nu)\{\frac{2}{l_{i-1}}\tau^{i-1}\cdot t\,w^{i-1} + 2(-\frac{1}{l_i}\tau^i\cdot t - \frac{1}{l_{i-1}}\tau^{i-1}\cdot t)w^i$$

$$+ \frac{2}{l_i}\tau^i\cdot t\,w^{i+1} + (\frac{3}{2}-\lambda_{i-1})\tau^{i-1}\cdot t\,w_1'^i - \frac{1}{2}\tau^i\cdot t\,w_1'^{i+1}$$

$$+ \frac{1}{2}\tau^{i-1}\cdot t\,w_2'^{i-1} - (\frac{3}{2}-\lambda_i)\tau^i\cdot t\,w_2'^i + (\lambda_{i-1}-1)\tau^{i-1}\cdot n\,w_{\nu_1}^i$$

$$- \tau^i\cdot n\,w_{\nu_1}^{i+1} + \tau^{i-1}\cdot n\,w_{\nu_2}^{i-1} - (\lambda_i-1)\tau^i\cdot n\,w_{\nu_2}^i\}\frac{1}{8\pi},$$

w_ν = normal derivative

The expressions

$$\tau^i\cdot n, \qquad \tau^i\cdot t$$

denote the scalar product between the tangent vector τ^i of element Γ_i and the vector n of the fundamental solution g_1 or the associated tangent vector $t = (t_1, t_2)^T = (-n_2, n_1)^T$.

6.7 Element matrices

At each smooth node on the boundary we formulate two coupling conditions, Eqs.(6.6) and (6.7). At corner points the second equation, (6.7), is formulated twice so that the number of equations is

$$\text{nodes} \times 2 + \text{corner points}$$

This is also the height of the element matrices, see Fig. 6.12. The coefficients a_{kj}^l and b_{kj}^l in the element matrix G^i have the following meaning

$$a_{kj}^l = \int\limits_{\Gamma_i} -\frac{\partial}{\partial\nu}g_j(y, x^k) * ds_y, \qquad b_{kj}^l = \int\limits_{\Gamma_i} g_j(y, x^k) * ds_y$$

where the lone star represents the pertinent local basis function.

The coefficients a_{i0}^l, b_{i0}^l, $a_{i+1,0}^l$, $b_{i+1,0}^l$ etc. are the singular integrals, that is, the integrals if the collocation point is the point x^i or x^{i+1}. These integrals were calculated in section 6.6 by integrating over the two elements that enclose the collocation point. Now we split these results again in contributions which refer to the single elements. To find these results simply look in the equations for the nodal degrees of freedom, $w^i, w_j'^i, M_i^j, V_i^j$ which refer to the pertinent element. The coefficients which multiply these nodal values go into the element

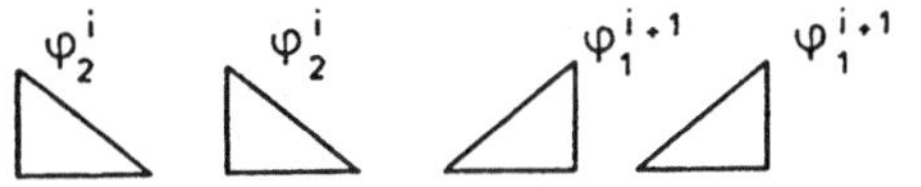

$$\begin{bmatrix} a_{10}^2 & b_{10}^2 & a_{10}^1 & b_{10}^1 \\ a_{11}^2 & b_{11}^2 & a_{11}^1 & b_{11}^1 \\ a_{20}^2 & b_{20}^2 & a_{20}^1 & b_{20}^1 \\ a_{21}^2 & b_{21}^2 & a_{21}^1 & b_{21}^1 \\ \cdot\cdot & \cdot\cdot & \cdot\cdot & \cdot\cdot \\ \\ a_{i0}^2 & b_{i0}^2 & a_{i0}^1 & b_{i0}^1 \\ a_{i1}^2 & b_{i1}^2 & a_{i1}^1 & b_{i1}^1 \\ a_{i+1.0}^2 & b_{i+1.0}^2 & a_{i+1.0}^1 & b_{i+1.0}^1 \\ a_{i+1.1}^2 & b_{i+1.1}^2 & a_{i+1.1}^1 & b_{i+1.1}^1 \\ \cdot\cdot & \cdot\cdot & \cdot\cdot & \cdot\cdot \\ \\ a_{K0}^2 & b_{K0}^2 & a_{K0}^1 & b_{K0}^1 \\ a_{K1}^2 & b_{K1}^2 & a_{K1}^1 & b_{K1}^1 \end{bmatrix} \begin{bmatrix} M_2^i \\ V_2^i \\ M_1^{i+1} \\ V_1^{i+1} \end{bmatrix}$$

K = number of the last node

Figure 6.12 The element matrix G^i

stiffness matrix. We first consider the singular entries of the element matrix G^i.

The collocation point is the node left (1st subscript: i)

1st IEQ (2nd subscript: 0)

$$a_{i0}^2 = a_{i0}^1 = 0, \qquad \alpha = \frac{1}{8\pi K},$$

$$b_{i0}^2 = \frac{1}{12}(\lambda_i - \frac{7}{12})l_i^3\alpha, \qquad b_{i0}^1 = \frac{1}{4}(\lambda_i - \frac{1}{4})l_i^3\alpha$$

2nd IEQ (2nd subscript: 1)

$$a_{i1}^2 = -\tau^i \cdot t\,l_i(1 - \lambda_i)\alpha, \qquad a_{i1}^1 = \tau^i \cdot t\,l_i\lambda_i\alpha,$$

$$b_{i1}^2 = -\tau^i \cdot n\,l_i^2\frac{1}{3}(\lambda_i - \frac{1}{3})\alpha, \qquad b_{i1}^1 = -\tau^i \cdot n\,l_i^2\frac{1}{3}(2\lambda_i + \frac{1}{3})\alpha$$

The collocation point is the node right (1st subscript: $i+1$)

1st IEQ (2nd subscript: 0)

$$a_{i+1,0}^2 = a_{i+1,0}^1 = 0,$$

$$b_{i+1,0}^2 = \frac{1}{4}(\lambda_i - \frac{1}{4})l_i^3\alpha, \qquad b_{i+1,0}^1 = \frac{1}{12}(\lambda_i - \frac{7}{12})l_i^3\alpha$$

2nd IEQ (2nd subscript: 1)

$$a_{i+1,1}^2 = \tau^i \cdot t\, l_i \lambda_i \alpha, \qquad a_{i+1}^1 = -\tau^i \cdot t\, l_i(1-\lambda_i)\alpha,$$

$$b_{i+1,1}^2 = -\tau^i \cdot n\, l_i^2 \frac{1}{3}(2\lambda_i + \frac{1}{3})\alpha, \qquad b_{i+1,1}^1 = \tau^i \cdot n\, l_i^2 \frac{1}{3}(\lambda_i - \frac{1}{3})\alpha$$

The coefficients c_{kj}^l and d_{kj}^l in the element matrix H^i, see Fig. 6.13, have the following meaning

$$
\begin{bmatrix}
c_{10}^1 & c_{10}^2 & c_{10}^3 & c_{10}^4 & d_{10}^5 & d_{10}^6 \\
c_{20}^1 & c_{20}^2 & c_{20}^3 & c_{20}^4 & d_{20}^5 & d_{20}^6 \\
\cdot\cdot & \cdot\cdot & \cdot\cdot & \cdot\cdot & \cdot\cdot & \cdot\cdot \\
 & & & & & \\
 & & & & & \\
 & & & & & \\
c_{i0}^1 & c_{i0}^2 & c_{i0}^3 & c_{i0}^4 & d_{i0}^5 & d_{i0}^6 \\
c_{i1}^1 & c_{i1}^2 & c_{i1}^3 & c_{i1}^4 & d_{i1}^5 & d_{i1}^6 \\
c_{i+1.0}^1 & c_{i+1.0}^2 & c_{i+1.0}^3 & c_{i+1.0}^4 & d_{i+1.0}^5 & d_{i+1.0}^6 \\
c_{i+1.1}^1 & c_{i+1.1}^2 & c_{i+1.1}^3 & c_{i+1.1}^4 & d_{i+1.1}^5 & d_{i+1.1}^6 \\
\cdot\cdot & \cdot\cdot & \cdot\cdot & \cdot\cdot & \cdot\cdot & \cdot\cdot \\
 & & & & & \\
\cdot\cdot & \cdot\cdot & \cdot\cdot & \cdot\cdot & \cdot\cdot & \cdot\cdot \\
c_{K1}^1 & c_{K1}^2 & c_{K1}^3 & c_{K1}^4 & d_{K1}^5 & d_{K1}^6
\end{bmatrix}
\begin{bmatrix}
w^i \\
w_2'^i \\
w^{i+1} \\
w_1'^{i+1} \\
w_{v2}^i \\
w_{v1}^{i+1}
\end{bmatrix}
$$

Figure 6.13 The ememt matrix H^i

$$c_{kj}^l = \int_{\Gamma_i} V_\nu(g_j(\boldsymbol{y}, \boldsymbol{x}^k)) * ds\boldsymbol{y}, \qquad d_{kj}^l = \int_{\Gamma_i} -M_\nu(g_j(\boldsymbol{y}, \boldsymbol{x}^k)) * ds\boldsymbol{y}$$

The coefficients c_{i0}^l, d_{i0}^l, $c_{i+1,0}$, $d_{i+1,0}^l$ etc. are the singular integrals, which are taken from section 6.6.

The collocation point is the node left (1st subscript: i)

1st IEQ (2nd subscript: 0)

$$c_{i0}^1 = 0.5\dot{c} = 0.5\frac{\Delta\varphi}{2\pi}, \qquad c_{i0}^2 = c_{i0}^3 = c_{i0}^4 = 0,$$

$$d_{i0}^5 = (\lambda_i(1+\nu)-1)l_i\frac{1}{8\pi} \qquad d_{i0}^6 = (\lambda_i(1+\nu)+\nu)l_i\frac{1}{8\pi}$$

2nd IEQ (2nd subscript: 1)

$$c_{i1}^1 = 4\beta(-\tau^i \cdot t\frac{1}{l_i} - \tau^{i-1} \cdot t\frac{1}{l_{i-1}})0.5, \qquad \beta = \frac{1+\nu}{8\pi},$$

$$c_{i1}^2 = -\beta(3-2\lambda_i)\tau^i \cdot t + \frac{1}{2}(-\dot{c}_1\nu_2^i + \dot{c}_2\nu_1^i),$$

$$c_{i1}^3 = 4\beta\,\tau^i \cdot t\frac{1}{l_i}, \qquad c_{i1}^4 = -\beta\,\tau^i \cdot t, \qquad d_{i1}^5 = -2\beta(\lambda_i-1)\tau^i \cdot n,$$

$$d_{i1}^6 = -2\beta\,\tau^i \cdot n$$

The terms ν_j^i are the components of the normal vector of the element Γ_i.

The collocation point is the node right (1st subscript: $i+1$)

1st IEQ (2nd subscript: 0)

$$c_{i+1,0}^1 = c_{i+1,0}^2 = c_{i+1,0}^4 = 0, \qquad c_{i+1,0}^3 = 0.5\,\dot{c},$$

$$d_{i+1,0}^5 = (\lambda_i(1+\nu)+\nu)l_i\frac{1}{8\pi}, \qquad d_{i+1,0}^6 = (\lambda_i(1+\nu)-1)l_i\frac{1}{8\pi}$$

2nd IEQ (2nd subscript: 1)

$$c_{i+1,1}^1 = 4\beta\,\tau^i \cdot t\frac{1}{l_i}, \qquad c_{i+1,1}^2 = \beta\,\tau^i \cdot t,$$

$$c_{i+1,1}^3 = 4\beta(-\tau^{i+1} \cdot t\frac{1}{l_{i+1}} - \tau^i \cdot t\frac{1}{l_i})0.5,$$

$$c_{i+1,1}^4 = \beta(3-2\lambda_i)\tau^i \cdot t + 0.5\,(-\dot{c}_1\nu_2^i + \dot{c}_2\nu_1^i),$$

$$d_{i+1,1}^5 = 2\beta\,\boldsymbol{\tau}^i \cdot \boldsymbol{n}\,\frac{1}{l_i}\,,$$

$$d_{i+1,1}^6 = 2\beta(\lambda_i - 1)\boldsymbol{\tau}^i \cdot \boldsymbol{n} + 0.5\,(\dot{c}_1\nu_1^i + \dot{c}_2\nu_2^i)$$

If the collocation point $\boldsymbol{x}^k$ is a corner point then we formulate the second integral equation twice so that a corner point is represented by three rows in the matrix $\boldsymbol{G}^i$

$$a_{k0}^l\,, \qquad b_{k0}^l\,, \qquad \cdots$$

$$a_{k1(l)}^l\,, \qquad b_{k1(l)}^l\,, \qquad \cdots$$

$$a_{k1(r)}^l\,, \qquad b_{k1(r)}^l\,, \qquad \cdots$$

and similarly in the matrix $\boldsymbol{H}^i$.

The influence integral of the deflection in the second equation

$$\int_\Gamma V_\nu(g_1(\boldsymbol{y}, \boldsymbol{x}^k))[w(\boldsymbol{y}) - w(\boldsymbol{x}^k)]\,ds_y =$$

$$= \int_{\Gamma_{i-1}\cup\Gamma_i} \cdots [w(\boldsymbol{y}) - w(\boldsymbol{x}^k)]\,ds_y + \int_{\Gamma_{\text{remainder}}} \cdots [w(\boldsymbol{y}) - w(\boldsymbol{x}^k)]\,ds_y$$

we split up into an integral over the two neighboring elements of the collocation point and the rest of the boundary. The integral over the two elements $\Gamma_{i-1}\cup\Gamma_i$ that enclose the collocation point is given in section 6.6. The integral over the remainder of the boundary can be split up into two integrals

$$\int_{\Gamma_{\text{remainder}}} V_\nu(g_1(\boldsymbol{y}, \boldsymbol{x}^k))w(\boldsymbol{y})\,ds_y - \int_{\Gamma_{\text{remainder}}} V_\nu(g_1(\boldsymbol{y}, \boldsymbol{x}^k))\,ds_y\,w(\boldsymbol{x}^k)$$

Like all non-singular integrals the first integral is calculated elementwise by numerical quadrature and the element contributions are stored in the element matrices $\boldsymbol{H}^i$, $i \neq k-1$, $i \neq k$. The second term is handled as follows. Recall that the calculation of the influence coefficients, the entries of the element matrices, is done by keeping the element fixed and evaluating successively the influence which the element layers have on the single collocation points. During this step we calculate for each node $\boldsymbol{x}^k$, which does not coincide with one of the two nodes of the element i, the regular integral

$$-\int_{\Gamma_i} V_\nu(g_1(\boldsymbol{y}, \boldsymbol{x}^k))\,ds_y$$

and add it to the influence vector of w^k and to that component whose row number is equal to the row number of the second integral equation at the node x^k.

The corner points contribute to the equations, too. To each corner point x^c belong two influence vectors, r^c and s^c, which represent the influence of the corner deflection and corner force F_c, respectively, on the collocation points x^k. The elements of these vectors are

$$r^c_{kj} = \begin{cases} F(g_j(y^c, x^k)) \\ 0 \end{cases} , \qquad s^c_{kj} = \begin{cases} g_j(y^c, x^k), & y^c \neq x^k, \\ 0, & y^c = x^k, \end{cases}$$

where the indices run over $k = 1, 2, \ldots, K$ and $j = 0, 1$ so that the height of the vectors is $K \times 2$. The influence of the term $w(x^k)$ in the expression

$$F(g_1(y^c, x^k))(w(y^c) - w(x^k))$$

requires a similar technique as explained before.

6.8 Degrees-of-freedom

The tangential derivative w' is not one of the Betti data and we therefore replace it by finite differences of the nodal deflections w^i,

$$w'^i = \frac{1}{2}(\frac{w^{i+1} - w^i}{l_i} + \frac{w^i - w^{i-1}}{l_{i-1}})$$

This substitution is done when the element matrix H^i is stored in the global matrix H. If the node is smooth then the column-vector which is associated with w'^i is

a) multiplied with the factor

$$\frac{1}{2}(\frac{1}{l_{i-1}} - \frac{1}{l_i})$$

and added to the column of w^i,

b) multiplied with the factor

$$-\frac{1}{2}\frac{1}{l_{i-1}}$$

and added to the column of w^{i-1},

c) multiplied with the factor

$$\frac{1}{2}\frac{1}{l_i}$$

and added to the column of w^{i+1}.

At corner points we add the column-vectors of w'^i to the columns of the normal derivatives. The relation between the tangential derivatives and the normal derivatives at a corner point are

$$\begin{bmatrix} w_1'^i \\ w_2'^i \end{bmatrix} = \begin{bmatrix} m_{11} & m_{12} \\ m_{21} & m_{22} \end{bmatrix} \begin{bmatrix} w_{\nu_1}^i \\ w_{\nu_2}^i \end{bmatrix},$$

where

$$m_{11} = -\frac{1}{D}[\nu_2^r \nu_2^l + \nu_1^r \nu_1^l], \qquad m_{12} = -\frac{1}{D}[(\nu_1^l)^2 + (\nu_2^l)^2],$$

$$m_{21} = -\frac{1}{D}[(\nu_1^r)^2 + (\nu_2^r)^2], \qquad m_{22} = -\frac{1}{D}[\nu_1^l \nu_1^r + \nu_2^l \nu_2^r]$$

The terms ν_i are the components of the normal vector on the left-hand side and right-hand side, respectively, and D is a determinant

$$D = \nu_1^l \nu_2^r - \nu_2^l \nu_1^r$$

Correspondingly we multiply at a corner point

the column of $w_1'^i$:

a) with m_{11} and add it to the column of $w_{\nu_1}^i$
b) with m_{12} and add it to the column of $w_{\nu_2}^i$

the column of $w_2'^i$:

a) with m_{21} and add it to the column of $w_{n_1}^i$
b) with m_{22} and add it to the column of $w_{n_2}^i$

After the elimination of w'^i four degrees-of-freedom remain at a smooth collocation point

$$w^i, \quad w_\nu^i, \quad M^i, \quad V^i$$

and at a corner point 8 such degrees

$$w^i, \quad w_{\nu_1}^i, \quad w_{\nu_2}^i, \quad M_1^i, \quad M_2^i, \quad V_1^i, \quad V_2^i, \quad F^i, \qquad 1 = \text{left}, \; 2 = \text{right}$$

At smooth points two of the four nodal values are prescribed so that the two integral equations suffice to determine the remaining two unknowns. At corner points we formulate three integral equations. If these and the boundary conditions alone do not suffice to determine the eight degrees-of-freedom then we can employ the continuity of the gradients ∇w and $\nabla\nabla w$ as side conditions. Such a typical argument is the following (hinged corner point): because the deflection is zero the tangential derivatives w_t are zero and therefore also the normal derivatives, etc. Often the situation is also reversed: only two degrees-of-freedom are undetermined or sometimes even only one, see Table 6.1.

Table 6.1

Nr.	BC		unknown terms	given terms
1	C + C		M_n , V_n	$w = w_n = 0$
2	H + H		w_n , V_n	$w = M_n = 0$
3	C + F		w , $w_n = \dfrac{\partial w}{\partial n}$	$M_n = V_n = 0$
4	C + H		M_n^l , V_n	$w = w_n = M_n^r = 0$
5	C + F		M_n^l , V_n^l	$w = w_n = M_n^r = V_n^r = 0$
6	H + F		w_n , V_n^l	$w = M_n = V_n^r = 0$
7	H + C		M_n^r , V_n	$w = w_n = M_n^l = 0$
8	F + C		M_n^r , V_n^r	$w = w_n = M_n^l = V_n^l = 0$
9	F + H		w_n , V_n^r	$w = M_n = V_n^l = 0$
1c	C + C		M_n^l , M_n^r , F	$w = w_n^l = w_n^r = V_n^l = V_n^r = 0$
2c	H + H		V_n^l , V_n^r , F	$w = w_n^l = w_n^r = M_n^l = M_n^r = 0$
3c	F + F		w , w_n^l , w_n^r	$M_n^l = M_n^r = V_n^l = V_n^r = F = 0$
4c	C + H		M_n^l , M_n^r , F	$w = w_n^l = w_n^r = V_n^l = V_n^r = 0$
5c	C + F		M_n^l , V_n^l , F	$w = w_n^l = w_n^r = M_n^r = V_n^r = 0$
6c	H + F		w_n^l , w_n^r , V_n^l	$M_n^l = M_n^r = F = w = V_n^r = 0$
7c	H + C		V_n^l , M_n^r , F	$w = w_n^l = w_n^r = M_n^l = V_n^r = 0$
8c	F + C		M_n^r , V_n^r , F	$w = w_n^l = w_n^r = M_n^l = V_n^l = 0$
9c	F + H		w_n^l , w_n^r , V_n^r	$M_n^l = M_n^r = F = w = V_n^l = 0$

6.9 The domain integrals

The influence of the distributed load

$$\int_{\Omega} g_i(y, x) p(y) \, d\Omega_y, \qquad i = 0, 1$$

can, in general, be expressed by an equivalent boundary integral. If the load p is constant then we proceed as in section 3.7: the integral of the fundamental solution $g_0(y, x)$ with respect to the Laplacian is

$$\frac{1}{8\pi K} \frac{r^4}{32} (2 \ln r - 1),$$

so that we obtain for the domain integral in Eq.(6.6) the representation ($p = 1$)

$$\frac{1}{8\pi K} \int_{\Omega} r^2 \ln r \, d\Omega_y = \frac{1}{8\pi K} \int_{\Gamma} \frac{\partial}{\partial \nu} \left[\frac{r^4}{32} (2 \ln r - 1) \right] ds_y$$

$$= \frac{1}{64\pi K} \int_{\Gamma} r^3 (2 \ln r - \frac{1}{2}) r_\nu \, ds_y$$

and for its normal derivative in Eq.(6.7) the expression

$$\frac{1}{8\pi K} \int_{\Omega} r(1 + 2\ln r) r_n \, d\Omega_y = \frac{1}{64\pi K} \int_{\Gamma} \{ r^2 [(6 \ln r + 0.5) r_n r_\nu$$

$$+ (2 \ln r - 0.5) r_\tau r_t] \} \, ds_y$$

If the load is the partial derivative $p(x) = \psi_{,i}(x)$ of a function $\psi(x)$ with the property

$$\Delta \psi(x) = k_0 = \text{constant},$$

then we use the substitution, see [71],

$$\frac{1}{8\pi K} \int_{\Omega} r^2 \ln r \, p(y) \, d\Omega_y = \int_{\Gamma} [r^2 \ln r \, \psi(y) \nu_i(y) + (\frac{\partial f(r)}{\partial \nu})_{,x_i} \psi(y)$$

$$- f(r)_{,x_i} \frac{\partial \psi}{\partial \nu}(y) - k_0 f(r) \nu_i(y)] \, ds_y,$$

$$f(r) = \frac{r^4}{32} (2 \ln r - 1),$$

We could as well split the deflection of the plate into a homogeneous and a particular solution

$$w = w_h + w_p ,$$

and determine only the homogeneous solution with boundary elements and later add the particular solution w_p. This is the approach in the program BE-PLATE-BENDING if the distributed load is linear. Because we cannot cut off a particular solution in the middle of the plate, we must require that the triangular load extends over the whole plate.

6.10 Actions on the boundary

The bending moment M_n and the Kirchhoffshear V_n being Betti data are the only actions on the boundary which are immediately available: they are either given or they are obtained by solving the collocation equations. All the other actions must be calculated separately.

To obtain the moment M_t at a free edge

$$M_t = -K(1 - \nu^2)w_{,tt}$$

we approximate the second order tangential derivative $w_{,tt}$ by finite differences. If the edge is clamped or hinged then the moment M_t is identical with M_n.

$$M_t = M_n$$

The twisting moment M_{nt} is zero on such edges while it is proportional to $w_{,nt}$ along a free edge

$$M_{nt} = -K(1 - \nu)w_{,nt} ,$$

so that we can approximate it by differentiating the slope w_n numerically.

For the shear force we use the formula

$$Q_n = V_n - \frac{d}{ds}M_{nt}$$

and approximate the tangential derivative of M_{nt} by finite differences.

Due to the approximate character of M_{nt} the difference between the two twisting moments at a corner

$$F = M_{nt}^r - M_{nt}^l \qquad \text{r = right, l = left}$$

will not be exactly the corner force F. But this is still an improvement over finite elements where one nodal value represents the sum of the two shear forces and the corner force.

6.11 Internal actions

The bending moments

$$M_{11} = -K(w,_{11} + \nu w,_{22}), \qquad M_{12} = -K(1 - \nu)w,_{12},$$

$$M_{22} = -K(w,_{22} + \nu w,_{11}),$$

are linear combinations of the second derivatives

$$w,_{11}, \quad w,_{12}, \quad w,_{22}$$

so that, instead of formulating three separate influence functions for the three bending moments, it is more efficient to formulate only one influence function for a second order directional derivative

$$\frac{\partial}{\partial m} \frac{\partial}{\partial n} w$$

and to substitute later for the two unit vectors

$$\boldsymbol{n} = \{n_1, n_2\}^T, \qquad \boldsymbol{m} = \{m_1, m_2\}^T$$

and their associated tangent vectors

$$\boldsymbol{t} = \{-n_2, n_1\}^T, \qquad \boldsymbol{p} = \{-m_2, m_1\}^T$$

the unit vectors $\boldsymbol{n} = \boldsymbol{e}_1$ and $\boldsymbol{m} = \boldsymbol{e}_1$, to calculate the derivative $w,_{11}$, etc.

We obtain the influence functions for these second order derivatives if we differentiate the first order derivative, the slope (6.7). The derivatives of the single kernels in Eq.(6.7) are

$$(g_0(\boldsymbol{y},\boldsymbol{x}))_{,n\,,m} = \frac{1}{8\pi K}[(3+2\ln r)r_m r_n + (1+2\ln r)r_t r_p],$$

$$(\frac{\partial}{\partial \nu}g_0(\boldsymbol{y},\boldsymbol{x}))_{,n\,,m} = \frac{1}{4\pi K r}[r_m(r_\nu r_n + r_t r_\tau) + r_p(r_n r_\tau + r_t r_\nu)],$$

$$(M_\nu(g_0(\boldsymbol{y},\boldsymbol{x})))_{,n\,,m} = \frac{1}{4\pi r^2}\{r_m[(1+\nu)r_n + 2(1-\nu)r_\nu r_\tau r_t]$$

$$- r_p[(1+\nu)r_t + 2(1-\nu)([r_\tau^2 - r_\nu^2]r_t - r_\tau r_n r_\nu)]\},$$

$$(V_\nu(g_0(\boldsymbol{y},\boldsymbol{x})))_{,n\,,m} = \frac{1}{4\pi r^3}\{2r_m\{[3-\nu-2(1-\nu)r_\tau^2](r_t r_\tau - r_\nu r_n)$$

$$+ 4(1-\nu)r_\nu^2 r_\tau r_t\} - r_p\{4(1-\nu)r_\nu r_\tau(r_\tau r_t - r_\nu r_n)$$

$$- 2[3-\nu-2(1-\nu)r_\tau^2](r_\nu r_t + r_\tau r_n)$$

$$+ 4(1-\nu)(2r_\nu r_\tau^2 r_t - r_\nu^3 r_t - r_\nu^2 r_\tau r_n)\}\}, \qquad (6.10)$$

$$(M_{\nu\tau}(g_0(\boldsymbol{y},\boldsymbol{x})))_{,n\,,m} = \frac{1-\nu}{4\pi r^2}\{-r_m r_t(r_\nu^2 - r_\tau^2)$$

$$+ r_p[-r_n(r_\nu^2 - r_\tau^2) + 4r_\nu r_\tau r_t]\}$$

In formulating Eq.(6.10) we assumed that the curvature $\kappa = \kappa(\boldsymbol{y})$ of the boundary is zero.
The derivatives of the distance r are

$$r_m = r_{,x_1}\,m_1 + r_{,x_2}\,m_2 = -r_{,1}\,m_1 - r_{,2}\,m_2,$$

$$r_p = r_{,x_1}\,p_1 + r_{,x_2}\,p_2 = r_{,1}m_2 - r_{,2}\,m_1$$

and the directional derivatives of the kernel

$$g_p = \frac{1}{64\pi K}[r^3(2\ln r - \frac{1}{2})r_\nu]$$

in the transformed domain integral are

$$(g_p)_{,n\,,m} = \frac{r}{64\pi K}\{r_m[r_n r_\nu(12\ln r + 7) + r_\tau r_t(4\ln r + 1)]$$

$$+ r_p[(r_t r_\nu + r_n r_\tau)(4\ln r + 1)]\}$$

For the shear forces

$$Q_1 = -K(w_{,111} + w_{,221}), \qquad Q_2 = -K(w_{,112} + w_{,222})$$

we need an influence function for the directional derivative of the third order

$$\frac{\partial}{\partial l}\frac{\partial}{\partial m}\frac{\partial}{\partial n}w$$

Hence, we must differentiate the kernels in Eq.(6.10) once more. This third direction we denote by $l = \{l_1, l_2\}^T$ and its associated tangent vector by $q = \{q_1, q_2\}^T = \{-l_2, l_1\}^T$.

$$(g_0)_{,n\,,m\,,l} = \frac{2}{8\pi K r}\{r_q[r_p r_n + r_m r_t] + r_l[r_m r_n + r_t r_p]\},$$

$$(\frac{\partial}{\partial \nu}g_0)_{,n\,,m\,,l} = \frac{1}{4\pi K r^2}[(r_\nu r_n + r_t r_\tau)(r_p r_q - r_l r_m) - (r_n r_\tau$$
$$+ r_t r_\nu)(r_l r_p + r_m r_q) + 2r_p r_q(r_t r_\tau - r_n r_\nu)],$$

$$(M_\nu(g_0))_{,n\,,m\,,l} = \frac{1}{4\pi r^3}[-2r_l\{r_m[(1+\nu)r_n + 2(1-\nu)r_\nu r_\tau r_t]$$
$$- r_p[(1+\nu)r_t + 2(1-\nu)([r_\tau^2 - r_\nu^2]r_t - r_t r_n r_\nu)]\}$$
$$+ r_q\{r_p[(1+\nu)r_n + 2(1-\nu)r_\nu r_\tau r_t] + r_m[(1+\nu)r_t$$
$$+ 2(1-\nu)[r_\tau^2 r_t - r_\nu^2 r_t - r_\nu r_\tau r_n]] + r_m[(1+\nu)r_t$$
$$+ 2(1-\nu)([r_t^2 - r_\nu^2]r_t - r_\tau r_\nu r_n)] - r_p[-(1+\nu)r_n$$
$$+ 2(1-\nu)(-5r_\tau r_\nu r_t - 2r_\tau^2 r_n + 2r_\nu^2 r_n)]\}],$$

$$(V_\nu(g_0))_{,n\,,m\,,l} = \frac{1}{4\pi}[\frac{-3}{r^4}r_l(2r_m b - r_p c)$$
$$+ \frac{1}{r^3}(\frac{2}{r}r_p r_q b + \frac{1}{r}r_m r_q c + 2r_m b_{,l} - r_p c_{,l})],$$

$$b = [3 - \nu - 2(1-\nu)r_\tau^2](r_\tau r_t - r_\nu r_n) + 4(1-\nu)r_\nu^2 r_\tau r_t,$$

$$c = 4(1-\nu)r_\nu r_\tau(r_\tau r_t - r_\nu r_n) - 2[3 - \nu - 2(1-\nu)r_\tau^2](r_\nu r_t + r_\tau r_n)$$
$$+ 4(1-\nu)(2r_\nu r_\tau^2 r_t - r_\nu^3 r_t - r_\nu^2 r_\tau r_n),$$

$$b_{,l} = 4(1-\nu)r_\tau r_\nu r_q \frac{1}{r}(r_\tau r_t - r_\nu r_n) - \frac{2}{3}[3 - \nu - 2(1-\nu)r_\tau^2](r_\nu r_t$$

$$+ r_\tau r_n)r_q + \frac{4(1-\nu)}{r}[2r_\tau^2 r_t - r_\nu^2 r_t - r_\nu r_\tau r_n]r_q r_\nu,$$

$$c_{,l} = \frac{4(1-\nu)}{r}[[r_\tau^2 - r_\nu^2](r_\tau r_t - r_\nu r_n) - 4r_\nu r_\tau(r_\nu r_t + r_\tau r_n)$$

$$+ \frac{d}{4(1-\nu)} + 2r_\tau^3 r_t - 8r_\nu^2 r_\tau r_t - 4r_\nu r_\tau^2 r_n + 2r_\nu^3 r_n]r_q,$$

$$d = -4[3 - \nu - 2(1-\nu)r_\tau^2](r_\tau r_t - r_\nu r_n),$$

$$(M_{\nu\tau}(g_0))_{,n,m,l} = \frac{1-\nu}{4\pi r^3}[e(2r_l r_m r_t - r_p r_q r_t + r_m r_n r_q)$$

$$- f[2r_l r_p + r_m r_q] - 4r_m r_t r_\nu r_\tau r_q + r_p f_{,l}],$$

$$e = r_\nu^2 - r_\tau^2, \qquad f = -r_n e + 4r_\nu r_\tau r_t,$$

$$f_{,l} = r_q[-5r_t r_\nu^2 + 3r_\tau^2 r_t - 8r_n r_\nu r_\tau]\frac{1}{r}$$

The third derivatives of the kernel g_p are

$$(g_p)_{,n,m,l} = \frac{1}{64\pi K}\{r_l h + r_p r_q\{r_n r_\nu[12\ln r + 7]$$

$$+ r_\tau r_t[4\ln r + 1]\} + r_m\{(r_t r_q r_\nu + r_n r_\tau r_q)(8\ln r + 6)$$

$$+ r_l(12r_n r_\nu + 4r_\tau r_t)\} - r_m r_q\{(r_t r_\nu + r_n r_\tau)[4\ln r + 1]\}$$

$$+ r_p\{2(r_t r_q r_\tau - r_n r_q r_\nu)[4\ln r + 1] + 4r_l(r_t r_\nu + r_n r_\tau)\}\},$$

where h denotes the auxiliary function

$$h = r_m[r_n r_\nu(12\ln r + 7) + r_\tau r_t(4\ln r + 1)]$$

$$+ r_p[(r_t r_\nu + r_n r_\tau)(4\ln r + 1)]$$

6.12 Internal supports and subdomain loads

If the plate is supported at a single point x^P then the influence of the support reaction P is considered by supplementing the influence functions (6.6) and (6.7) with the functions

$$g_i(\boldsymbol{x}^P, \boldsymbol{x})P, \qquad i = 0 \quad \text{in Eq.(6.6)}, \qquad i = 1 \quad \text{in Eq.(6.7)}$$

To determine the support reaction P we require that the deflection is zero at $\boldsymbol{x}^P$,

$$w(\boldsymbol{x}^P) = 0 = \int_{\Gamma} \left[g_0(\boldsymbol{y}, \boldsymbol{x}^P) V_\nu(w)(\boldsymbol{y}) - \frac{\partial}{\partial \nu} g_0(\boldsymbol{y}, \boldsymbol{x}^P) M_\nu(w)(\boldsymbol{y}) \right.$$

$$- V_\nu(g_0(\boldsymbol{y}, \boldsymbol{x}^P)) w(\boldsymbol{y}) + M_\nu(g_0(\boldsymbol{y}, \boldsymbol{x}^P)) \frac{\partial w}{\partial \nu}(\boldsymbol{y}) \Big] \, ds_{\boldsymbol{y}}$$

$$+ \int_{\Omega} g_0(\boldsymbol{y}, \boldsymbol{x}^P) p(\boldsymbol{y}) \, d\Omega_{\boldsymbol{y}} + g_0(\boldsymbol{x}^P, \boldsymbol{x}^P) P$$

$$+ \sum_c \{ g_0(\boldsymbol{y}^c, \boldsymbol{x}^P) F(w)(\boldsymbol{y}^c) - w(\boldsymbol{y}^c) F(g_0)(\boldsymbol{y}^c, \boldsymbol{x}^P) \} \qquad (6.11)$$

This equation is added to the system $\boldsymbol{Hu} = \boldsymbol{Gt} + \boldsymbol{d}$ and simultaneously we append a dof u (= the deflection at the support) and a dof t (= support reaction) to the vectors $\boldsymbol{u}$ and $\boldsymbol{t}$.

If the point support is an elastic support with stiffness c, then the deflection $w(\boldsymbol{x}^P)$ and the support reaction P are coupled by

$$w(\boldsymbol{x}^P) c - P = 0$$

If we let the deflection $w(\boldsymbol{x}^P)$ be the dependent unknown and P be the primary unknown then we replace the deflection $w(\boldsymbol{x}^P)$ on the left-hand side of Eq.(6.11) by P/c.

The influence of the support reactions $p_s(\boldsymbol{y})$ of a line support γ is represented by a line integral

$$\int_{\gamma} g_0(\boldsymbol{y}, \boldsymbol{x}) p_s(\boldsymbol{y}) \, ds_{\boldsymbol{y}}$$

In the program BE-PLATE-BENDING such a line support is subdivided into elements Γ_i and the unknown support reactions are approximated by a series of linear functions

$$p_s(\boldsymbol{x}) = p_s(\boldsymbol{x}^j) \varphi_j(\boldsymbol{x})$$

so that the integral over γ becomes a sum of integrals over the single support elements.

$$\sum_i \int_{\Gamma_i} g_0(\boldsymbol{y}, \boldsymbol{x}) \varphi_j(\boldsymbol{y}) \, ds_{\boldsymbol{y}} \, p^j, \qquad p^j = p_s(\boldsymbol{x}^j)$$

and the condition that the deflection of the plate is zero at the nodes of the

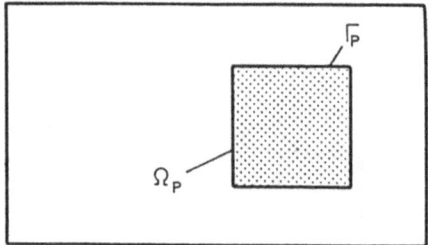

Figure 6.14 Subdomain load

support supplies the additional equations to determine the nodal values p^j of the support reactions.

The influence of a (constant) subdomain load, see Fig 6.14, is transformed into an equivalent boundary integral over the boundary Γ_p of the subdomain Ω_p,

$$\frac{p}{8\pi K} \int\limits_{\Omega_p} r^2 \ln r \, d\Omega y = \frac{p}{64\pi K} \int\limits_{\Gamma_p} r^3 (2\ln r - \frac{1}{2}) r_\nu \, ds y$$

This means that we encircle each subdomain with an extra set of boundary elements.

6.13 Examples

The first example is a rectangular plate with an opening, see Fig. 6.15. The plate is clamped on two sides and hinged on the two opposite sides. The outer boundary was subdivided into 26 boundary elements and the edge of the opening into eight elements. The agreement between the BE-results and FE-results is very good, see Fig. 6.16.

The second example is a floor plate which is supported internally by 16 columns, see Fig. 61.7a. Figure 6.17 b contains a comparison between the computed bending moments and the distribution of the bending moments as recommended by the German building code. A student obtained these results with only 16 boundary elements, four boundary elements on each side (the author would not have dared to do this). This plate demonstrates the ease with which the BE-method solves such plate problems. Because the stress singularities caused by the single columns are, so to speak, built into the computer code no special measures need be taken as in the FE-method where a mesh-refinement would be necessary to capture the high stress peaks.

Let us use this occasion to make a remark regarding singular stresses. Engineers often argue that stress singularities are mathematical artefacts because

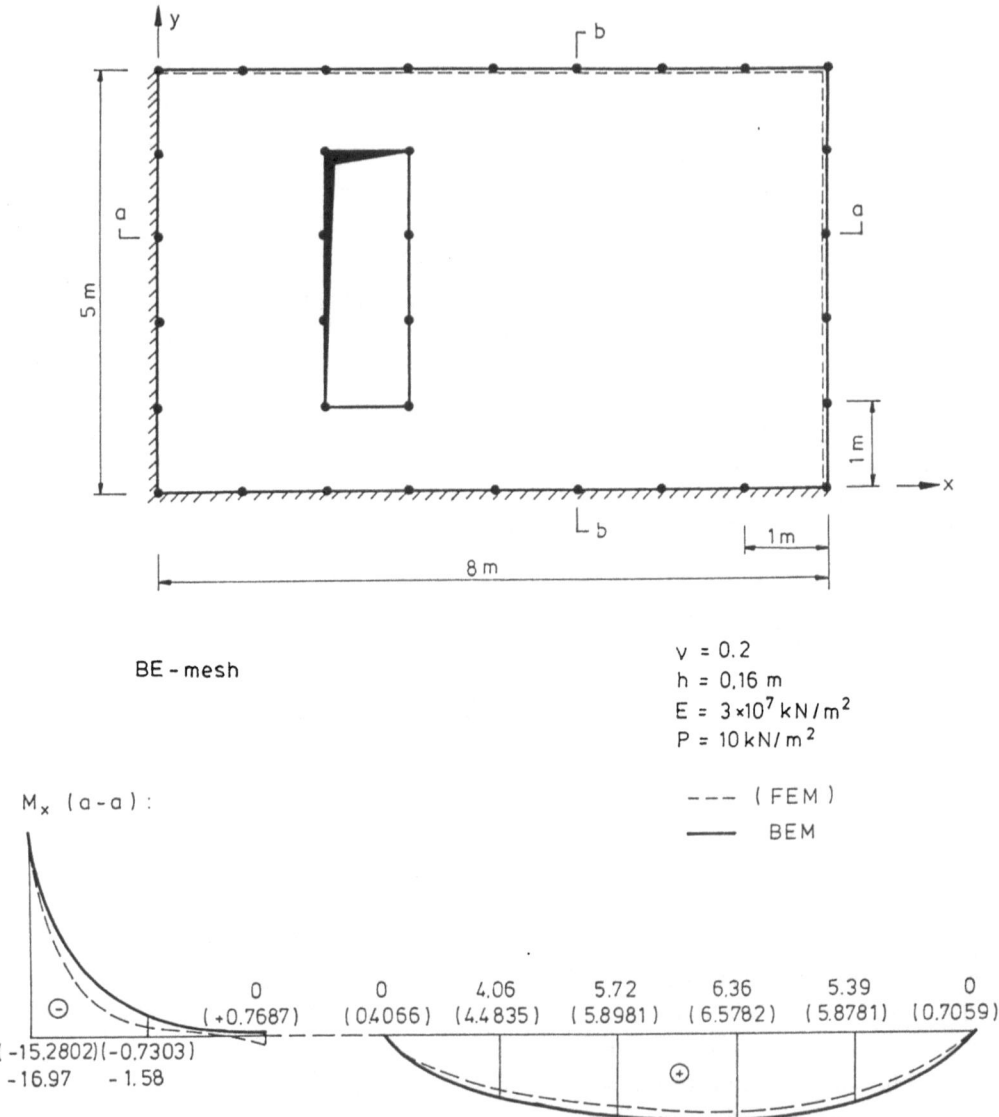

Figure 6.15 Plate with an opening under uniform loads and the bending moments in the cross-section a–a

"in reality" the underlying mechanical theory and so the differential equation is no longer valid. This might be true, but it is also true that you must model these high stress peaks as well as possible (even if you believe that they do not occur) to obtain an accurate solution at points *away* from the singularity, [72].

The third example is a skew floor plate which was solved with ADINA, see Fig. 6.18, and with the program BE-PLATE-BENDING, see Fig. 6.19. The

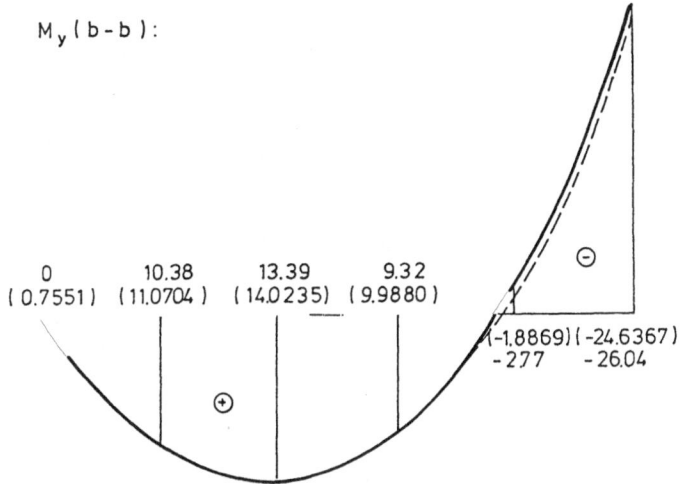

M$_y$(b–b):

0   10.38   13.39   9.32
(0.7551) (11.0704) (14.0235) (9.9880)

(−1.8869)(−24.6367)
−277   − 26.04

Figure 6.16 The bending moments in the cross-section b–b

agreement between the two methods is (as usual) very good. Notable deviations only occur at the skew-angled hinged corner points, see Fig. 6.20, where the exact solution becomes singular, see Table 6.2. This causes oscillations and peaks in the support forces of the two approximate solutions. Note also that the BE-program outputs at each corner point three values: the value of the Kirchhoff shear V_n on the left and on the right-hand side and, separately, the corner force F while the FE-program outputs only one nodal force F which combines the action of the three forces. This habit of the FE-solution is also clearly seen at the intersection of the interior supports, see Fig. 6.20, point A. While the support force of the BE-solution changes from plus to minus and stays that way up to the intersection, the FE-solution changes its sign abruptly from minus to plus to reach the positive level at the intersection.

The fourth example is a continuous hinged plate, see Fig. 6.21. In the first field the thickness of the plate is $h = 15cm$ and in the second $h = 20cm$. To solve this problem we made use of the technique outlined in section 3.13. We formulated the influence functions for the two subdomains separately, multiplied each equation with the stiffness K of the pertinent domain, and added the two equations, so that with the interface conditions

$$w_a - w_b = 0, \qquad \frac{\partial w_a}{\partial n} + \frac{\partial w_b}{\partial n} = 0, \qquad M_a + M_b = 0, \qquad V_a + V_b = f,$$

and the interface properties of the two fundamental solutions

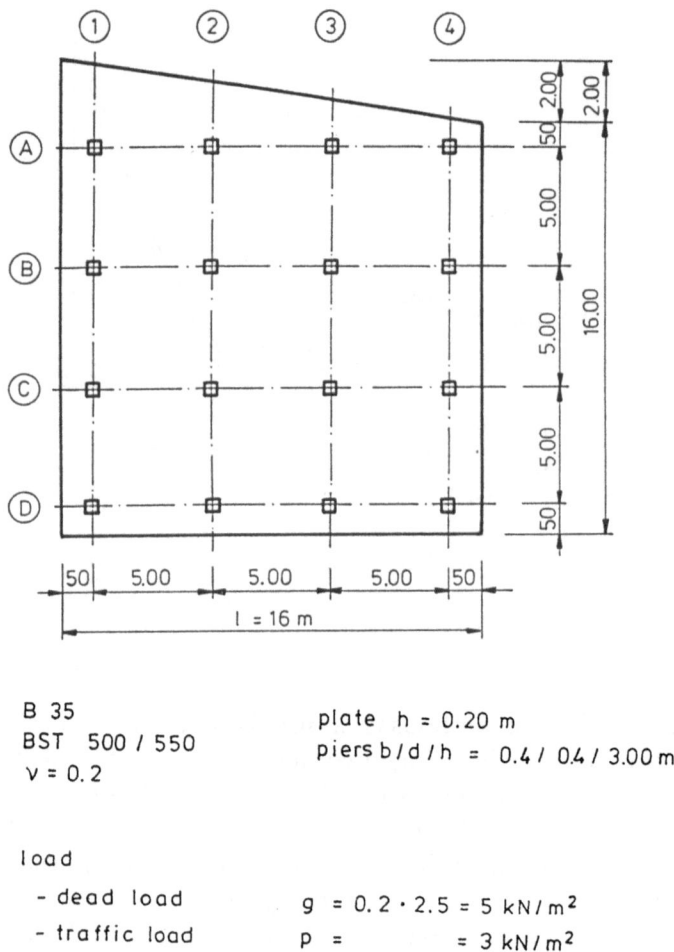

B 35
BST 500 / 550
ν = 0.2

plate h = 0.20 m
piers b / d / h = 0.4 / 0.4 / 3.00 m

load
 - dead load
 - traffic load

$g = 0.2 \cdot 2.5 = 5 \text{ kN} / m^2$

$p = = 3 \text{ kN} / m^2$

$\Sigma = g + p = 8 \text{ kN} / m^2$

a

Figure 6.17a Plate that rests on 16 piers

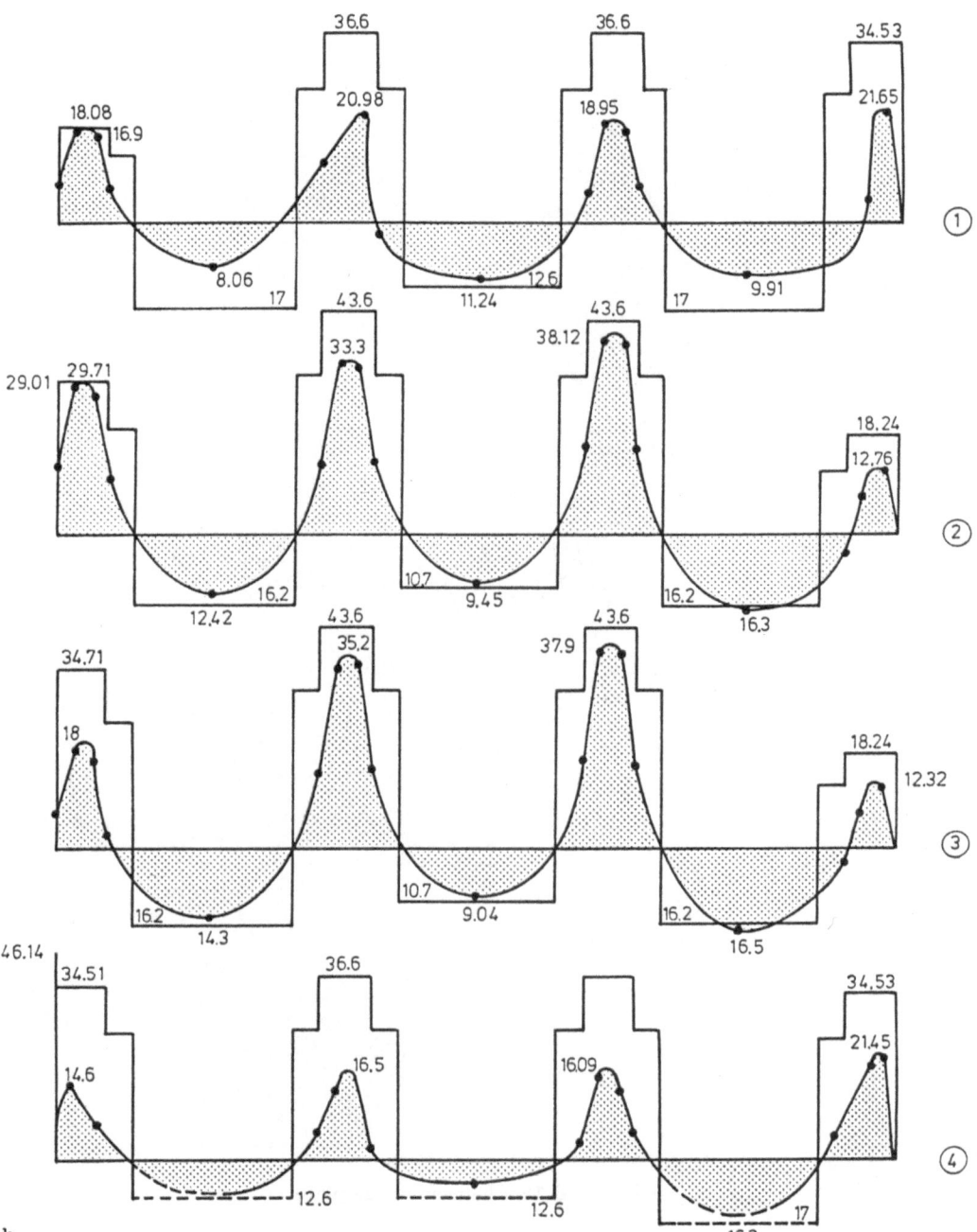

Figure 6.17b Distribution of bending moments

b

286

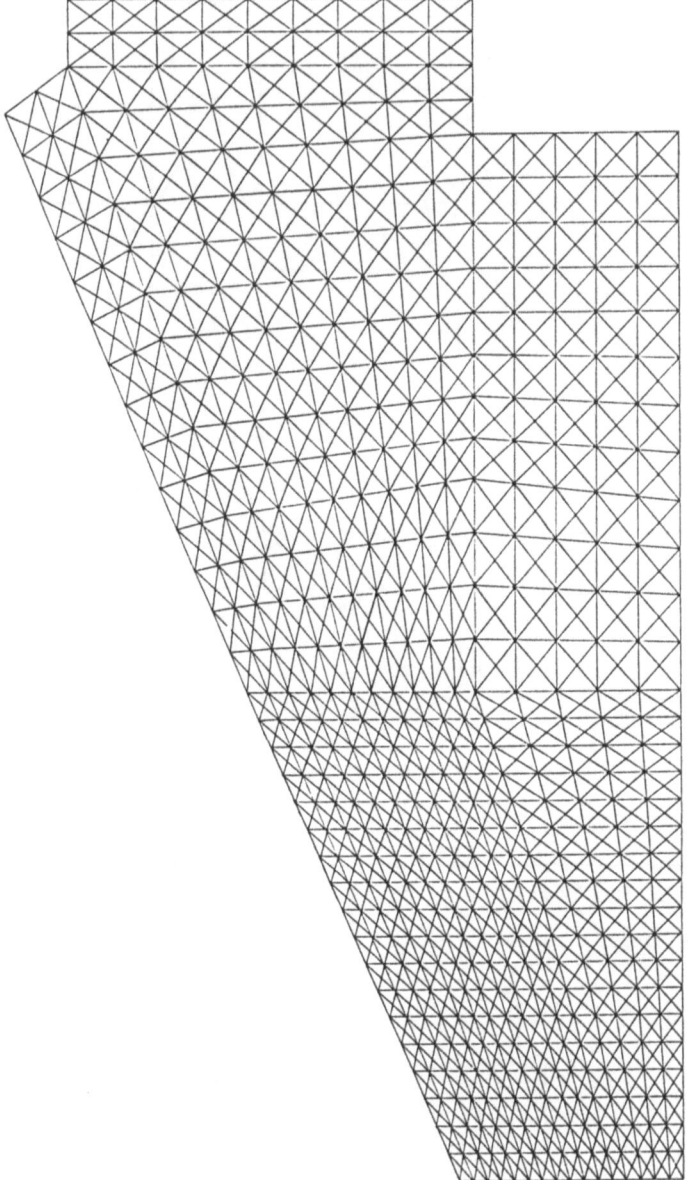

Figure 6.18 FE-mesh of the plate

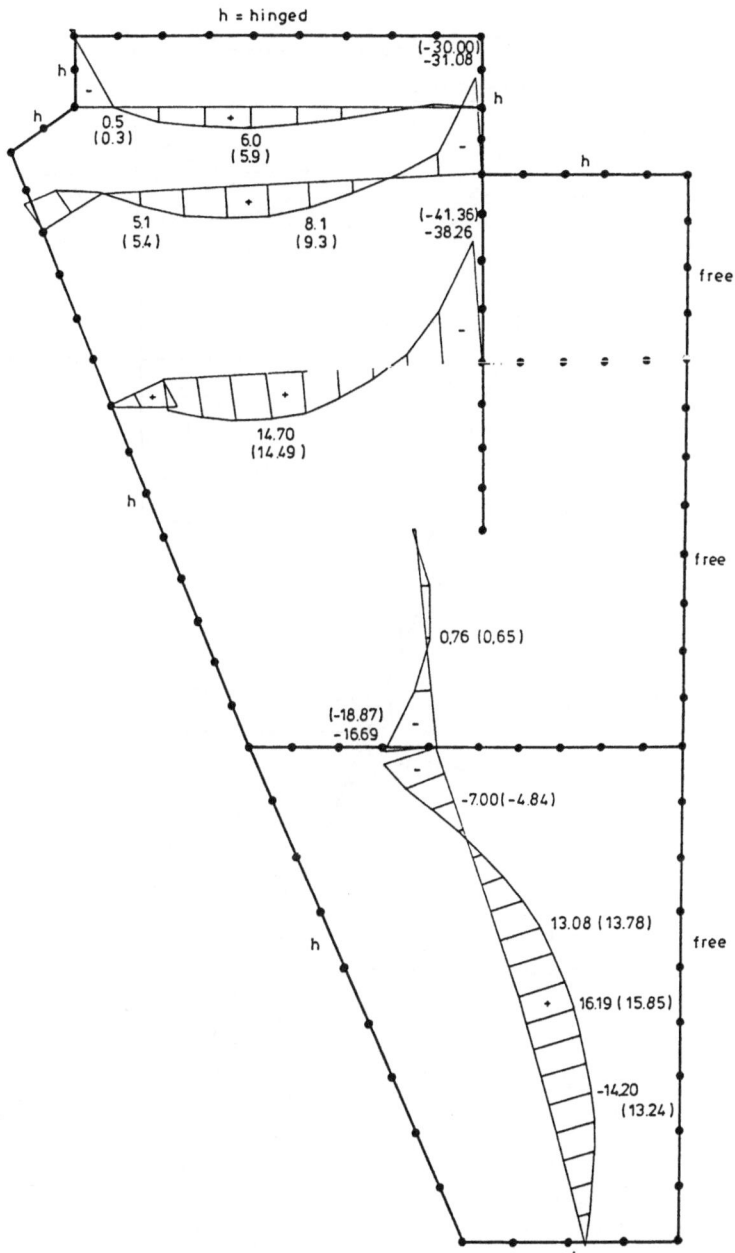

Figure 6.19 BE-mesh and distribution of bending moments

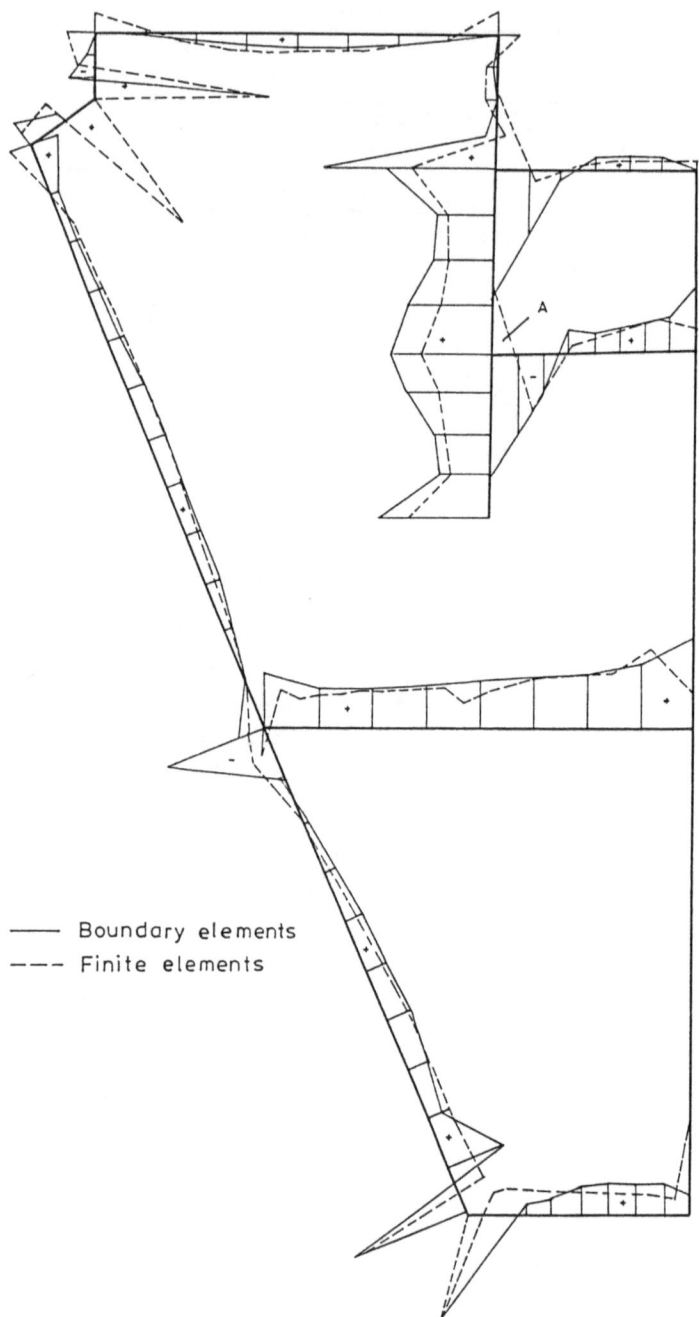

Figure 6.20 The support forces

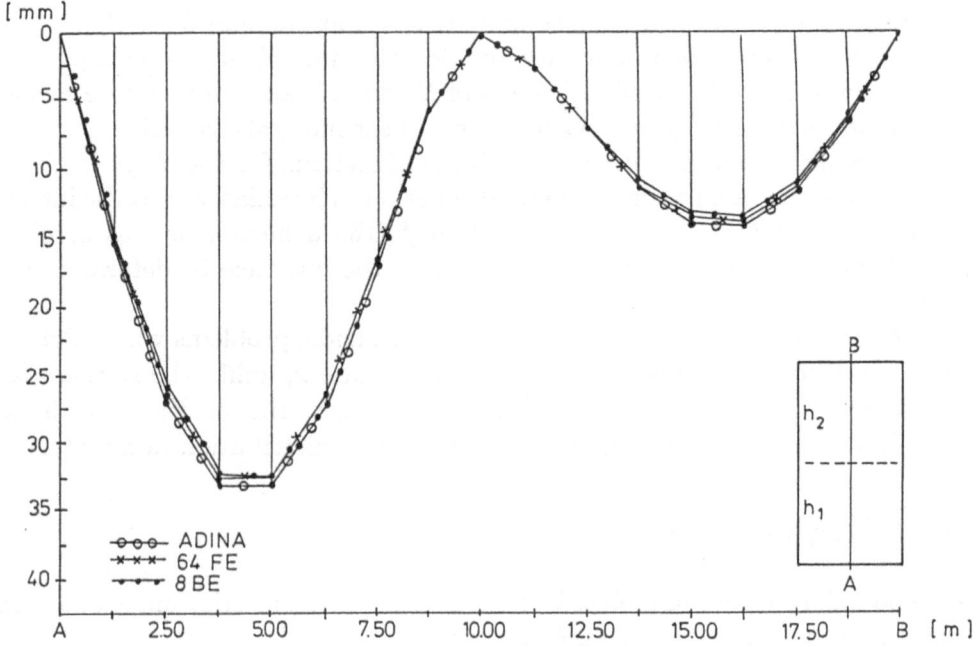

Figure 6.21 Deflection along the center line of a continuous plate with different heights

$$K_a g_0^a = K_b g_0^b \qquad K_a \frac{\partial}{\partial \nu_b} g_0^a = -K_b \frac{\partial}{\partial \nu_a} g_0^b$$

$$V_\nu(g_0^a) = -V_\nu(g_0^b) \qquad M_\nu(g_0^a) = M_\nu(g_0^b)$$

the sum of these two equations became

$$(K_a c_a(\boldsymbol{x}) + K_b c_b(\boldsymbol{x})) w(\boldsymbol{x}) = \int_\Gamma K[g_0(\boldsymbol{y}, \boldsymbol{x}) V_\nu(w)(\boldsymbol{y}) - \frac{\partial}{\partial \nu} g_0(\boldsymbol{y}, \boldsymbol{x}) M_\nu(w)(\boldsymbol{y})$$

$$- V_\nu(g_0(\boldsymbol{y}, \boldsymbol{x})) w(\boldsymbol{y}) + M_\nu(g_0(\boldsymbol{y}, \boldsymbol{x})) \frac{\partial w}{\partial \nu}(\boldsymbol{y})] \, ds_{\boldsymbol{y}} + \int_\Omega K g_0(\boldsymbol{y}, \boldsymbol{x}) p(\boldsymbol{y}) \, d\Omega_{\boldsymbol{y}}$$

$$+ \int_{\Gamma_i} [M_\nu(g_0^a(\boldsymbol{y}, \boldsymbol{x}))(K_a - K_b) \frac{\partial w}{\partial \nu}(\boldsymbol{y}) - V_\nu(g_0^a(\boldsymbol{y}, \boldsymbol{x}))(K_a - K_b) w(\boldsymbol{y})$$

$$+ K_a g_0^a(\boldsymbol{y}, \boldsymbol{x}) f(\boldsymbol{y})] ds_{\boldsymbol{y}} + \sum_c K[g_0(\boldsymbol{y}^c, \boldsymbol{x}) F(w)(\boldsymbol{y}^c) - w(\boldsymbol{y}^c) F(g_0)(\boldsymbol{y}^c, \boldsymbol{x})]$$

where f denotes the unknown support reaction at the interface. So as not to have to write down each boundary and domain integral twice we simply wrote Γ instead of Γ_a and Γ_b and Ω instead of Ω_a and Ω_b and we therefore made no distinction between K_a and K_b in the pertinent integrals as well.

A similar combination of the influence functions for the slope in the two subdomains rendered a second integral equation. The unknowns at the interface were the slope and the support reaction f (the deflection w was zero) and additional collocation points were placed at the interface to determine these functions.

Note that this coupling technique cannot handle problems where Poisson's ratio ν is different for the two subdomains because ν, unlike the stiffness K, is not simply a multiplier of the differential equation. But on the other hand ν does have only a marginal influence on the stress distribution in a plate.

6.14 Singularities

The boundary values of a plate become singular if the interior angle of a corner point exceeds the values in Table 6.2, see [73].

Table 6.2 critical angles for moments and shear forces

Boundary condition	Moments	Kirchhoff shear
c/c	180°	126°
c/h	129°	90°
c/f	95°	52°
h/h	90°	60°
h/f	90°	51°
f/f	180°	78°

c = clamped, f = free, h = hinged.

Exceptions are only the corner points in table 6.3 at which the stresses remain bounded.

Table 6.3 exceptional points

Boundary condition	
h/h	90°
h/h	180°
f/f	180°
h/f	90°

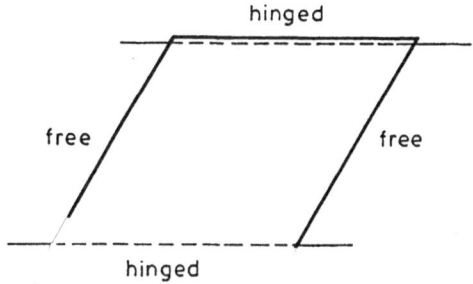

Figure 6.22 Two-skew bridge

Note that at any skew-angled hinged corner whose interior angle exceeds 60° the support forces become singular! This is the most severe deficiency of the Kirchhoff plate bending theory; it causes a lot of trouble for finite elements and boundary elements.

The two-skew bridge in Fig. 6.22 was chosen as a bench mark to check the behavior of different *advanced* FE-models; simple FE-models have serious problems with such skew bridges because at the wide-angled corner points the bending moment parallel to the support (M_x) and the support forces become infinite.

All the FE-solutions used uniform meshes so that we considered an BE-solution and an FE-solution "equal" if the meshes coincided on the boundary. Based on this assumption it was possible to assign to each BE-solution an equivalent number of FE degrees-of-freedom. The plots in Fig. 6.23a and 6.23b show that the BE-solution converges faster than the medium FE-solution. This tendency was also observed at other stress points. However, the rate of convergence was about the same with regard to the deflections.

An interesting point is the distribution of the shear force Q_2 along the support. The BE-shear force is plotted in Fig. 6.23c. While along the main part of the support the shear force follows the "exact" solution it begins to oscillate near the corner point. No effort was made to depict these oscillations graphically.

On first guess we would assume that such strong oscillations have a very negative effect on the solution. But this is not confirmed by the results in the interior. The shear force Q_2 in the interior follows the "exact" solution, see Fig. 6.23d, very closely.

In this respect we should also see that the Betti data w, $\partial w/\partial n$, M_n and V_n, which are obtained by solving the discrete coupling condition on the boundary are *not* the boundary values of the BE-solution, see section 1.7. The BE-solution is calculated by integrating products of Betti data × kernel functions over the boundary. By this integration oscillations of the Betti data

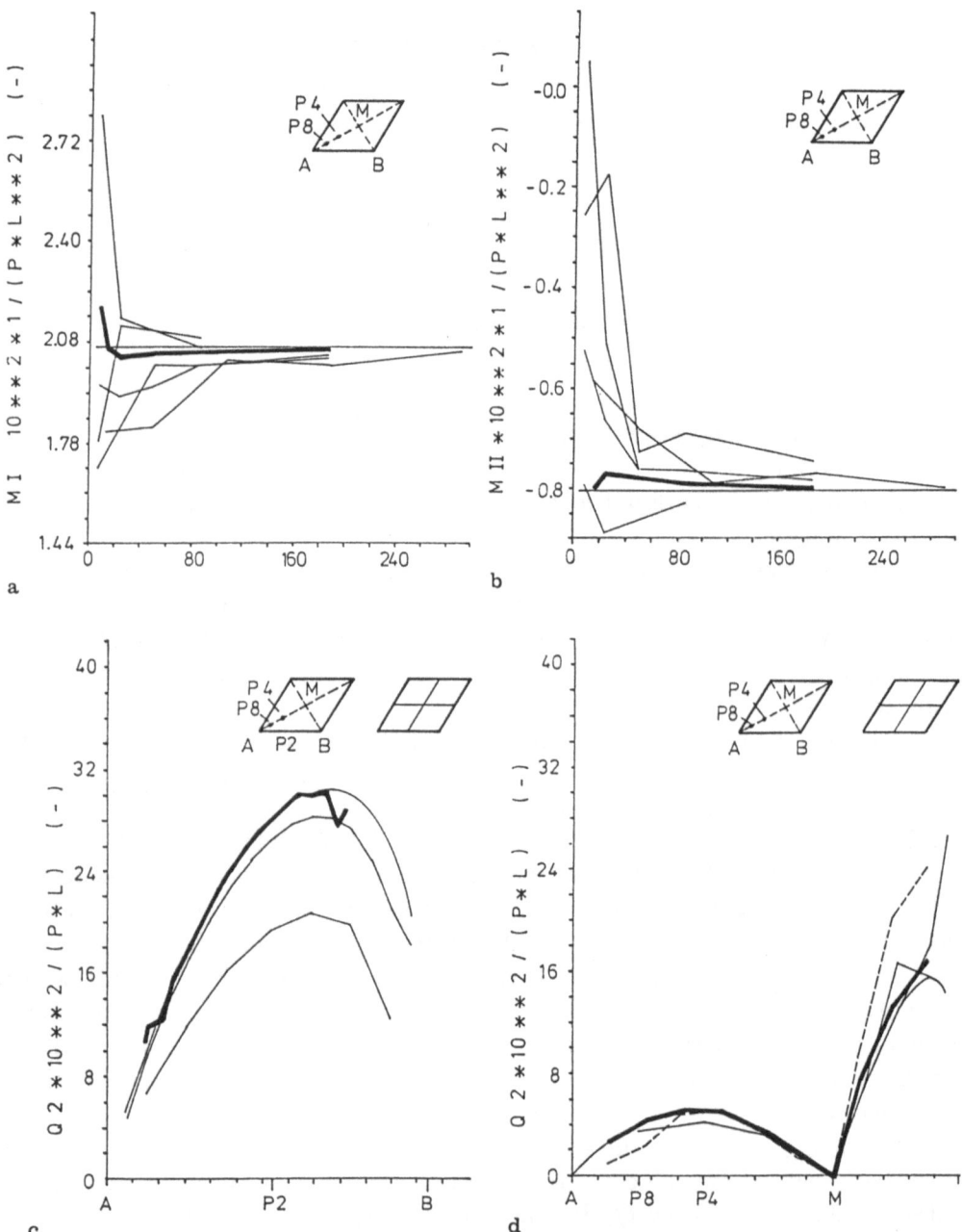

Figure 6.23 a-d Results for the two-skew bridge: **a b** Convergence of the principal moment M_I and M_{II} at the point $P8$; **c** Distribution of the shear force Q_2 along the support (maximum number of elements); **d** Distribution of the shear force Q_2 between the points $A - M - B$

are dampened. At a certain distance from the boundary it is no longer possible to say whether the boundary layers oscillate or whether they are smooth. The influence felt is the same. This is known as St. Venant's principle. Oscillations are a clear hint of numerical difficulties. But on the other hand we should not be too scared if we see oscillations. If oscillating data are filtered by influence functions then the peaks are levelled off. An impressive example for this fact is the shear force Q_2 in Fig. 6.23d and the FE-solution in section 1.12. Though the fourth derivatives $K\Delta\Delta w$ of the FE-solution oscillate enormously near the point supports, the bending moments in the same area are smooth because the oscillating fourth derivatives are integrated twice.

6.15 Influence surfaces

In this book the word influence function is a synonym for integral representations while in engineering applications it often has a narrower meaning. A typical engineering example is an influence line for the bending moment at a section of a continuous beam which shows the variation in the bending moment at this section as a unit transverse load traverses the beam. The calculation of such influence functions, or rather *influence surfaces*, for plates is the topic of this section.

According to Betti's principle, see Fig. 6.24,

$$P_L w_R(\boldsymbol{\xi}, \boldsymbol{x}) = P_R w_L(\boldsymbol{x}, \boldsymbol{\xi})$$

is the influence surface for the quantity $\partial^i w(\boldsymbol{\xi})$ equal to the deflection surface $w(\boldsymbol{\xi}, \boldsymbol{x})$, if the singularity ∂^j that is conjugated to ∂^i, $i + j = 3$, acts at the point $\boldsymbol{\xi}$.

The influence function for the deflection at $\boldsymbol{\xi}$ is the deflection surface of the plate if a concentrated force $P = 1$ acts at $\boldsymbol{\xi}$, that is if the Kirchhoff shear at $\boldsymbol{\xi}$ has the property

$$\lim_{\varepsilon \to 0} \int_{\Gamma_{N\varepsilon}(\boldsymbol{\xi})} V_n(\boldsymbol{x}, \boldsymbol{\xi}) \hat{w}(\boldsymbol{x}) \, ds_{\boldsymbol{x}} = \hat{w}(\boldsymbol{\xi}) \qquad \text{for all } \hat{w} \in C^4(\bar{\Omega})$$

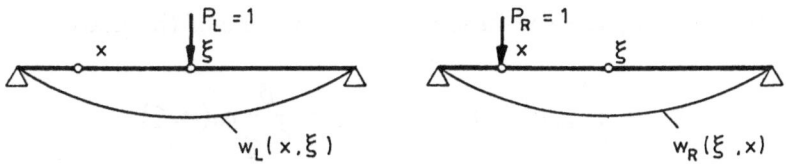

Figure 6.24 An application of Betti's principle

If the circle

$$\Gamma_{N_\epsilon}(\xi) = \{x \in \Omega \mid |x - \xi| = \epsilon\}$$

shrinks to a mere point ξ the work that is done by the Kirchhoff shear on acting through the virtual deflection $\hat{w}$ must tend to $\hat{w}(\xi)$. Because $\hat{w} = 1$ is also a virtual deflection (boundary conditions play no role in this context) this is equivalent to saying that the integral of the Kirchhoff shear tends to 1,

$$\lim_{\epsilon \to 0} \int_{\Gamma_{N_\epsilon}(\xi)} V_n(x, \xi)\, ds_x = 1, \qquad (\hat{w} = 1)$$

According to the same principle the influence surface for the slope $(\partial w / \partial f)$ in the direction of the vector f is the deflection surface if a concentrated couple $M_f = 1$ acts at the point ξ

$$\lim_{\epsilon \to 0} \int_{\Gamma_{N_\epsilon}(\xi)} M_n(x, \xi)\frac{\partial \hat{w}}{\partial n}(x)\, ds_x = \frac{\partial \hat{w}}{\partial f}(\xi) \qquad \text{for all } \hat{w} \in C^4(\bar{\Omega}),$$

The influence surface for the bending moment M_f is the deflection surface of the plate if the slope exhibits a peculiar bend

$$\lim_{\epsilon \to 0} \int_{\Gamma_{N_\epsilon}(\xi)} \frac{\partial w}{\partial n}(x, \xi) M_n(\hat{w})\, ds_y = M_f(\hat{w})(\xi) \qquad \text{for all } \hat{w} \in C^4(\bar{\Omega}),$$

and the influence surface for the Kirchhoff shear V_f is the deflection surface if a dislocation occurs at the point ξ

$$\lim_{\epsilon \to 0} \int_{\Gamma_{N_\epsilon}(\xi)} w(x, \xi) V_n(\hat{w})(x)\, ds_x = V_f(\hat{w})(\xi) \qquad \text{for all } \hat{w} \in C^4(\bar{\Omega})$$

These surfaces have the form

$$w_j(x, \xi) = g_j(x, \xi) + w_{R_j}(x),$$

where the functions g_j are the fundamental solutions of the plate

$$g_0 = \frac{1}{8\pi K} r^2 \ln r, \qquad g_1 = \frac{\partial}{\partial f_\xi} g_0(x, \xi),$$

$$g_2 = M_{f_\xi}(g_0(x, \xi)), \qquad g_3 = V_{f_\xi}(g_0(x, \xi))$$

The vector f now plays the role of the normal vector n across a cut. The subscript ξ is to indicate that we must differentiate with respect to the coordinates ξ_i.

The functions $w_{R_j}(x)$ are (unknown) regular, homogeneous solutions of the plate equation which must be added to the fundamental solutions to satisfy the boundary conditions.

If we formulate with the fundamental solution $g_0(y, x)$ and $g_1(y, x)$ respectively and such a deflection surface $w_j(y, \xi)$ Green's second identity (6.5)

$$B(g_0(y, x), w_j(y, \xi))$$

and

$$B(g_1(y, x), w_j(y, \xi)),$$

then we obtain after taking the limit (both points, x as well as the point ξ, must now be excluded from the domain by cuts with radius ε and η respectively) the same Eqs.(6.6) and (6.7) as before, only that now the domain integrals

$$\int_\Omega g_i K \Delta \Delta w \, d\Omega y, \qquad i = 0, 1$$

in Eqs.(6.6) and (6.7) must be replaced by the functions

$$s_j(\xi, x) = \left. \partial_y^j g_0(y, x) \right|_{y=\xi} \qquad \text{in Eq.(6.6)},$$

$$\frac{\partial}{\partial n_x} s_j(\xi, x) = \left. \frac{\partial}{\partial n_x} \partial_y^j g_0(y, x) \right|_{y=\xi} \qquad \text{in Eq.(6.7)}$$

Note that the function g_0 that is differentiated here is the first argument in the identity $B(g_0, w_j)$ and that the index j corresponds to the index j of the influence function w_j.

If we denote the tangent vector which is associated with the direction $f = \{f_1, f_2\}^T$ with $g = \{g_1, g_2\}^T = \{-f_2, f_1\}^T$, then these functions are:

Unit force

$$s_0(\xi, x) = \frac{1}{8\pi K} r^2 \ln r, \qquad \frac{\partial}{\partial n_x} s_0(\xi, x) = \frac{1}{8\pi K} r r_n (1 + 2 \ln r)$$

Unit couple

$$s_1(\boldsymbol{\xi}, \boldsymbol{x}) = -\frac{1}{8\pi K} r r_f (1 + 2\ln r),$$

$$\frac{\partial}{\partial n_x} s_1(\boldsymbol{\xi}, \boldsymbol{x}) = -\frac{1}{8\pi K} [(r_n r_f + r_g r_t)(1 + 2\ln r) + 2 r_f r_n]$$

Unit bend

$$s_2(\boldsymbol{\xi}, \boldsymbol{x}) = -\frac{1}{8\pi} [2(1 + \nu)\ln r + (3 + \nu) r_f^2 + (1 + 3\nu) r_g^2],$$

$$\frac{\partial}{\partial n_x} s_2(\boldsymbol{\xi}, \boldsymbol{x}) = -\frac{1}{4\pi r} [(1 + \nu) r_n + 2(1 - \nu) r_g r_f r_t]$$

Unit dislocation

$$s_3(\boldsymbol{\xi}, \boldsymbol{x}) = -\frac{1}{4\pi r} [2 r_f + (1 - \nu)(r_f - \kappa r)(r_f^2 - r_g^2)],$$

$$\frac{\partial}{\partial n_x} s_3(\boldsymbol{\xi}, \boldsymbol{x}) = \frac{1}{4\pi r^2} [r_n \{2 r_f + (1 - \nu) r_f (r_f^2 - r_g^2)\}$$
$$- r_t \{2 r_g + (1 - \nu)[r_g(r_f^2 - r_g^2) + 4 r_f^2 r_g]\}]$$

We easily recognize the rule according to which these expressions are formulated

$$s_i = \partial_{y(f)}^i g_0(\boldsymbol{y}, \boldsymbol{x})\big|_{y=\xi} \quad ,$$

and, therefore, it is quite simple to derive the corresponding terms for the influence surface of the shear force Q_f

$$\bar{s}_3(\boldsymbol{\xi}, \boldsymbol{x}) = Q_f(g_0) = \frac{1}{2\pi r} r_f,$$

$$\frac{\partial}{\partial n_x} \bar{s}_3(\boldsymbol{\xi}, \boldsymbol{x}) = -\frac{1}{2\pi r^2}(r_n r_f - r_g r_t)$$

and the twisting moment M_{fg},

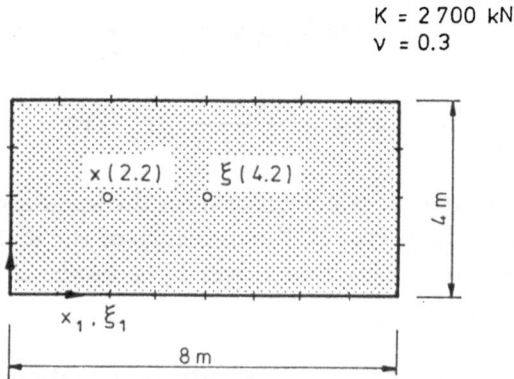

$$K = 2\,700 \text{ kNm}$$
$$v = 0.3$$

$x\,(2.2)$ $\xi\,(4.2)$

$x_1 \cdot \xi_1$

8 m

Figure 6.25 The plate and the boundary element mesh

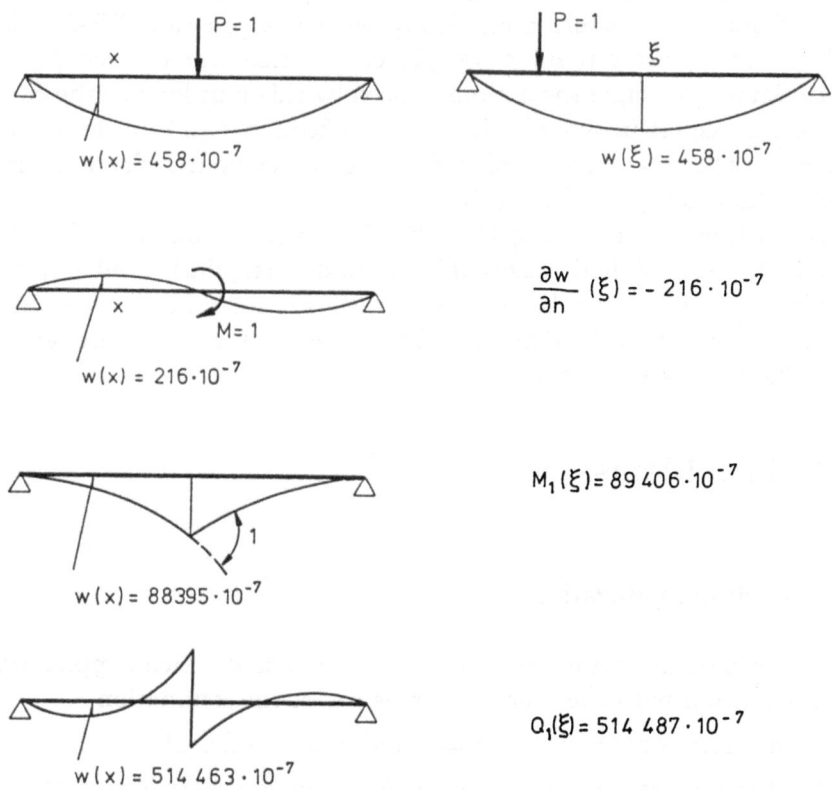

$P = 1$

x

$w(x) = 458 \cdot 10^{-7}$

$P = 1$

ξ

$w(\xi) = 458 \cdot 10^{-7}$

x $M = 1$

$w(x) = 216 \cdot 10^{-7}$

$$\frac{\partial w}{\partial n}(\xi) = -216 \cdot 10^{-7}$$

$w(x) = 88395 \cdot 10^{-7}$

$$M_1(\xi) = 89\,406 \cdot 10^{-7}$$

$w(x) = 514\,463 \cdot 10^{-7}$

$$Q_1(\xi) = 514\,487 \cdot 10^{-7}$$

Figure 6.26 Numerical check of Betti's principle

$$s_4(\boldsymbol{\xi}, \boldsymbol{x}) = M_{fg}(g_0) = -\frac{(1-\nu)}{4\pi} r_f r_g \,,$$

$$\frac{\partial}{\partial n_{\boldsymbol{x}}} s_4(\boldsymbol{\xi}, \boldsymbol{x}) = -\frac{(1-\nu)}{4\pi} r_t (r_g^2 - r_f^2)$$

The further treatment does not differ from the standard boundary element approach. The role of the vector $\boldsymbol{d}$, the influence of the distributed load in the equation $\boldsymbol{Hu} = \boldsymbol{Gt} + \boldsymbol{d}$, is now taken over by a vector $\boldsymbol{d}'$ whose components are the following terms

$$s_j(\boldsymbol{\xi}, \boldsymbol{x}^k) \qquad \text{1st integral equation}\,,$$

$$\frac{\partial}{\partial n} s_j(\boldsymbol{\xi}, \boldsymbol{x}^k) \qquad \text{2nd integral equation}$$

If the discrete coupling condition is solved for the unknown nodal values then the influence surface is obtained from Eq.(6.6).

A hinged rectangular plate, see Fig. 6.25, served as a test example to check the validity of Betti's principle numerically with the program BE-PLATE-BENDING. At the quarter point $\boldsymbol{x}$ we placed a concentrated force $P = 1$. According to Betti's principle the actions caused by this unit load at the centre point $\boldsymbol{\xi}$ should be equal in magnitude to the deflection $w(\boldsymbol{x})$ at the quarter point if a conjugated singularity is placed at the center point. The numerical results confirm the theory, see Fig. 6.26.

The ease with which these singular deflection surfaces can be modelled by the boundary element method makes this method particularly well suited to calculate influence surfaces for bridge structures. Figure 6.27 shows such an influence surface for a two-skew bridge. The agreement with experimentally obtained results, [74], is very good.

6.16 Special problems

6.16.1 Plates on elastic foundations

The simplest and best known model is the *Winkler model* which supplements the Kirchhoff equation with the reaction forces $c\,w$ of the foundation

$$K\Delta\Delta w + c\,w = p\,, \qquad c = \text{foundation coefficient}$$

The associated fundamental solutions can be found in the contribution of Puttonen and Varpasuo, [75], who also considered the application of the BEM to the Pasternak model

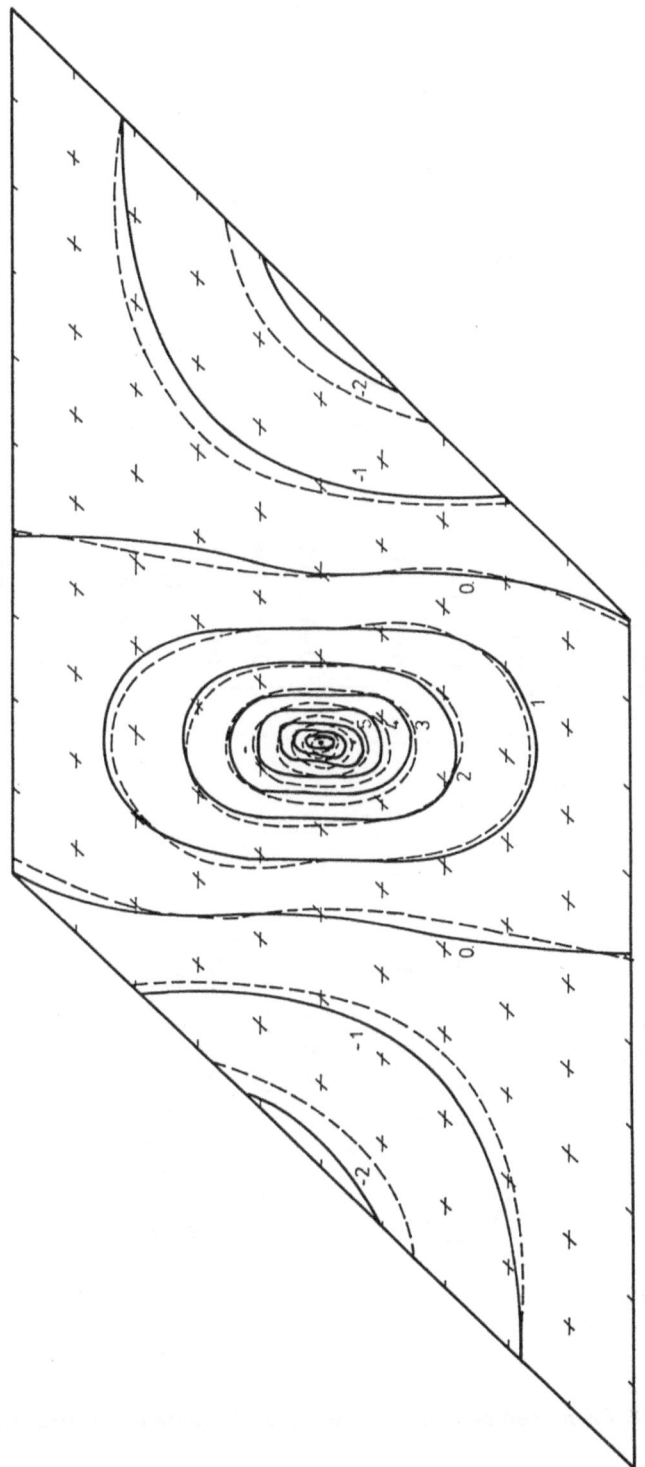

Figure 6.27 Influence surface for the bending moment M_n of a two-skew bridge

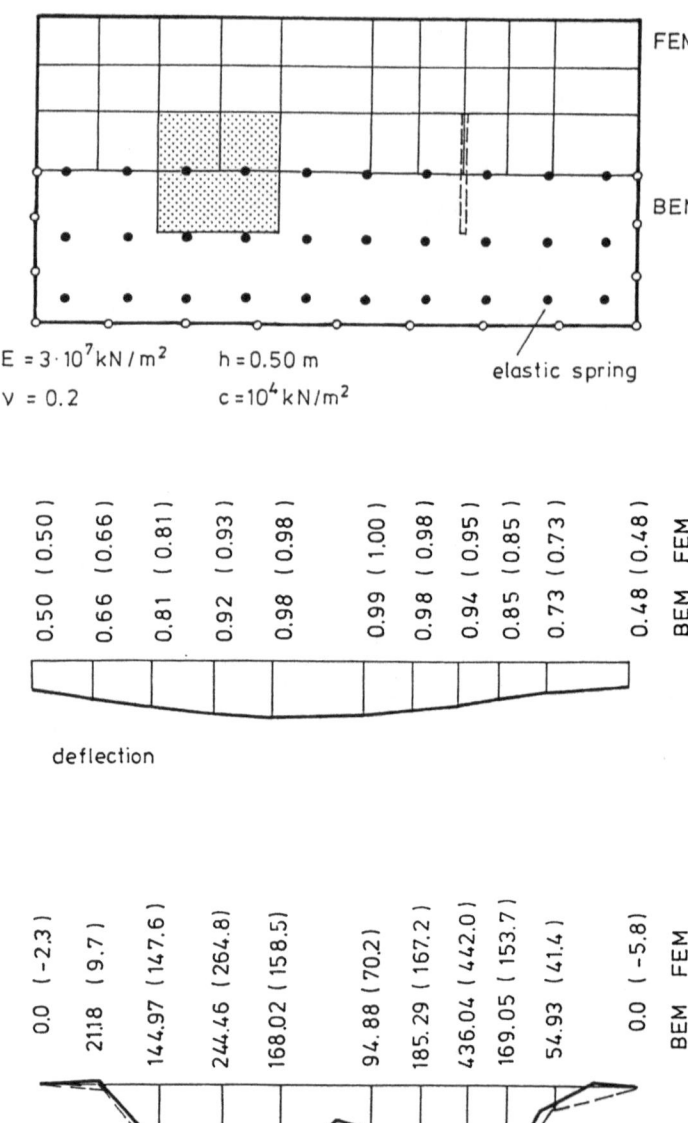

$E = 3 \cdot 10^7 \, kN/m^2$ $h = 0.50 \, m$

$v = 0.2$ $c = 10^4 \, kN/m^2$

elastic spring

Figure 6.28 FE-modelling and BE-modelling of a plate on an elastic foundation

$$KΔΔw + cw - GΔw = p, \qquad G = \text{shear modulus of the surface},$$

and the model that is based on the interaction between the plate and the elastic halfspace (Boussinesq).

In the program BE-PLATE-BENDING plates on elastic foundations are treated according to the Winkler model and the influence of the elastic foundation is simply modelled by a series of elastic springs. Figure 6.28 shows that the BE-solution so obtained is in good agreement with an FE-solution.

6.16.2 Plate buckling

The second-order theory of plate bending is governed by the differential equation

$$KΔΔw - (N_1 w_{,1})_{,1} - (N_{12} w_{,1})_{,2} - (N_2 w_{,2})_{,2} - (N_{21} w_{,2})_{,1} = p$$

where

$$N_1 = h\,\sigma_{11}, \qquad N_{12} = h\,\sigma_{12} = h\,\sigma_{21} = N_{21}, \qquad N_2 = h\,\sigma_{22}$$

are the normal stresses multiplied with the plate thickness h. For particular cases as

$$N_1 = c, \qquad N_{12} = N_{21} = N_2 = 0,$$

Kitahara has found the fundamental solutions, [29, p. 211], and he has calculated the buckling load by finding the lowest zero of the determinant, which is a transcendental function of the load parameter λ, with a search algorithm.

Costa and Brebbia reduced the same problem to an algebraic eigen-value problem. But for this they had also to discretize the interior, see [76].

Exercises

1. Two couples, M_1 and M_2, are applied at the end points of a hinged beam, see Fig. 6.29a. Show your friends how you obtain the internal distribution of the bending moment $M(x)$ with a ruler and a pencil.
2. Calculate the deflection at the centre of the beam in Fig. 6.29d by the principle of virtual forces and by Betti's principle

$$1 \times w(x) = \int_0^l \frac{M_0(y,x)M(y)}{EI}\,dy \qquad 1 \times w(x) = \int_0^l G_0(y,x)p(y)\,dy$$

where G_0 is the Green's function (2.25).

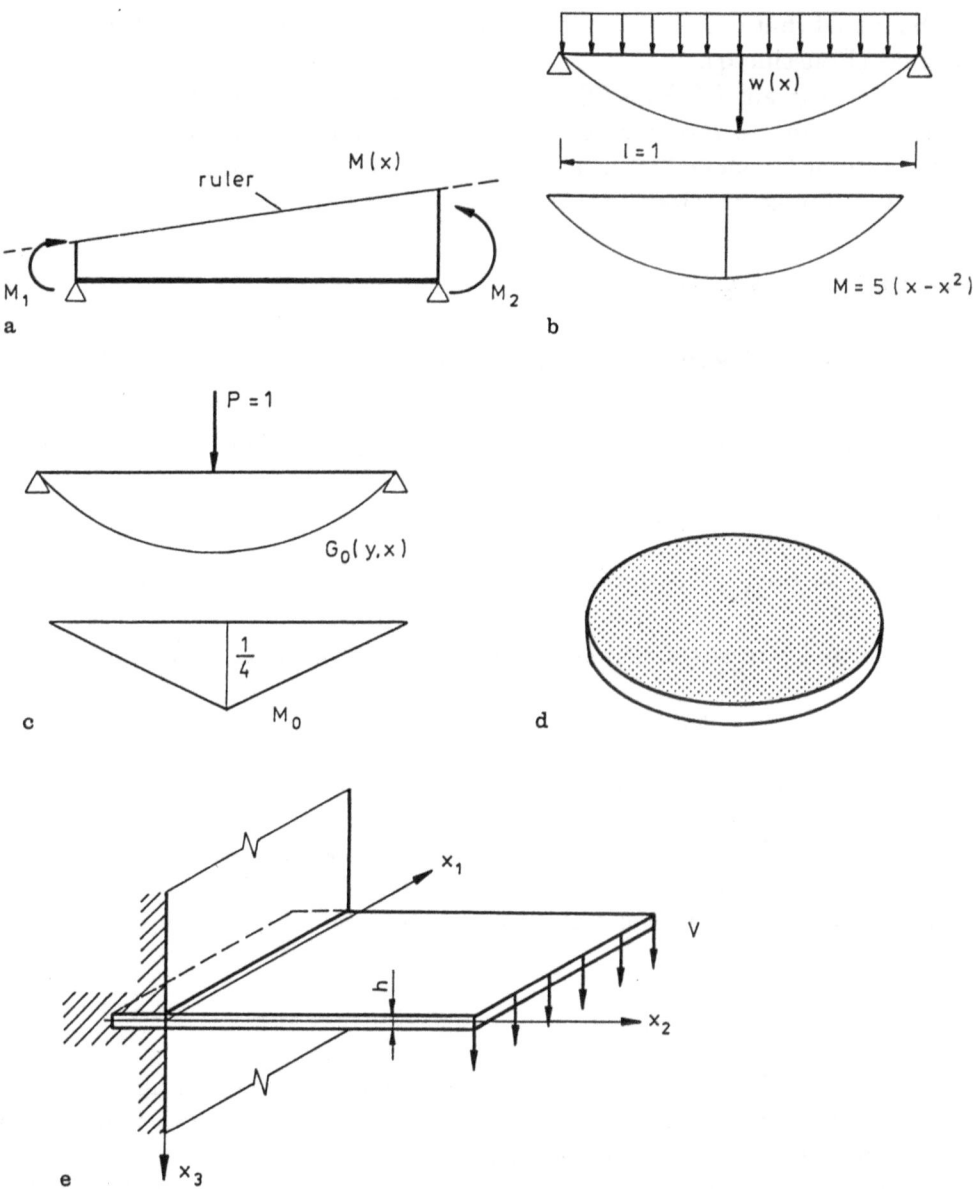

Figure 6.29 a-e Exercises: **a** a linear function is determined by its boundary values; **b** distributed load; **c** unit-dummy load; **d** circular plate; **e** cantilever beam

3. Calculate the deflection at the centre of the circular plate in Fig. 6.29c by Betti's principle

$$1 \times w(\boldsymbol{x}) = \int_{\Omega} G_0(\boldsymbol{y}, \boldsymbol{x}) p(\boldsymbol{y}) \, d\Omega_{\boldsymbol{y}}$$

where $G_0(\boldsymbol{y}, \boldsymbol{x})$ is the Green's function (6.1) and the load $p(\boldsymbol{y})$ is some constant p_0. The solution is $w(\boldsymbol{x}) = (p_0(5 + \nu)R^4)/(64K(1 + \nu))$.

4. In the BEM we use Betti's principle $B(w, g_0[\boldsymbol{x}]) = 0$ to find the nodal values of the unknown boundary functions. Show that also the equilibrium conditions are an application of Betti's principle: the reciprocal exterior work between the plate solution w and any rigid-body motion must be the same, $B(w, a + b_1 x_1 + b_2 x_2) = 0$.

5. Using the program BE-PLATE-BENDING calculate the deflection of a cantilever beam, see Fig. 6.29e, by modelling the beam as a plate with Poisson's ratio $\nu = 0$.

6. Using the program BE-PLATE-BENDING verify Betti's principle numerically as in Fig. 6.25, 26.

7. Calculate the derivatives

$$\frac{\partial}{\partial n} \frac{\partial}{\partial m} M_\nu(g_0), \qquad \frac{\partial}{\partial n} \frac{\partial}{\partial m} V_\nu(g_0).$$

8. Check the singular integrals in section 6.6.

9. Show that

$$\int_{\Gamma} V_\nu(g_0(\boldsymbol{y}, \boldsymbol{x})) ds_{\boldsymbol{y}} = c(\boldsymbol{x}), \qquad \int_{\Gamma} V_\nu(g_1(\boldsymbol{y}, \boldsymbol{x})) ds_{\boldsymbol{y}} = 0.$$

(Hint: apply (6.6) and (6.7) to a translation $w = 1$). Which result do you obtain if $w = ax_1 + bx_2$ is a rotation?

10. Show that at a corner point of a clamped plate all the boundary values are zero.

11. The influence which the boundary values exert on the interior points depends on the kernels in the influence functions. Which of these boundary values has the most delicate kernels, the kernels with the strongest singularity? Which boundary conditions are therefore numerically the most difficult?

12. Let Γ_e be a boundary element on a free edge, so that the deflection w is non-zero. Let Γ_e have length 1. Assume that the interior point $\boldsymbol{x}$ lies on the normal vector that passes through the centre of the element. Place the interior point closer and closer to Γ_e and trace the variation of the kernel function $V_\nu(g_0)$ over the length of the element at each step. Experiment with different quadrature formulas to calculate the integral of the kernel function.

13. A hinged square plate under a uniform load

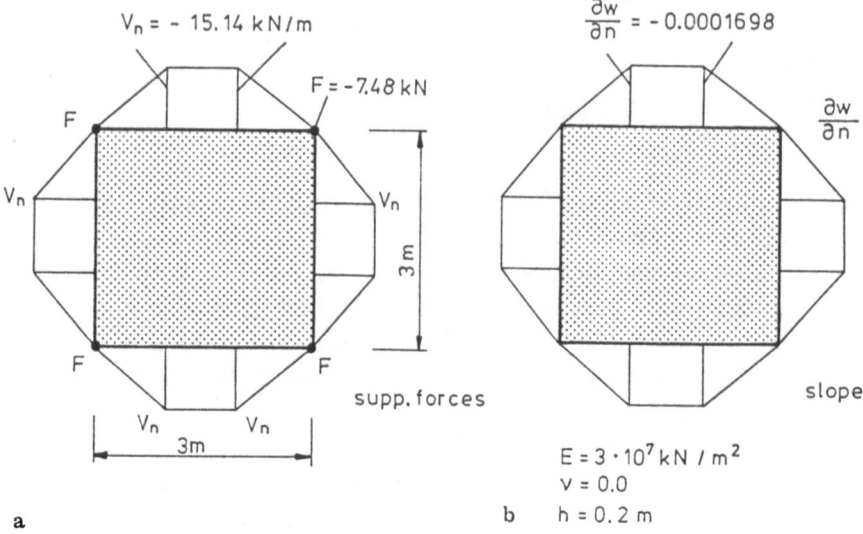

Figure 6.30 a-b Exercises: a-b distribution of support forces and slope for a plate under uniform load

$$K\Delta\Delta w = 10 \quad \mathrm{kN/m^2} \qquad w = M_n = 0$$

was discretized with three boundary elements on each side. The distribution of the slope and the Kirchhoff shear is seen in Fig. 6.30. The four corner forces were $F = -7.48\,\mathrm{kN}$ each. Substitute these values and the data $w = M_n = 0$ into the influence function for the deflection

$$w(\boldsymbol{x}) = \int_{\Gamma} [g_0(\boldsymbol{y},\boldsymbol{x})V_\nu(w)(\boldsymbol{y}) + M_\nu(g_0(\boldsymbol{y},\boldsymbol{x}))\frac{\partial w}{\partial \nu}(\boldsymbol{y})]\,ds_{\boldsymbol{y}}$$

$$+ \int_{\Omega} g_0(\boldsymbol{y},\boldsymbol{x})p(\boldsymbol{y})\,d\Omega_{\boldsymbol{y}} + \sum_{c} g_0(\boldsymbol{y}^c,\boldsymbol{x})F(\boldsymbol{y}^c)$$

$$E = 3 * 10^7\,\mathrm{kN/m^2} \qquad \nu = 0.0 \qquad h = 0.2\,\mathrm{m}$$

and calculate the deflection at the centre of the plate. (Hint: transform the domain integral into an equivalent boundary integral and make use of the symmetry of the problem). Compare your solution with the exact solution $w = 0.0487pl^4/(Eh^3)$. (Note: the program BE-PLATE-BENDING outputs support forces as in Fig. 6.30a as positive).

7 Boundary elements and finite elements

Both methods have their strong points

FEM	BEM
element library	reduction of dimension
robust	higher precision
variable coefficients	exterior problems

so that a coupling of the two methods or, as it was phrased, a *Marriage à la Mode*, [77], should benefit from both. The coupling will usually be done by reformulating the coupling conditions of the boundary data of the BE-domain as a stiffness matrix and to couple this stiffness matrix with the stiffness matrices of the neighboring finite elements.

But before we discuss this procedure in more detail let us first concentrate on the underlying theory.

7.1 Theory

The stiffness matrix of one-dimensional problems

$$K u = f$$

is obtained by multiplying the coupling condition

$$H u = G f$$

from the left with the matrix G^{-1}.

In the case of a two-dimensional problem as a membrane

$$G^{-1} H u = t, \tag{7.1}$$

the vector on the right side is the vector of the tractions at the nodes but not the vector of the equivalent nodal forces. These terms are work terms: the equivalent nodal force f_i is the work done by the traction $t = t_j \psi_j$ on acting through the deflection φ_i

$$f_i = \int_\Gamma \varphi_i \, t \, ds = \int_\Gamma \varphi_i \, \psi_j \, ds \, t_j$$

To obtain the vector of the equivalent nodal forces f we, therefore, have to multiply Eq.(7.1) from the left with *Gram's matrix*

$$F = [\int_\Gamma \varphi_i \, \psi_j \, ds]$$

so that

$$FG^{-1}Hu = Ft = f$$

(In one-dimensional problems the matrix F is the unit matrix.)

The entry F_{ij} in Gram's matrix is the work done by the unit force ψ_j on acting through the unit deflection φ_i. If we choose for ψ_j and φ_i hat-functions as indicated in Fig. 7.1 then F becomes a tridiagonal matrix with the elements

$$\frac{1}{6} l_i , \qquad \frac{1}{3}(l_i + l_{i+1}), \qquad \frac{1}{6} l_{i+1} , \qquad \text{(row } i\text{)}$$

where l_i denotes the length of the element to the left, (i), and to the right, $(i+1)$, of the node i.

However, the BE-matrix

$$K = FG^{-1}H, \tag{7.2}$$

has none of the properties exactly

(Kernel) $\qquad Ku^\circ = o, \quad (u^\circ = \text{vector of a rigid-body motion}),$

(Equ.) $\quad u^{\circ T} Ku = 0,$

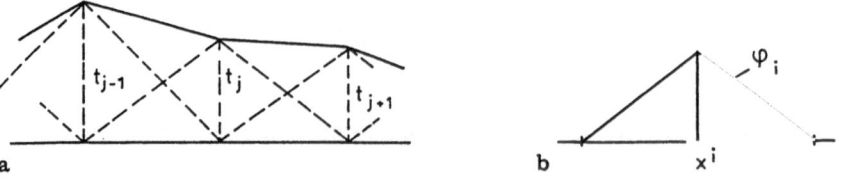

Figure 7.1 a-b. Virtual work on the boundary: a the boundary forces; b the virtual displacements

(Sym.) $\boldsymbol{u}^T \boldsymbol{K} \hat{\boldsymbol{u}} = \hat{\boldsymbol{u}}^T \boldsymbol{K} \boldsymbol{u}$,

(Pos.def.) $\boldsymbol{u}^T \boldsymbol{K} \boldsymbol{u} > 0$, $\boldsymbol{u} \neq \boldsymbol{u}^\circ$,

which distinguish a normal stiffness matrix.

In particular the matrix is not exactly symmetric. The reason for this is that the piecewise polynomials

$$u = u_i \varphi_i, \qquad t = t_j \psi_j \tag{7.3}$$

satisfy the coupling condition only at the collocation points so that, strictly speaking, they are not compatible, not the boundary values of the *same* elastic state of the membrane. To see this more clearly let us assume that the opposite is true:

(i) The fact that two pairs of vectors $\boldsymbol{u}, \boldsymbol{t}$ and $\hat{\boldsymbol{u}}, \hat{\boldsymbol{t}}$ satisfy the discrete coupling condition $\boldsymbol{H}\boldsymbol{u} = \boldsymbol{G}\boldsymbol{t}$ and $\boldsymbol{H}\hat{\boldsymbol{u}} = \boldsymbol{G}\hat{\boldsymbol{t}}$, respectively, would imply that

(ii) the corresponding boundary functions

$$u = u_i \varphi_i, \qquad t = t_j \psi_j, \qquad \hat{u} = \hat{u}_i \varphi_i, \qquad \hat{t} = \hat{t}_j \psi_j$$

are the boundary values of two homogeneous deflections u and $\hat{u}$ (homogeneous = no distributed loads in the interior, $-N\Delta u = 0$).

If this were true then the Betti data of these two deflections should satisfy Betti's principle

$$\int_\Gamma u \hat{t} \, ds = \int_\Gamma t \hat{u} \, ds, \qquad (W_{1,2} = W_{2,1})$$

But this implies that the matrix $\boldsymbol{K}$ is symmetric. Namely the left side is

$$\int_\Gamma u \hat{t} \, ds = \boldsymbol{u}^T \boldsymbol{F} \hat{\boldsymbol{t}} = \boldsymbol{u}^T \boldsymbol{F} \boldsymbol{G}^{-1} \boldsymbol{H} \hat{\boldsymbol{u}} = \boldsymbol{u}^T \boldsymbol{K} \hat{\boldsymbol{u}}$$

and the right side

$$\int_\Gamma t \hat{u} \, ds = \boldsymbol{t}^T \boldsymbol{F} \hat{\boldsymbol{u}} = \boldsymbol{u}^T \boldsymbol{H}^T \boldsymbol{G}^{-1T} \boldsymbol{F} \hat{\boldsymbol{u}} = \boldsymbol{u}^T \boldsymbol{K}^T \hat{\boldsymbol{u}},$$

and this only matches if $\boldsymbol{K} = \boldsymbol{K}^T$.

At this point one is tempted to simply continue with the matrix

$$\tilde{\boldsymbol{K}} = \frac{1}{2}(\boldsymbol{K} + \boldsymbol{K}^T)$$

A manipulation which seemingly can even be justified: the potential energy of a membrane which is acted upon by boundary forces $\bar{t}$ only

$$\Pi_1(u) = \frac{1}{2} E(u, u) - \int_\Gamma \bar{t} u \, ds$$

can, with Green's first identity,

$$E(u, u) = \int_\Omega -N \Delta u \, u \, d\Omega + \int_\Gamma t u \, ds = \int_\Gamma t u \, ds$$

be transformed into boundary integrals only.

$$\Pi_1(u) = \frac{1}{2} \int_\Gamma t u \, ds - \int_\Gamma \bar{t} u \, ds$$

If we replace t and u by the piecewise polynomials (7.3) then $\Pi_1(u)$ becomes

$$\Pi_1(u) = \frac{1}{2} u^T F G^{-1} H u - u^T F \bar{t},$$

and the minimum condition

$$\frac{\partial \Pi_1}{\partial u_i} = 0, \qquad i = 1, 2, \ldots, n$$

leads to the very same equation as before:

$$\frac{1}{2}(K + K^T) u = Ft$$

So that the symmetrization seems justified. But this derivation suffers from the same defect as the simple symmetrization: The 'derivation' rests basically on the equation

$$E(u, u) = \int_\Gamma t u \, ds \qquad (\text{first identity, } \Delta u = 0)$$

and this equation is only true if the two functions u and t are the boundary values of the *same* domain function, the deflection u in the energy integral on the left-hand side. But this assumption does not hold true for the BE-solution. Otherwise the matrix K would be symmetric all by itself.

Unlike FE-methods where the right-hand side, f, of a stiffness matrix

$$Ku = f$$

is the vector of the equivalent nodal forces of the very same function $u = u_i\varphi_i$ whose termwise energy products constitute the elements of K, the right-hand side, f, of a BE-stiffness matrix in no way at least mathematically speaking is associated with the u-vector on the left side. It is a separate independent quantity. No differentiation or integration by parts will get us from u_i to t_i.

Genuine symmetric matrices are obtained if we formulate the coupling condition with the Green's function G_0 of the *'clamped plate problem'* (all displacement terms on the boundary are zero). We apply this technique in the following to a Kirchhoff plate.

A BE-stiffness matrix formulates a relation between the nodal displacements and the equivalent nodal edge forces under the side-condition that there are no distributed loads in the interior, $K\Delta\Delta w = 0$. From a mathematical point of view this mapping $u \to Ku$ corresponds to a boundary-value problem where certain deflections w and rotations $\partial w/\partial n$ are prescribed on the boundary of the plate. As only displacements are prescribed the potential energy consists only of the domain integral of the strain energy

$$\Pi_1(w) = \frac{1}{2}E(w, w)$$

and so it can be expressed by an equivalent boundary integral

$$\Pi_1(w) = \frac{1}{2}E(w, w) = \frac{1}{2}\int_\Gamma (V_n w - M_n \frac{\partial w}{\partial n})\, ds \tag{7.4}$$

(For simplicity we neglected the corner terms.)

Having established this transformation to the boundary let us now focus on the four coupling conditions between the boundary values of a plate.

$$\frac{1}{2}\begin{bmatrix} \partial^0 w \\ \partial^1 w \\ \partial^2 w \\ \partial^3 w \end{bmatrix} = \int_\Gamma \begin{bmatrix} \partial_y^0 \partial_x^0 g_0 & \partial_y^1 \partial_x^0 g_0 & \partial_y^2 \partial_x^0 g_0 & \partial_y^3 \partial_x^0 g_0 \\ \partial_y^0 \partial_x^1 g_0 & \partial_y^1 \partial_x^1 g_0 & \partial_y^2 \partial_x^1 g_0 & \partial_y^3 \partial_x^1 g_0 \\ \partial_y^0 \partial_x^2 g_0 & \partial_y^1 \partial_x^2 g_0 & \partial_y^2 \partial_x^2 g_0 & \partial_y^3 \partial_x^2 g_0 \\ \partial_y^0 \partial_x^3 g_0 & \partial_y^1 \partial_x^3 g_0 & \partial_y^2 \partial_x^3 g_0 & \partial_y^3 \partial_x^3 g_0 \end{bmatrix} \begin{bmatrix} +\partial^3 w \\ -\partial^2 w \\ +\partial^1 w \\ -\partial^0 w \end{bmatrix} ds_y$$

$$+ \int_\Omega \begin{bmatrix} \partial_x^0 g_0 \\ \partial_x^1 g_0 \\ \partial_x^2 g_0 \\ \partial_x^3 g_0 \end{bmatrix} p\, d\Omega_y ,$$

If we would derive these conditions not with the fundamental solution g_0 but with the deflection $G_0 = g_0 + w_r$, the deflection of a clamped plate loaded with a concentrated force $\hat{P} = 1$ (i.e. we add to g_0 a function w_r so that $g_0 + w_r$ satisfy the boundary conditions of the clamped plate) then the third and the fourth equation would become

$$\frac{1}{2} M_n = \int_\Gamma (\partial_y^2 \partial_x^2 G_0 \frac{\partial w}{\partial n} - \partial_y^3 \partial_x^2 G_0 w)\, ds_y \,,$$

$$\frac{1}{2} V_n = \int_\Gamma (\partial_y^2 \partial_x^3 G_0 \frac{\partial w}{\partial n} - \partial_y^3 \partial_x^3 G_0 w)\, ds_y$$

If we substitute these two influence functions for M_n and V_n into Eq.(7.4) and if we approximate the displacement terms on the boundary by the same set of functions

$$w = \delta_i \varphi_i \,, \qquad \frac{\partial w}{\partial n} = \varepsilon_i \varphi_i \,,$$

then the potential energy becomes a quadratic form of the nodal variables

$$\Pi_1(w) = \frac{1}{2} [\delta, \varepsilon] \begin{bmatrix} A & B \\ B^T & C \end{bmatrix} \begin{bmatrix} \delta \\ \varepsilon \end{bmatrix} \,,$$

that is the stiffness matrix

$$a_{ij} = \int_\Gamma \int_\Gamma \partial_y^3 \partial_x^3 G_0 \varphi_i\, ds_y\, \varphi_j\, ds_x \,,$$

$$b_{ij} = \int_\Gamma \int_\Gamma \partial_y^3 \partial_x^2 G_0 \varphi_i\, ds_y\, \varphi_j\, ds_x = \int_\Gamma \int_\Gamma \partial_y^2 \partial_x^3 G_0 \varphi_j\, ds_y\, \varphi_i\, ds_x \,,$$

$$c_{ij} = \int_\Gamma \int_\Gamma \partial_y^2 \partial_x^2 G_0 \varphi_i\, ds_y\, \varphi_j\, ds_x$$

is symmetric.

Instead of considering the energy we could also obtain symmetric matrices if we formulate the coupling conditions with Green's function and solve these with Galerkin's method. Let us explain this by considering the coupling condition of the boundary values of a membrane.

If we would replace in the second coupling condition

$$\frac{1}{2} \frac{\partial u}{\partial n} = \int_\Gamma (\partial_y^0 \partial_x^1 g_0 \frac{\partial u}{\partial n} - \partial_y^1 \partial_x^1 g_0 u)\, ds_y$$

the fundamental solution g_0 by Green's function G_0 then in the process of the derivation of the condition the first boundary integral would vanish and we would be left with the expression

$$\frac{1}{2}\frac{\partial u}{\partial n} = -\int_{\Gamma} \partial_y^1 \partial_x^1 G_0 u \, ds_y$$

The kernel of the integral operator is symmetric and, therefore, the solution of this integral equation with Galerkin's method would render a symmetric stiffness matrix.

However, all this must remain theory because, in general, we do not know the Green's function G_0 that corresponds to homogeneous displacement boundary conditions ("clamped plate"). But the foregoing discussion was only intended to clarify the issue, to make clear that it is not the fault of the boundary element method if the matrices are unsymmetric but rather our fault because we use free-space Green's function, fundamental solutions, instead of as it would be correct Green's functions.

When we multiply the collocation equations from the left with Gram's matrix F

$$FG^{-1}Hu = Ft = f$$

this can then be interpreted as the solution of the integral equations with Galerkin's method; not of the original equations but certain projections of these equations onto the set of basis functions. So why not solve the integral equations from the start with Galerkin's method? This would be a perfect match with the finite element method which, too, is a Galerkin method. One could then use the same test functions on the interface, [78], [79], and if we choose on each part of the boundary that integral equation whose free term is conjugated to the unknown boundary value, see section 1.8, then we would even obtain a symmetric matrix. Such a symmetric coupling was proposed by Costabel [80] and Polizzotto [81].

To explain this coupling procedure let us consider the boundary-value problem, see Fig. 7.2,

$$-\Delta u = f \qquad \text{on} \quad \Omega_{FE}$$

$$u = 0 \qquad \text{on} \quad \Gamma_{FE}$$

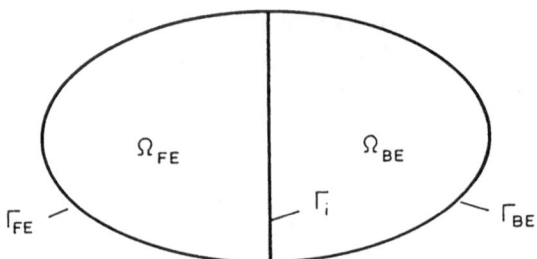

Figure 7.2 The FE-domain and the BE-domain

$$-\Delta u = 0 \qquad \text{on} \quad \Omega_{BE}$$

$$u = \bar{u} \qquad \text{on} \quad \Gamma_{BE}$$

$$u - u = 0 \qquad \text{on} \quad \Gamma_i \qquad \text{(the interface)}$$

$$t + t = 0 \qquad \text{on} \quad \Gamma_i$$

Assume that the traction t across the interface is known. Then it can be treated as an exterior force for the FE-domain so that the potential energy becomes

$$\Pi(u) = \frac{1}{2} E(u, u) - \int_{\Omega} f u \, d\Omega - \int_{\Gamma_i} t u \, ds$$

Galerkin's method would then lead to

$$E(u, \hat{u}) - \int_{\Gamma_i} t \hat{u} \, ds = \int_{\Omega} f \hat{u} \, d\Omega \qquad \text{for all} \quad \hat{u}$$

where $\hat{u}$ are test functions that satisfy the homogeneous geometric boundary conditions.

The system of integral equations for the Betti data of the BE-domain reads

$$c_0(\boldsymbol{x}) u(\boldsymbol{x}) = \int_{\Gamma} [\, g_0 t - \frac{\partial}{\partial \nu} g_0 u] ds_{\boldsymbol{y}}$$

$$c_1(\boldsymbol{x}) u(\boldsymbol{x}) = \int_{\Gamma} [\, g_1 t - \frac{\partial}{\partial \nu} g_1 u] ds_{\boldsymbol{y}}$$

If we divide these equations by the characteristic functions and if we place all terms on one side then this system of integral equations can be written more concisely as

$$\left[(\boldsymbol{I} - \boldsymbol{M}) \begin{pmatrix} u \\ t \end{pmatrix} \right] = \begin{pmatrix} 0 \\ 0 \end{pmatrix}$$

where $\boldsymbol{M}$ is an array (2×2) of integral operators.

Galerkin's method would then consist of

$$\int_\Gamma \left[(\boldsymbol{I} - \boldsymbol{M}) \begin{pmatrix} u \\ t \end{pmatrix} \right]^T \begin{pmatrix} \hat{t} \\ \hat{u} \end{pmatrix} ds = 0 \qquad \text{for all} \quad \hat{t}, \hat{u}$$

where $\hat{t}$ and $\hat{u}$ are test functions which are zero outside of Γ_2 (the traction boundary) and Γ_1 (the deflection boundary).

Now let us combine the Galerkin procedure for the two methods. The unknowns are the deflection u in the FE-domain and the traction t on the edge of the BE-domain.

The test for u would consist of

$$E(u, \hat{u}) - \int_{\Gamma_i} t\hat{u}\, ds = \int_\Omega f\hat{u}\, d\Omega \qquad \text{for all} \quad \hat{u}$$

and the test for t as

$$\int_\Gamma \left[(\boldsymbol{I} - \boldsymbol{M}) \begin{pmatrix} u \\ t \end{pmatrix} \right]^T \begin{pmatrix} \hat{t} \\ \hat{u} \end{pmatrix} ds = 0 \qquad \text{for all} \quad \hat{u}, \hat{t}$$

where $u = u_{FE}$ on Γ_i and $u = \bar{u}$ on Γ_{BE}. The test functions $\hat{u}$ are zero on the boundary Γ_{BE}; this is the boundary of the BE-domain minus the interface while no such restriction applies to the test functions $\hat{t}$ because the traction boundary is the whole boundary of the BE-domain. The interface Γ_i is simultaneously a traction boundary as a deflection boundary. The minus sign which multiplies the virtual traction $\hat{t}$ is a mathematical oddity. Without the minus the bilinear form would not be positive definite, see [80]. If we add these equations then we obtain the Galerkin equations

$$E(u, \hat{u}) - \int_{\Gamma_i} t\hat{u}\, ds + \int_{\Gamma_{BE} + \Gamma_i} \left[(\boldsymbol{I} - \boldsymbol{M}) \begin{pmatrix} u \\ t \end{pmatrix} \right]^T \begin{pmatrix} -\hat{t} \\ \hat{u} \end{pmatrix} ds_y$$

$$= \int_\Omega f\hat{u}\, d\Omega - \int_\Gamma \left[(\boldsymbol{I} - \boldsymbol{M}) \begin{pmatrix} \bar{u} \\ 0 \end{pmatrix} \right]^T \begin{pmatrix} -\hat{t} \\ \hat{u} \end{pmatrix} ds_y$$

which render a symmetric system matrix. The proof uses the same technique as in section 1.8, see [80]

7.2 Practice

To keep things simple, we demonstrate the coupling of a BE-domain with a FE-domain by considering a beam which is "discretized with boundary elements" and which connects at its right end with a continuous beam which is discretized with finite elements, see Fig. 7.3. The continuous beam on the right is to represent the FE-domain and the beam on the left the BE-domain. The coupling matrix is the matrix which formulates the connection between the end actions M_r, Q_r and the end displacements w_r, w'_r at that side of the left beam which connects with the continuous beam.

"Discretized by boundary elements" simply means that we formulate the connection between the displacement terms and force terms of the left beam in the spirit of the boundary element method, see section 2.2.

$$
\begin{bmatrix} H_1 & H_2 & H_3 & H_4 \end{bmatrix}
\begin{bmatrix} w_l \\ w'_l \\ w_r \\ w'_r \end{bmatrix}
=
\begin{bmatrix} G_1 & G_2 & G_3 & G_4 \end{bmatrix}
\begin{bmatrix} M_l \\ Q_l \\ M_r \\ Q_r \end{bmatrix}
+
\begin{bmatrix} d_1 \\ d_2 \\ d_3 \\ d_4 \end{bmatrix}
$$

The vector d represents the influence of the distributed load. The vector $G^{-1}d$ would be the vector of the end fixing forces.

At the left end of the beam the deflection and the moment are prescribed

$$ w_l = \bar{w}_l = 0, \qquad M_l = \bar{M}_l = 0 $$

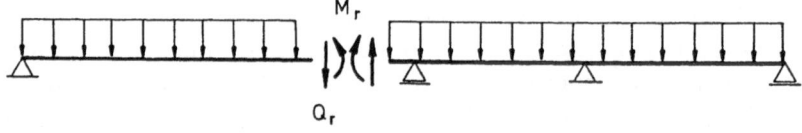

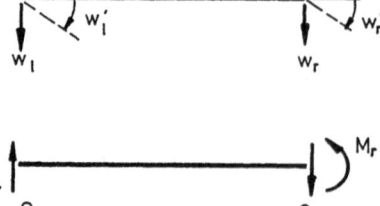

Figure 7.3 A continuous beam

so that, if we put all unknown terms on one side, the following system of equations results

$$
\begin{bmatrix} -G_2 & H_2 & -G_3 & -G_4 & H_3 & H_4 \end{bmatrix} \begin{bmatrix} Q_l \\ w_l' \\ M_r \\ Q_r \\ w_r \\ w_r' \end{bmatrix}
$$

$$
= \begin{bmatrix} d_1 \\ d_2 \\ d_3 \\ d_4 \end{bmatrix} + \begin{bmatrix} -H_1 & G_1 \end{bmatrix} \begin{bmatrix} \bar{w}_l \\ \bar{M}_l \end{bmatrix} =: \begin{bmatrix} r_1 \\ r_2 \\ r_3 \\ r_4 \end{bmatrix} \qquad (7.5)
$$

If, for ease of notation, we replace the first four columns by a block A and the columns H_3 and H_4 by a block B then we can write

$$
A_{(4\times4)} \begin{bmatrix} Q_l \\ w_l' \\ M_r \\ Q_r \end{bmatrix} + B_{(4\times2)} \begin{bmatrix} w_r \\ w_r' \end{bmatrix} = \begin{bmatrix} r_1 \\ r_2 \\ r_3 \\ r_4 \end{bmatrix}
$$

and on multiplying this equation from the left with A^{-1} we obtain the desired result

$$
A^{-1} B \begin{bmatrix} w_r \\ w_r' \end{bmatrix} = - \begin{bmatrix} Q_l \\ w_l' \\ M_r \\ Q_r \end{bmatrix} + \begin{bmatrix} \bar{r}_1 \\ \bar{r}_2 \\ \bar{r}_3 \\ \bar{r}_4 \end{bmatrix},
$$

because the lower block K within the matrix $A^{-1} B$ is the 2×2 coupling matrix. The vector $\bar{r}$ is the vector $A^{-1} r$,

$$
[K] \begin{bmatrix} w_r \\ w_r' \end{bmatrix} = - \begin{bmatrix} M_r \\ Q_r \end{bmatrix} + \begin{bmatrix} \bar{r}_3 \\ \bar{r}_4 \end{bmatrix} \qquad (7.6)
$$

The drawback of this so-called *global coupling procedure* is that the whole matrix A must be inverted. Beer, [82], and Li et al., [83], therefore, have proposed a *local coupling procedure*. To understand this modification let us recall how we solve a system of equations as $Ax = b$ by Gauss' method: first we add to the system matrix A the right-hand side, $[A, b]$, and we then bring the array A

into a triangular shape, $[A, b] \to [\tilde{A}, \tilde{b}]$, and finally we solve the system $\tilde{A}x = \tilde{b}$ by back substitution. We proceed now as in Gauss' method: we supplement the matrix $[A, B]$ in Eq.(7.5) with the right-hand side $[A, B, r]$ and we then bring the array A into a triangular shape

$$Q_l \quad w'_l \quad M_r \quad Q_r \qquad w_r \quad w'_r \qquad \text{r.S.}$$

$$
\begin{array}{cccccccc}
* & * & * & * & \vdots & * & * & \vdots & * \\
0 & * & * & * & \vdots & * & * & \vdots & * \\
0 & 0 & * & * & \vdots & * & * & \vdots & * \\
0 & 0 & 0 & * & \vdots & * & * & \vdots & *
\end{array}
$$

and, finally, we apply the same elementary operations to the lower triangular matrix in the last two rows to transform it into a unit matrix.

$$Q_l \quad w'_l \quad M_r \quad Q_r \qquad w_r \quad w'_r \qquad \text{r.S.}$$

$$
\begin{array}{cccccccc}
* & * & * & * & \vdots & * & * & \vdots & * \\
0 & * & * & * & \vdots & * & * & \vdots & * \\
0 & 0 & 1 & 0 & \vdots & \diamond & \diamond & \vdots & \diamond \\
0 & 0 & 0 & 1 & \vdots & \diamond & \diamond & \vdots & \diamond
\end{array}
$$

By this manipulation the matrix to the right of the 2×2 unit matrix becomes the coupling matrix because the last two rows

$$
\begin{bmatrix} M_r \\ Q_r \end{bmatrix} + \begin{bmatrix} \diamond & \diamond \\ \diamond & \diamond \end{bmatrix} \begin{bmatrix} w_r \\ w'_r \end{bmatrix} = \begin{bmatrix} \bar{r}_3 \\ \bar{r}_4 \end{bmatrix}
$$

are identical with Eq.(7.6).

Hence, the idea behind the local coupling is to invert only that part of A which belongs to the interface.

The extension to higher dimensional problems, as for example the plate in Fig. 7.4, is straightforward. Assume that along the boundary Γ_I of the BE-domain the traction vector t^I is prescribed. The coupling condition between the boundary data of the BE-domain

$$
[H_I \quad H_{II}] \begin{bmatrix} u^I \\ u^{II} \end{bmatrix} = [G_I \quad G_{II}] \begin{bmatrix} t^I \\ t^{II} \end{bmatrix}
$$

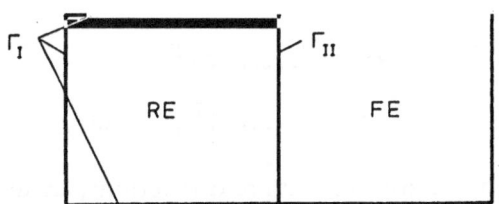

Figure 7.4 The coupling between a BE-domain and an FE-domain

then becomes

$$[H_I \quad -G_{II} \quad H_{II}] \begin{bmatrix} u^I \\ t^{II} \\ u^{II} \end{bmatrix} = [G_I][t^I],$$

and, as above, the Gaussian algorithm allows us to derive the coupling matrix

$$[K]u^{II} = t^{II} + \begin{bmatrix} r^I \\ r^{II} \end{bmatrix} \tag{7.7}$$

between the nodal displacements u^{II} and the tractions t^{II} on the interface Γ_{II}.

To obtain equivalent nodal forces this equation must be multiplied with a matrix $\tilde{F}$ whose entries are the elements of the Gram's matrix F.

The external work of the tractions t_i on the boundary on acting through virtual displacements u_i is

$$\int_\Gamma (u_1 t_1 + u_2 t_2)\, ds$$

Considering the approximation of the displacements

$$u_1 = u_1^j \varphi_j(\boldsymbol{x}), \qquad u_2 = u_2^j \varphi_j(\boldsymbol{x})$$

and tractions

$$t_1 = t_1^j \psi_j(\boldsymbol{x}), \qquad t_2 = t_2^j \psi_j(\boldsymbol{x})$$

it follows

$$\int_\Gamma (u_1 t_1 + u_2 t_2)\, ds = u_1 F t_1 + u_2 F t_2$$

where the matrix F is the Gram's matrix of the basis functions

$$F_{ij} = \int_\Gamma \varphi_i \psi_j\, ds,$$

and the vectors

$$\boldsymbol{u_1} = \{u_1^1, u_1^2, \ldots, u_1^n\}^T,$$

$$\boldsymbol{t_1} = \{t_1^1, t_1^2, \ldots, t_1^n\}^T, \qquad \text{etc.},$$

are the vectors of nodal displacements and tractions, u_1^j and t_1^j, respectively.

7.3 Experience

Tullberg and Bolteus, [84], employed seven different methods to obtain a stiffness matrix from the equation

$$\boldsymbol{H u} = \boldsymbol{G t}$$

This high number of versions is simply due to the many different strategies they employed to satisfy the equilibrium conditions and to get symmetric matrices. The equilibrium condition was satisfied either algebraically by simply distributing the defects onto the single columns, or by adding the equilibrium condition

$$\int_{\Gamma} \boldsymbol{t} \cdot (\boldsymbol{a} + \boldsymbol{b} \times \boldsymbol{x}) \, ds \qquad (\boldsymbol{a} + \boldsymbol{b} \times \boldsymbol{x}) = \text{rigid-body motion}$$

explicitly as a side condition to the system of equations. Figure 7.5 contains the results obtained for a cantilever plate. The best result renders method M1. This is simply the matrix

$$\boldsymbol{K} = \boldsymbol{F G^{-1} H},$$

obtained if you do not try to "improve" the results.

Similar observations were made in [83] where the coupling of boundary elements and finite elements was applied to three-dimensional problems of elasticity. The results showed clearly that the best results are to expected if no artificial symmetrization is tried. The deviations are considerable on a coarse mesh but they decrease if the mesh is refined as then the asymmetry also diminishes.

BE-stiffness matrices are matrices which represent the elastic properties of a domain in terms of boundary actions alone. Such matrices, naturally, can also be obtained with finite elements by eliminating the interior degrees-of-freedom. Li et al. do something similar. They eliminate from the FE-stiffness matrix all internal and external degrees of freedom which do not refer to the interface. They call this procedure *bi-condensation* because in the end the elastic properties of the two domains are only described by the displacements and actions on the interface.

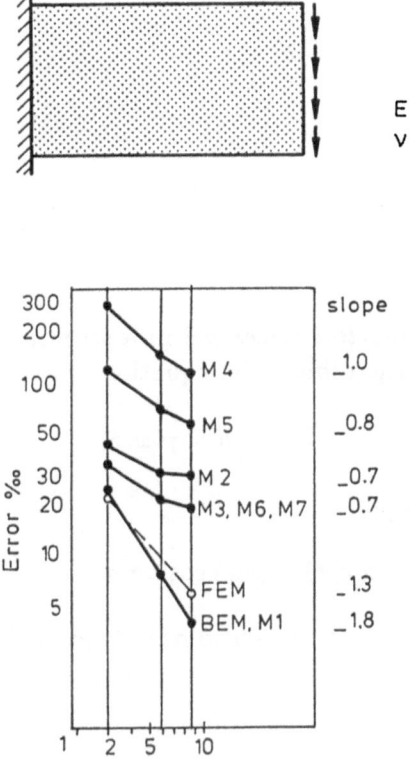

Figure 7.5 The results of Tullberg and Bolteus

8 Harmonic oscillations

Dynamical loads cause *inertial forces* $\rho\ddot{u}$ in a structure. These forces appear on the left-hand side of the differential equation

$$Du + \rho\ddot{u} = p(x,t)$$

If the excitation is harmonic

$$p(x,t) = p(x)\cos(\omega t + \varphi),$$

then the response of the structure is also harmonic. This important case is the topic of this chapter.

8.1 Rods

The differential equation of the vibrating rod

$$-EAu''(x,t) + \mu\ddot{u}(x,t) = p(x)\cos(\omega t + \varphi), \qquad \mu = \rho A, \quad A = \text{area},$$

becomes, after a separation of the variables,

$$u(x,t) = u(x)\cos(\omega t + \varphi)$$

a differential equation

$$-EAu''(x) - \mu\omega^2 u(x) = p(x)$$

for the amplitude $u(x)$ alone.

To this differential equation belong the identities

$$G(\hat{u}, u) = \int_0^l (-EA\hat{u}'' - \mu\omega^2\hat{u})u\, dx + [\hat{N}u]_0^l - \int_0^l (EA\hat{u}'u' - \mu\omega^2\hat{u}u)\, dx = 0$$

and

$$B(\hat{u}, u) = \int\limits_0^l (-EA\hat{u}'' - \mu\omega^2\hat{u})u\, dx + [\hat{N}u - \hat{u}N]_0^l - \int\limits_0^l \hat{u}(-EAu'' - \mu\omega^2 u)\, dx$$

By a proper choice of the constants a and b in the homogeneous solution

$$u(x) = a\sin(\frac{\mu\omega^2}{EA}\, x) + b\cos(\frac{\mu\omega^2}{EA}\, x)$$

we can construct two solutions φ_1 and φ_2 which correspond to unit end displacements

$$\varphi_1(0) = 1\,, \quad \varphi_1(l) = 0\,, \qquad \varphi_2(0) = 0\,, \quad \varphi_2(l) = 1\,,$$

With these two unit functions we can derive, as in section 2.1, coupling conditions between the end displacements u_i and end actions f_i of an amplitude $u \in C^2[0, l]$,

$$Ku = f + p\,,$$

where

$$K_{ij} = E(\varphi_i, \varphi_j) = \int\limits_0^l (EA\varphi_i'\varphi_j' - \mu\omega^2\varphi_i\varphi_j)\, dx\,,$$

$$p_i = \int\limits_0^l p(x)\varphi_i(x)\, dx\,, \qquad p(x) = -EAu''(x) - \mu\omega^2 u(x)$$

The terms p_i are the negative end fixing forces. By rearranging this coupling condition we can obtain a transfer matrix and, therefore, the matrix displacement method can be applied to harmonic oscillations as well.

8.2 Beams

The differential equation of the vibrating beam

$$EIw^{IV} + \mu\ddot{w} = p(x)\cos(\omega t + \varphi)\,, \qquad \mu = \rho A\,,$$

becomes after a separation of the variables

$$w(x, t) = w(x)\cos(\omega t + \varphi)$$

a differential equation for the amplitude

$$EIw^{IV}(x) - \mu\omega^2 w(x) = p(x)$$

To this differential equation belong the identities

$$G(\hat{w}, w) = \int_0^l (EI\hat{w}^{IV} - \mu\omega^2 \hat{w})w\, dx + [\hat{Q}w - \hat{M}w']_0^l$$

$$- \int_0^l (EI\hat{w}''w'' - \mu\omega^2 \hat{w}w)\, dx = 0$$

and

$$B(\hat{w}, w) = G(\hat{w}, w) - G(w, \hat{w}) = 0.$$

The general homogeneous solution is

$$w(x) = a_1 \cos(\lambda x) + a_2 \sin(\lambda x) + a_3 \cosh(\lambda x) + a_4 \sinh(\lambda x)$$

where

$$\lambda = \left(\frac{\mu\omega^2}{EI}\right)^{1/4}$$

By an appropriate choice of the integration constants we can obtain four solutions ψ_i which correspond to unit end displacements

$$\psi_1(0) = 1, \qquad \psi_1'(0) = \psi_1(l) = \psi_1'(l) = 0, \qquad \text{etc.}$$

The energy products of these functions constitute the elements of the stiffness matrix K. This matrix formulates a coupling condition between the end displacements u_i and end actions f_i of a smooth amplitude $w \in C^4[0, l]$,

$$Ku = f + p \tag{8.1}$$

where

$$K_{ij} = E(\psi_i, \psi_j) = \int_0^l (EI\psi_i''\psi_j'' - \mu\omega^2 \psi_i\psi_j)\, dx,$$

$$p_i = \int_0^l p(x)\psi_i(x)\, dx, \qquad p(x) = EIw^{IV} - \mu\omega^2 w$$

The terms p_i are the negative end fixing forces. By a simple rearrangement of Eq.(8.1) we can derive the transfer matrix and, therefore, the matrix-displacement method is complete.

8.3 Elastic plates and bodies

The vibrations v of an isotropic, homogeneous linear elastic body satisfy the differential equations

$$-\mu \Delta v - \frac{\mu}{1 - 2\nu} \nabla \operatorname{div} v + \rho \ddot{v} = b(x, t)$$

In elastodynamics the elastic constants ν and μ are often replaced by the constants

$$c_2 = \left(\frac{\mu}{\rho}\right)^{1/2}, \qquad c_1 = \left(\frac{2\mu}{\rho}(\frac{1 - \nu}{1 - 2\nu})\right)^{1/2},$$

the isochoric velocity $c_2 (= c_s)$, or s-wave velocity, and the irrotational velocity $c_1 (= c_p)$, or p-wave velocity so that the equations take the form

$$-c_2^2 \Delta v - (c_1^2 - c_2^2) \nabla \operatorname{div} v + \ddot{v} = \frac{1}{\rho} b(x, t) \tag{8.2}$$

Harmonic excitations

$$\frac{1}{\rho} b(x, t) = p_1(x) \cos \omega t + p_2(x) \sin \omega t$$

and harmonic displacement fields

$$v(x, t) = v_1(x) \cos \omega t + v_2(x) \sin \omega t$$

can be considered the real parts of complex-valued functions

$$\frac{1}{\rho} b(x, t) = \Re \left\{ p(x) e^{-i\omega t} \right\}, \qquad v(x, t) = \Re \left\{ u(x) e^{-i\omega t} \right\},$$

where

$$p(x) = p_1(x) + i\, p_2(x), \qquad u(x) = v_1(x) + i\, v_2(x)$$

If we substitute these expressions into Eq.(8.2) then we obtain the following system of equations for the complex-valued amplitude

$$-c_2^2 \Delta u - (c_1^2 - c_2^2) \nabla \operatorname{div} u - \omega^2 u = p(x) \tag{8.3}$$

To this system belong the identities

$$G(\hat{u}, u) = \int_\Omega (-L\hat{u} - \omega^2\hat{u}) \cdot u \, d\Omega + \int_\Gamma \tau(\hat{u}) \cdot u \, ds - E(\hat{u}, u)$$

$$+ \int_\Omega \omega^2 \hat{u} \cdot u \, d\Omega = 0,$$

$$B(\hat{u}, u) = G(\hat{u}, u) - G(u, \hat{u}) = 0$$

The operator $-L$ is the operator of the static problem and $E(\hat{u}, u)$ is the associated energy product.

The fundamental solutions of Eq.(8.3) are

$$U_{ij} = \frac{1}{4\pi\rho c_2^2}[\psi\,\delta_{ij} - \chi\,r_{,i}\,r_{,j}]$$

where

$$\psi = \frac{1}{r}\left[e^{-i\omega r/c_2}\left(1 - \frac{c_2^2}{\omega^2 r^2} + \frac{c_2}{i\omega r}\right) - \frac{c_2^2}{c_1^2}\left(-\frac{c_1^2}{\omega^2 r^2} + \frac{c_1}{i\omega r}\right)e^{-i\omega r/c_1}\right],$$

$$\chi = \left(-\frac{3c_2^2}{\omega^2 r^2} + \frac{3c_2}{i\omega r} + 1\right)\frac{e^{-i\omega r/c_2}}{r} - \frac{c_2^2}{c_1^2}\left(-\frac{3c_1^2}{\omega^2 r^2} + \frac{3c_1}{i\omega r} + 1\right)\frac{e^{-i\omega r/c_1}}{r},$$

and the components of the associated traction vectors are

$$T_{ij} = \frac{1}{4\pi}[\left(\frac{d\psi}{dr} - \frac{1}{r}\chi\right)(\delta_{ij}r_\nu + r_{,j}\,\nu_i) - \frac{2}{r}\chi(\nu_j r_{,i} - 2r_{,i}\,r_{,j}\,r_\nu)$$

$$- 2\frac{d\chi}{dr}r_{,i}\,r_{,j}\,r_\nu + \left(\frac{c_1^2}{c_2^2} - 2\right)\left(\frac{d\psi}{dr} - \frac{d\chi}{dr} - \frac{2}{r}\chi\right)r_{,i}\,\nu_j]$$

If we formulate with the three fundamental solutions $g_0^i = \{U_{ij}\}$ the limit

$$\lim_{\varepsilon \to 0} B(g_0^i, u)_{\Omega_\varepsilon} = 0,$$

then we obtain three influence functions for the complex-valued amplitude

$$C_{ij}(x)u_j(x) = \int_\Gamma [U_{ij}(y, x)t_j(y) - T_{ij}(y, x)u_j(y)]\,ds_y$$

$$+ \int_\Omega U_{ij}(y, x)p_j(y)\,d\Omega_y.$$

The terms $C_{ij}(x)$ are the characteristic functions of the static problem.

The 2-D fundamental solutions are given in [54, p. 365]. For applications we refer to [85], where two-dimensional problems are solved with the Laplace transform. The numerical technique is nearly identical because a Laplace transformation leads to the same system of differential equations as the separation of the variables.

8.4 Kirchhoff plates

The response of a plate to a harmonic excitation

$$K\Delta\Delta v + \rho\ddot{v} = b(\boldsymbol{x}, t) = p_1(\boldsymbol{x})\cos\omega t + p_2(\boldsymbol{x})\sin\omega t,$$

$$\rho = \text{specific mass} \times \text{plate thickness},$$

is a harmonic oscillation

$$v(\boldsymbol{x}, t) = v_1(\boldsymbol{x})\cos\omega t + v_2(\boldsymbol{x})\sin\omega t$$

If we fomulate as before

$$v(\boldsymbol{x}, t) = \Re\left\{w(\boldsymbol{x})e^{-i\omega t}\right\}, \qquad b(\boldsymbol{x}, t) = \Re\left\{p(\boldsymbol{x})e^{-i\omega t}\right\}$$

where

$$w(\boldsymbol{x}) = w_1(\boldsymbol{x}) + i\,w_2(\boldsymbol{x}), \qquad p(\boldsymbol{x}) = p_1(\boldsymbol{x}) + i\,p_2(\boldsymbol{x}),$$

then the differential equation for the complex-valued amplitude $w(\boldsymbol{x})$ becomes

$$K\Delta\Delta w(\boldsymbol{x}) - \rho\omega^2 w(\boldsymbol{x}) = p(\boldsymbol{x}) \tag{8.4}$$

Associated with this equation are the identities

$$G(\hat{w}, w) = \int_{\Omega}(K\Delta\Delta\hat{w} - \rho\omega^2\hat{w})w\,d\Omega + \int_{\Gamma}(\hat{V}_n w - \hat{M}_n\frac{\partial w}{\partial n})\,ds$$

$$+ \sum_{e}[\hat{F}(\boldsymbol{x}^e)w(\boldsymbol{x}^e) - \hat{w}(\boldsymbol{x}^e)F(\boldsymbol{x}^e)] - E(\hat{w}, w) + \int_{\Omega}\rho\omega^2\hat{w}w\,d\Omega = 0,$$

$$B(\hat{w}, w) = G(\hat{w}, w) - G(w, \hat{w}) = 0$$

The integral $E(\hat{w}, w)$ is the energy product of the static problem.

The fundamental solution of the differential equation (8.4) is

$$g_0(\lambda r) = icJ_0(\lambda r) + cY_0(\lambda r) + dK_0(\lambda r)$$

where

$$\lambda = \frac{\omega^2}{\rho K}, \qquad c = \frac{1}{8\lambda^2}, \qquad d = \frac{1}{4\pi\lambda^2}$$

The functions J_0, Y_0 and K_0 are Bessel functions.

If we formulate with the deflection w and the fundamental solution g_0 and $g_1 = \partial g_0/\partial n$ the limit

$$\lim_{\varepsilon \to 0} B(g_i, w)_{\Omega_\varepsilon} = 0, \qquad i = 0, 1,$$

then we obtain two influence functions for the deflection $w(\boldsymbol{x})$ and the slope $\partial w(\boldsymbol{x})/\partial n$, which are very similar to those of the static problem so that we do not list them here. The interested reader will find these equations in [86].

8.5 Natural frequencies

The determination of the natural frequencies and natural modes of elastic systems can be considered a particular application of the coupling condition between the Betti data of the system.

From a mathematical point of view the natural modes are the non-trivial solutions of fully homogeneous boundary-value problems as

$$Du + \lambda u = 0 \qquad + \quad m \text{ homogeneous boundary conditions } \partial^i u = 0$$

The natural modes are subject to the same coupling conditions as the other amplitude functions.

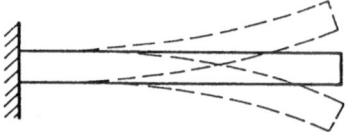

Figure 8.1 Vibrating beam

In the case of an oscillating cantilever beam as in Fig. 8.1 the Betti data of any natural mode must satifsfy the system $\boldsymbol{Ku} = \mathbf{o}$ where

$$\boldsymbol{K} = EI\frac{\lambda}{s}\frac{1}{(1 - Cc)}\begin{bmatrix} \frac{\lambda^2}{s^2}(Cs + Sc) & -\frac{\lambda}{s}Ss & -\frac{\lambda^2}{s^2}(S + s) & -\frac{\lambda}{s}(C - c) \\ * & Cs - Sc & \frac{\lambda}{s}(C - c) & S - s \\ * & * & \frac{\lambda^2}{s^2}(Cs + Sc) & \frac{\lambda}{s}Ss \\ * & * & * & Cs - Sc \end{bmatrix}$$

$$c = \cos\lambda, \qquad s = \sin\lambda, \qquad C = \cosh\lambda, \qquad S = \sinh\lambda$$

is the symmetric stiffness matrix of the oscillating beam, see section 8.2. The (2×2) array in the lower right half is the reduced stiffness matrix K_r. The full matrix reduces to this matrix if we cancel all those columns and rows which correspond to zero degrees-of-freedom. For non-trivial nodal displacements u_3 and u_4 to exist the determinant of the reduced stiffness matrix must be zero,

$$\det(K_r(\lambda)) = 1 + \cos \lambda \cosh \lambda = 0$$

The zeros of this transcendental function are the natural frequencies of the cantilever beam and the associated eigenvectors are the coefficients of the natural mode $w(x)$ with respect to the four normalized, homogeneous solutions ψ_i

$$w(x) = 0\,\psi_1(x) + 0\,\psi_2(x) + u_3\,\psi_3(x) + u_4\,\psi_4(x)$$

This example, simple as it is, illustrates the principal difficulty when it comes to the determination of the natural frequencies or critical loads: Because the fundamental solutions are *transcendental functions* of the parameter λ,

$$\cdots \quad \sin \lambda \quad \cdots \quad \cos \lambda \,,$$

the elements of the stiffness matrix too are transcendental functions and, therefore, also the determinant itself. But to find the zeros of complicated transcendental functions is not an easy task. In particular when you consider large frame structures where the reduced stiffness matrix easily exceeds the size 100×100.

The finite element method solves this problem by replacing the exact stiffness matrix by an approximation

$$\tilde{K} = K_0 + \lambda K_1 \,,$$

whose elements

$$\tilde{K}_{ij} = E(\varphi_i, \varphi_j) = \int_0^l (EI\varphi_i'' \varphi_j'' - \mu\omega^2 \varphi_i \varphi_j)\, dx$$

are the energy products of the four homogeneous solutions φ_i of the simple beam-equation $EIw^{IV} = 0$.

Because of the particular choice of basis functions the matrix $\tilde{K}$ consists of the stiffness matrix of the simple beam-equation, K, and the so-called *consistent mass matrix* K_1. (One can show that these two matrices are the first two terms of the Taylor series of K.)

Because the entries of both matrices do not depend on λ the determinant becomes a polynomial in λ. The price we have to pay for this transformation of a transcendental determinant into a polynomial determinant is that we must subdivide each beam into single finite elements and to thus raise the number of degrees-of-freedom.

All this holds true for higher dimensional problems as well. Their fundamental solutions, too, are transcendental functions of the parameter λ, therefore also the complex-valued matrices $\boldsymbol{G}$ and $\boldsymbol{H}$

$$\boldsymbol{H}(\lambda)\,\boldsymbol{u} = \boldsymbol{G}(\lambda)\,\boldsymbol{t}\,,$$

and, hence, a search-algorithm is required to find the zeros of the determinant. For a detailed review of all the relevant questions we refer the reader to Kitahara's monograph, [29].

8.6 Helmholtz equation (membrane)

The response of a stretched membrane to harmonic excitations

$$-\Delta v + \rho v = b(\boldsymbol{x}, t) = p_1(\boldsymbol{x})\cos\omega t + p_2(\boldsymbol{x})\sin\omega t$$

are solutions of the form

$$v(\boldsymbol{x}, t) = v_1(\boldsymbol{x})\cos\omega t + v_2(\boldsymbol{x})\sin\omega t$$

If we write

$$b(\boldsymbol{x}, t) = \Re\left\{p(\boldsymbol{x})\mathrm{e}^{-i\omega t}\right\}\,, \qquad p(\boldsymbol{x}) = p_1(\boldsymbol{x}) + i\,p_2(\boldsymbol{x})\,,$$

and

$$v(\boldsymbol{x}, t) = \Re\left\{u(\boldsymbol{x})\mathrm{e}^{-i\omega t}\right\}\,, \qquad u(\boldsymbol{x}) = u_1(\boldsymbol{x}) + i\,u_2(\boldsymbol{x})\,,$$

then this leads to the *Helmholtz equation*

$$-\Delta u(\boldsymbol{x}) - \lambda u(\boldsymbol{x}) = p(\boldsymbol{x})\,, \qquad \lambda = \rho\omega^2$$

for the complex-valued amplitude $u(\boldsymbol{x})$. Associated with this differential equations is the identity

$$G(\hat{u}, u) = \int_\Omega (-\Delta\hat{u} - \lambda\hat{u})u\,d\Omega + \int_\Gamma \frac{\partial\hat{u}}{\partial n}u\,ds$$

$$-\int_\Omega \nabla\hat{u}\cdot\nabla u\,d\Omega + \int_\Omega \lambda\hat{u}\,u\,d\Omega = 0$$

If we formulate the second identity $B(\hat{u}, u) = G(\hat{u}, u) - G(u, \hat{u}) = 0$ with the fundamental solutions

$$g_0(\boldsymbol{y}, \boldsymbol{x}) = \frac{-i}{2}H_0^{(1)}(\lambda r) \qquad \text{(2-D)} \text{ (Hankel-function)}\,,$$

$$g_0(y, x) = -\frac{1}{4\pi r} e^{i\lambda r} \qquad \text{(3-D)}$$

$$g_1(y, x) = \frac{\partial}{\partial n_x} g_0(y, x)$$

then we obtain the influence function for $u(x)$,

$$c(x)u(x) = \int_\Gamma [g_0(y, x)\frac{\partial u}{\partial \nu}(y) - \frac{\partial}{\partial \nu}g_0(y, x)u(y)]\, ds_y$$

$$+ \int_\Omega g_0(y, x)(-\Delta u(y) - \lambda u(y))\, d\Omega_y \qquad (8.5)$$

and the normal derivative

$$c_j(x)u_{,j}(x) = \int_\Gamma [g_1(y, x)\frac{\partial u}{\partial \nu}(y) - \frac{\partial}{\partial \nu}g_1(y, x)(u(y) - u(x))]\, ds_y$$

$$+ \int_\Omega g_1(y, x)(-\Delta u(y) - \lambda u(y))\, d\Omega_y, \qquad (8.6)$$

The characteristic functions are the functions of the static case, see chapter 3.

8.6.1 The radiation resistance of a rotating engine

For an application we consider a problem in acoustics, [87]. The cover of a rotating engine will excite the surrounding air to harmonic oscillations which we perceive as noise. A measure for the generated noise is the sound power and the radiation resistance of the air. To quantify these terms we model the surrounding air as a frictionless, compressible fluid where sound waves propagate in agreement with the impulse balance

$$p_{,i} + \rho v_i = 0, \qquad p(x, t) = \text{sound pressure},$$

$$v(x, t) = \text{velocity},$$

$$\rho = \text{density of the fluid},$$

and according to the law that connects the sound pressure with the volume change

$$\dot{p} = -K v_{i,i} \qquad K = \text{modulus of compression}$$

If we introduce the scalar-valued potential $\Phi(x, t)$

$$v_i = \Phi_{,i} \qquad p = -\rho \dot{\Phi},$$

then these equations reduce to a single equation, *the wave equation,*

$$\Phi_{,ii} - \frac{1}{c^2}\ddot{\Phi} = 0, \qquad c = \left(\frac{K}{\rho}\right)^{1/2} = \text{speed of sound.}$$

The normal derivative of the potential Φ on the surface Γ, the cover of the engine

$$\frac{\partial\Phi}{\partial n} = \Phi_{,i}\,n_i = v_i n_i = v$$

is the normal component of the surface speed. We denote this scalar-valued function by v. Its value on the cover of the engine is given. Because the engine rotates v is a periodic function with respect to time t and, after an initial phase, the surrounding air will oscillate with the same frequency as the cover of the engine.

Hence, we are looking for the complex-valued velocity potential Φ, which satisfies in the exterior region the equations

$$\Delta\Phi + k^2\Phi = 0, \qquad k = \frac{\omega}{c} = \frac{2\pi}{\lambda}, \qquad \lambda = \text{wave length}\,,$$

$$\frac{\partial\Phi}{\partial n} = v \qquad \text{on } \Gamma\,,$$

and *Sommerfeld's radiation condition*

$$\lim_{r\to\infty}\left(\frac{\partial\Phi}{\partial n} - ik\Phi\right)r = 0, \qquad r = |y - x|$$

This condition excludes incoming waves and guarantees the unique solvability of the problem. Because the BE-solution is the influence function (8.5) the satisfaction of the coupling conditions guarantees also that the radiation condition is satisfied.

The boundary values of the unknown potentials Φ can either be determined by solving the first coupling condition (8.5)

$$\frac{1}{2}\Phi(x) + \int_{\Gamma}\frac{\partial}{\partial\nu}g_0(y,x)\Phi(y)\,ds_y = \int_{\Gamma}g_0(y,x)v(y)\,ds_y$$

or by solving the second condition (8.6)

$$\int_{\Gamma}\frac{\partial}{\partial\nu}g_1(y,x)[\Phi(y) - \Phi(x)]\,ds_y = -\frac{1}{2}v(x) + \int_{\Gamma}g_1(y,x)v(y)\,ds_y$$

If we introduce the symbols

$$I\Phi = \Phi,$$

$$Lv = \int_\Gamma g_0(y, x)v(y)\, ds_y,$$

$$N\Phi = \int_\Gamma \frac{\partial}{\partial\nu} g_1(y, x)[\Phi(y) - \Phi(x)]\, ds_y,$$

$$M\Phi = \int_\Gamma \frac{\partial}{\partial\nu} g_0(y, x)\Phi(y)\, ds_y,$$

for the integral operators then the coupling condition can also be written as

$$(\frac{1}{2}I + M)\Phi = Lv \tag{8.7}$$

$$N\Phi = (-\frac{1}{2}I + M^T)v \tag{8.8}$$

In the following we shall make use of *Fredholm's alternative* and to this end we need yet the coupling conditions between the Betti data of the interior domain.

Equations (8.7) and (8.8) are the coupling conditions for the boundary data of the exterior domain, the normal vector on Γ points into the interior domain. The situation is reversed if we formulate the coupling conditions for the boundary data of the interior domain. We then must multiply M and v with (-1) and N with $(-1)(-1) = 1$, (N contains the normal vector twice) so that the coupling conditions between the boundary data of the interior domain read

$$(-\frac{1}{2}I + M)\Phi = Lv,$$

$$N\Phi = (\frac{1}{2}I + M^T)v$$

Fredholm's alternative is best explained with vectors and matrices.

Recall that a scalar as $\hat{x}^T A x$ is invariant to transpositions so that the following expression is an identity

$$B(x, \hat{x}) = \hat{x}^T A x - x^T A^T \hat{x} = 0$$

Simply by substituting the solution x into this equation we learn that the system of equations

$$Ax = b$$

is solvable if and only if the right-hand side b is orthogonal to the eigen solutions $\hat{x}$,

$$A^T\hat{x} = o\,,$$

of the adjoint matrix. If the adjoint (= transposed) matrix has only the eigen solution $\hat{x} = o$, then the equation $Ax = b$ is uniquely solvable. But if the adjoint matrix has n non-trivial, linearly independent eigen solutions then the equation $Ax = b$ also has n such eigen solutions and the equation therefore is only up to n vectors x_i uniquely solvable.

This is Fredholm's alternative; in our context it is a corollary of Betti's principle.

In classical mechanics the adjoint operator is the same operator as A, and the eigen solutions of the adjoint operator are the rigid-body-motions. Therefore, Fredholm's alternative is identical with the equilibrium conditions. To any set of exterior forces exists an equilibrium position u if and only if the scalar product between the exterior forces and the rigid-body-motions is zero, that is if the exterior forces are in equilibrium.

Simply by adopting the technique applied to the matrix A we can now formulate with the two integral operators of the exterior problem two identities,

$$B(\varPhi, \hat{\varPhi}) = \int_\Gamma (\frac{1}{2}I + M)\varPhi\,\hat{\varPhi}\,ds - \int_\Gamma \varPhi\,(\frac{1}{2}I + M^T)\hat{\varPhi}\,ds = 0\,,$$

$$B(\varPhi, \hat{\varPhi}) = \int_\Gamma N\varPhi\,\hat{\varPhi}\,ds - \int_\Gamma \varPhi\,N\hat{\varPhi}\,ds = 0\,, \qquad (N = N^T)\,,$$

and we conclude that according to Fredholm's alternative the integral equation (8.7) is uniquely solvable if the adjoint homogeneous problem

$$(\frac{1}{2}I + M^T)v = 0$$

only has the trivial solution $v = 0$. But this condition is just the coupling condition for the boundary data $\varPhi = 0, v$ of an interior problem. At some frequencies k_{Di}, the *natural frequencies*, standing waves appear in the interior which satisfy these conditions.

The same holds true for the second integral equation (8.8). The adjoint integral equation

$$N\varPhi = 0$$

formulates the coupling condition between the boundary data $\varPhi, v = 0$ of the interior domain. At some particular frequencies, k_{Ni}, non-trivial solutions also exist.

According to Fredholm's alternative the integral equations (8.7) and (8.8) of the exterior problem are therefore not uniquely solvable at the critical wave numbers k_{Di} and k_{Ni}; the BE-matrix becomes singular at these frequencies. To overcome this difficulty we could couple the two integral equations,

$$\left\{\frac{1}{2}I + M + \alpha N\right\}\Phi = \left\{L - \frac{\alpha}{2}I + \alpha M^T\right\}v$$

As long as the coupling constant α is purely imaginary this coupling is successful because then the compound integral equation has a unique solution for all wave numbers, see [88]. In application α may be neither too small nor too large because then one of the two integral equations dominates.

The example of a harmonically oscillating ("breathing") sphere with the surface velocity

$$v(\boldsymbol{x}, t) = v_0\, e^{-i\omega t}, \qquad v_0 = \text{constant}$$

will demonstrate this more clearly. The pertinent potential

$$\Phi(\boldsymbol{x}, t) = \Phi(r = a)e^{-i\omega t}, \qquad a = \text{ radius of the sphere}$$

is constant on the sphere

$$\Phi(r = a) = \frac{v_0 a(1 + ika)}{1 - k^2 a^2} = \text{constant}$$

The critical wave numbers of the interior problem with homogeneous Dirichlet conditions $\Phi = 0$ is

$$k_{Di} = \pi, \quad 2\pi, \quad 3\pi, \quad 4\pi, \quad \ldots$$

and the same numbers for the interior problem with homogeneous Neumann conditions $v = 0$ are

$$k_{Ni} = 4.4935 \qquad 7.7150 \qquad 10.904 \qquad \ldots$$

Figure 8.2a contains a comparison between the approximate potential Φ_h (10 constant ring elements) and the exact potential Φ for a coupling constant $\alpha = 0$. Clearly we see that the numerical solution becomes unstable near the critical wave numbers k_{Di}. At these points the system matrix is singular. If we choose as coupling constant $\alpha = ia$, see Fig. 8.2b, then the solution is stable near the critical wave numbers k_{Di} but it becomes unstable near the wave numbers k_{Ni} because for large wave numbers k the operator N dominates.

Parameter studies demonstrated, as predicted in [89], that $\alpha = i/k$ renders the best solution, see Fig. 8.2c. The agreement with the exact solution over the whole range is then very good.

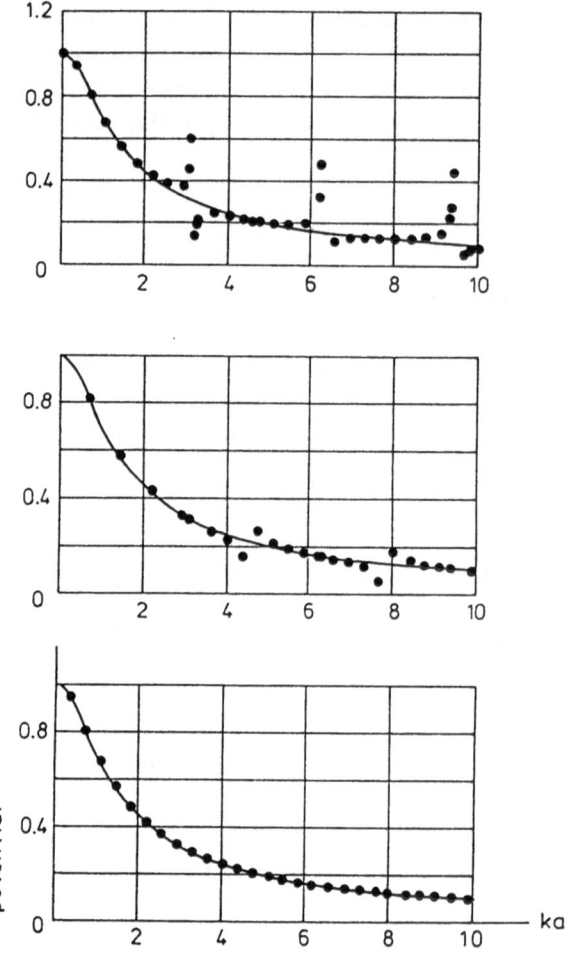

Figure 8.2 The influence of the coupling parameter α on the numerical solution (Akyol [87])

Axisymmetric bodies allow the two boundary functions Φ and v to expand into a symmetric Fourier series with respect to the zero meridian ($\varphi = 0$).

$$v(\boldsymbol{x}) = v_m(s) \cos m\varphi, \qquad \Phi(\boldsymbol{x}) = \Phi_m(s) \cos m\varphi$$

The coordinate s is the arc-length on the contour, see Fig. 8.3. If the axis of symmetry does not coincide with the zero meridian then we can rotate, termwise, the zero meridian because the integral equation over the surface Γ of the body reduces for each term to an integral equation on the contour S

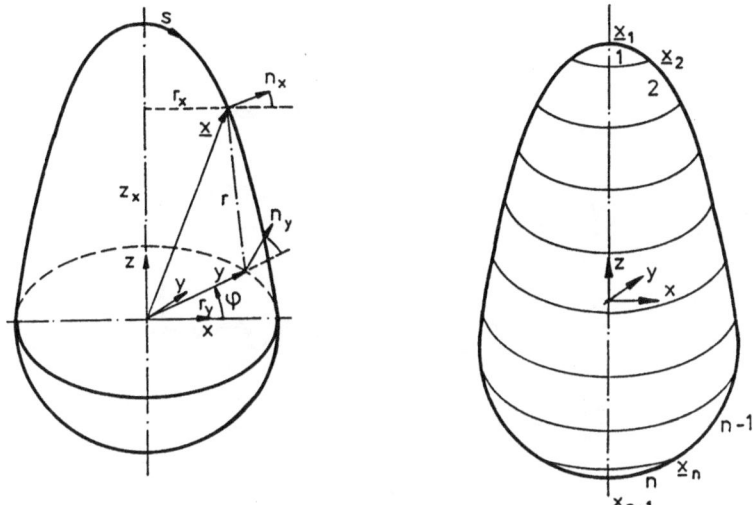

Figure 8.3 Discretization of an axisymmetric body

$$\frac{1}{2}\Phi_m(s) + 2\int_0^{s_0}\int_0^{\pi}[\frac{\partial g_0}{\partial \nu} + \alpha\frac{\partial}{\partial \nu}g_1]\Phi_m(\sigma)\cos m\varphi \; r \, d\varphi \, d\sigma$$

$$= 2\int_0^{s_0}\int_0^{\pi}[g_0 + \alpha g_1]v_m(\sigma)\cos m\varphi \; r \, d\varphi \, d\sigma - \frac{\alpha}{2}v(s)$$

The circumferential integrals are of the type

$$\int_0^{2\pi}\frac{1}{r^n}e^{ikr}\cos^l\varphi\cos m\varphi \, d\varphi, \quad m = 1, 2, \ldots; \quad n = 1, 2, \ldots, 5; \quad l = 0, 1, 2;$$

and, therefore, the oscillations are increasing rapidly with the wave number k. In Fig. 8.4 the real parts of the integrands are plottedd versus a normalized wave number $\kappa = kr_0$ (r_0 normalizing quantity).

At high wave numbers all numerical quadrature formulas fail and we must use asymptotic expansions. At small wave numbers and medium range wave numbers numerical quadrature is applicable. Experience teaches us that the *trapezoidal rule*

$$\int_0^{2\pi}y(\varphi)\,d\varphi = \frac{h}{2}(y_0 + 2y_1 + \cdots + 2y_{n-1} + y_n)$$

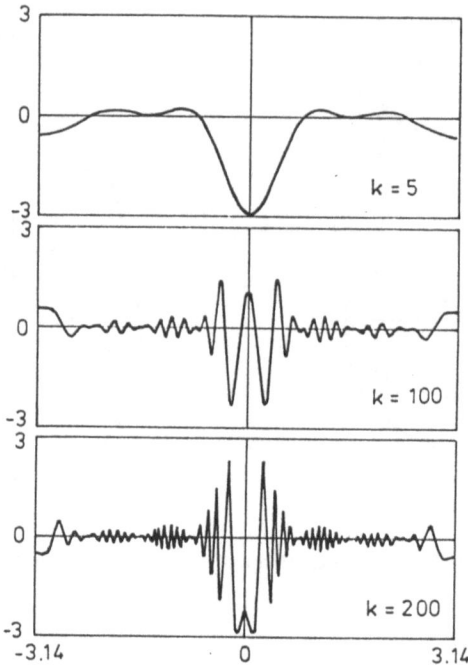

Figure 8.4 The oscillations of the integrand are increasing with the wave number k

$$-\frac{h^2}{12}(y_n' - y_0') + \frac{h^4}{720}(y_n^{(3)} - y_0^{(3)}) - \frac{h^6}{30\,240}(y_n^{(5)} - y_0^{(5)}) + \cdots$$

is superior to Gaussian quadrature when it comes to the integration of periodic functions because the error terms cancel each other.

The acoustic behavior of a pulsating cylinder whose surface velocity is locally constant and zero on the two faces was investigated by different authors with semi-analytic or numerical methods. Table 8.1 compares the results of these authors with Akyol's results for a cylinder with the ratio $h/a = 2$.

Table 8.1. Radiation resistance of a pulsating cylinder

ka	Copley	Will./Mor.	Sandman	Fenlon	Shenderov	Akyol
1	0.621	0.633	0.733	0.653	0.729	0.7297
2	0.758	0.822	0.880	0.780	0.881	0.8672
3	0.900	0.944	0.974	0.870	/	0.9762

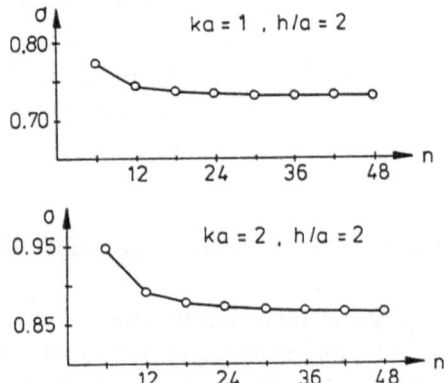

Figure 8.5 The convergence of the radiation resistance σ

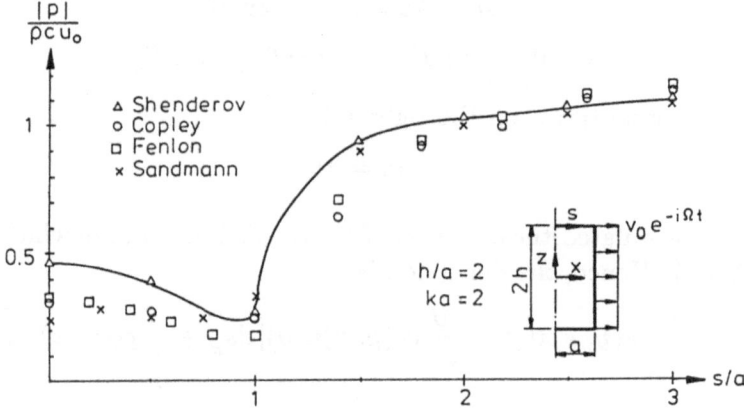

Figure 8.6 The surface pressure

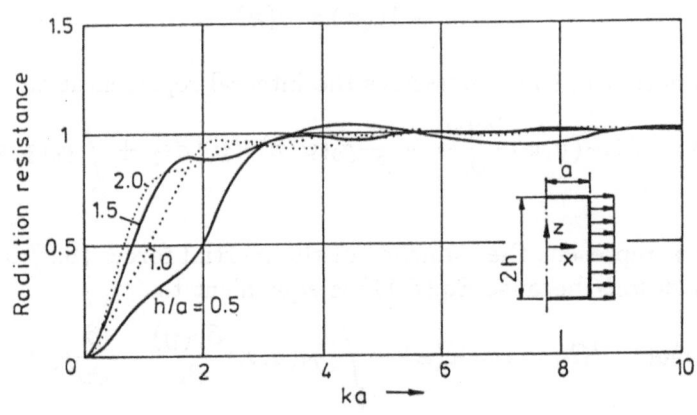

breathing cylinder

Figure 8.7 Connection between radiation resistance and size of the cylinder, $k =$ wave number

The good convergence of the boundary element method is illustrated in Fig. 8.5. In Fig. 8.6 the calculated surface pressure is compared with the results of other authors. Figure 8.7 depicts the relation between the geometry of the cylinder and the radiation resistance of the cylinder.

8.7 Algebraization of the eigenvalue problem

Nardini and Brebbia, [90], reduced the determination of the natural frequencies of a plate by the so-called *dual reciprocity boundary element method* to an algebraic eigenvalue problem. We illustrate this technique by considering the eigenvalue problem of a stretched membrane.

$$-\Delta u - \lambda u = 0, \qquad \text{on } \Omega$$

$$u = 0 \qquad \text{on } \Gamma_1, \qquad t = 0 \qquad \text{on } \Gamma_2$$

If we put all inertial terms on the right side

$$-\Delta u = \lambda u,$$

then we have formulated the equation of a stretched membrane which is loaded by forces $\lambda u(\boldsymbol{x})$. Hence, the deflection is

$$c(\boldsymbol{x})u(\boldsymbol{x}) = \int_{\Gamma} [g_0(\boldsymbol{y}, \boldsymbol{x})\, t(\boldsymbol{y}) - \frac{\partial}{\partial \nu} g_0(\boldsymbol{y}, \boldsymbol{x}) u(\boldsymbol{y})]\, ds_{\boldsymbol{y}} + \int_{\Omega} g_0(\boldsymbol{y}, \boldsymbol{x})\lambda u(\boldsymbol{y})\, d\Omega_{\boldsymbol{y}}$$

(8.9)

Let us assume that the function $u(\boldsymbol{x})$ is also the right-hand side of a function $v(\boldsymbol{x})$

$$-\Delta v(\boldsymbol{x}) = u(\boldsymbol{x}),$$

(8.10)

that is a function v, which possesses the integral representation

$$c(\boldsymbol{x})v(\boldsymbol{x}) = \int_{\Gamma} [g_0(\boldsymbol{y}, \boldsymbol{x})\frac{\partial v(\boldsymbol{y})}{\partial \nu} - \frac{\partial}{\partial \nu} g_0(\boldsymbol{y}, \boldsymbol{x})v(\boldsymbol{y})]\, ds_{\boldsymbol{y}} + \int_{\Omega} g_0(\boldsymbol{y}, \boldsymbol{x})u(\boldsymbol{y})\, d\Omega_{\boldsymbol{y}}$$

(8.11)

We can then represent the influence of the inertial forces (8.9) by $c(\boldsymbol{x})v(\boldsymbol{x}) +$ *boundary integrals* because Eq.(8.11) is equivalent to

$$\int_{\Omega} g_0(\boldsymbol{y}, \boldsymbol{x})u(\boldsymbol{y})\, d\Omega_{\boldsymbol{y}} = c(\boldsymbol{x})v(\boldsymbol{x}) - \int_{\Gamma} [g_0(\boldsymbol{y}, \boldsymbol{x})\frac{\partial v(\boldsymbol{y})}{\partial \nu} - \frac{\partial}{\partial \nu} g_0(\boldsymbol{y}, \boldsymbol{x})v(\boldsymbol{y})]\, ds_{\boldsymbol{y}}$$

If we substitute this into Eq.(8.9) and if we put the point $\boldsymbol{x}$ on the boundary, then Eq.(8.9) becomes

$$\dot{c}(\boldsymbol{x})u(\boldsymbol{x}) = \int_\Gamma [g_0(\boldsymbol{y}, \boldsymbol{x})\, t(\boldsymbol{y}) - \frac{\partial}{\partial \nu} g_0(\boldsymbol{y}, \boldsymbol{x})u(\boldsymbol{y})]\, ds_{\boldsymbol{y}} + \lambda\{\dot{c}(\boldsymbol{x})v(\boldsymbol{x})$$

$$- \int_\Gamma [g_0(\boldsymbol{y}, \boldsymbol{x})\frac{\partial v(\boldsymbol{y})}{\partial \nu} - \frac{\partial}{\partial \nu} g_0(\boldsymbol{y}, \boldsymbol{x})v(\boldsymbol{y})]\, ds_{\boldsymbol{y}}\} \qquad (8.12)$$

If we approximate the deflection and the traction on the boundary with functions $\varphi_i(\boldsymbol{x})$ and $\psi_i(\boldsymbol{x})$

$$u(\boldsymbol{x}) = u_i\varphi_i(\boldsymbol{x}), \qquad t(\boldsymbol{x}) = t_i\psi_i(\boldsymbol{x})$$

and the domain function with functions $d_i(\boldsymbol{x})$

$$v(\boldsymbol{x}) = v_i d_i(\boldsymbol{x}),$$

then Eq.(8.12) becomes, after collocating at K boundary points,

$$\boldsymbol{H}\boldsymbol{u} = \boldsymbol{G}\boldsymbol{t} + \lambda\boldsymbol{A}\boldsymbol{v} \qquad (8.13)$$

where

$$a_{ij} = \dot{c}(\boldsymbol{x}^i)d_j(\boldsymbol{x}^i) - \int_\Gamma [g_0(\boldsymbol{y}, \boldsymbol{x}^i)\frac{\partial d_j(\boldsymbol{y})}{\partial \nu} - \frac{\partial}{\partial \nu} g_0(\boldsymbol{y}, \boldsymbol{x}^i)d_j(\boldsymbol{y})]\, ds_{\boldsymbol{y}}$$

In agreement with Eq.(8.10) we now substitute for the boundary values of u at the collocation points

$$u(\boldsymbol{x}^i) = v_j(-\Delta d_j)(\boldsymbol{x}^i),$$

so that

$$\boldsymbol{u} = \boldsymbol{B}\boldsymbol{v}, \qquad b_{ij} = (-\Delta d_j)(\boldsymbol{x}^i) \qquad (8.14)$$

If the number of domain basis functions d_j agrees with the number of collocation points on the boundary then $\boldsymbol{B}$ is a quadratic matrix. If, in addition, the nodal values d_j are linearly independent vectors then (8.14) can be solved for the vector $\boldsymbol{v}$ and can be substituted into (8.13).

$$\boldsymbol{H}\boldsymbol{u} = \boldsymbol{G}\boldsymbol{t} + \lambda\boldsymbol{D}\boldsymbol{u}, \qquad \boldsymbol{D} = \boldsymbol{A}\boldsymbol{B}^{-1}$$

Because of the boundary conditions the vectors $\boldsymbol{u}_1$ and $\boldsymbol{t}_2$ are zero and, therefore, the vector $\boldsymbol{t}_1$ can be eliminated from the equation

$$\begin{bmatrix} H_{11} & H_{12} \\ H_{21} & H_{22} \end{bmatrix}\begin{bmatrix} \boldsymbol{o} \\ \boldsymbol{u}_2 \end{bmatrix} = \begin{bmatrix} G_{11} & G_{12} \\ G_{21} & G_{22} \end{bmatrix}\begin{bmatrix} \boldsymbol{t}_1 \\ \boldsymbol{o} \end{bmatrix} + \lambda\begin{bmatrix} D_{11} & D_{12} \\ D_{21} & D_{22} \end{bmatrix}\begin{bmatrix} \boldsymbol{o} \\ \boldsymbol{u}_2 \end{bmatrix}$$

We so obtain the algebraic eigenvalue problem

$$[(G_{11}G_{21}^{-1}H_{22} - H_{12}) - \lambda(D_{22} - D_{12})]u_2 = o$$

This technique can be applied to all similar problems, see e.g. [91].

A possible choice for the domain functions $-\Delta d_j$ are the functions, [92],

$$f_j = c - r(x^j, y) \qquad j = 1, 2 \ldots K \tag{8.15}$$

where r is the distance from the collocation point x^j on the boundary and c is the maximum distance between two points in the domain Ω. By centering the functions f_j at the collocation points the number of functions d_j and the number of collocation points is automatically the same. The constant c is chosen such that the functions f_j never become negative.

In 2-D elasticity the domain functions $d^j = \{d_1^j, d_2^j\}$ are vector-valued functions. If we characterize the single field d^j by its action $Ld^j = f^j$ (the right-hand side) where

$$f^j = \begin{bmatrix} c - r \\ 0 \end{bmatrix} \qquad or \qquad f^j = \begin{bmatrix} 0 \\ c - r \end{bmatrix}$$

and where $r = r(x^j, y)$ has the same meaning as above then the components of d^j are (in the first case = horizontal load)

$$d_1^j = (\frac{1 - 2\nu}{5 - 4\nu}c - \frac{r}{30(1 - \nu)})y_1 y_2 - \frac{9 - 10\nu}{90(1 - \nu)}r^3$$

$$d_2^j = (\frac{1 - 2\nu}{5 - 4\nu}c - \frac{r}{30(1 - \nu)})y_1 y_2 - \frac{9 - 10\nu}{90(1 - \nu)}r^3$$

Figure 8.8 shows the first four natural modes of an elastic plate with openings. This problem was first solved by Nardini, [93], and later also by Latz [92] with the functions (8.15), see Table 8.2.

Table 8.2

Frequency	Latz BEM 58 nodes	Nardini BEM 58 nodes	Nardini FEM 559 nodes
1	0.372	0.331	0.330
2	1.243	1.143	1.130
3	1.264	1.217	1.214
4	2.123	1.883	1.901
5	2.840	2.538	2.445
6	3.159	2.967	2.924
7	3.465	3.226	3.165
8	3.815	3.623	3.534

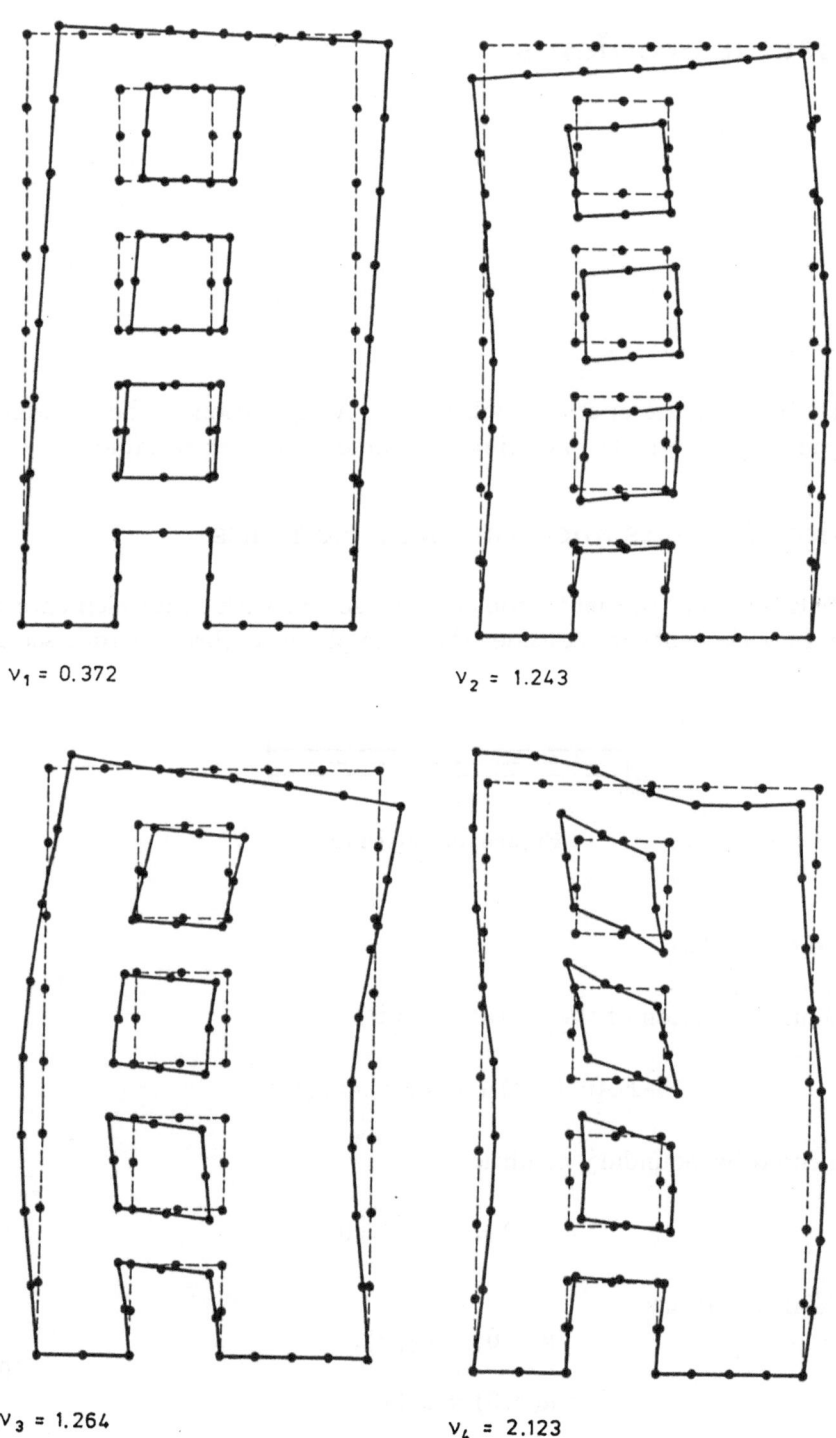

$v_1 = 0.372$

$v_2 = 1.243$

$v_3 = 1.264$

$v_4 = 2.123$

Figure 8.8 Vibrations of a plate

9 Transient problems

Transient vibrations are aperiodic vibrations. A separation of the variables is therefore no longer possible. The time t becomes a further variable.

9.1 Finite elements and boundary elements

To demonstrate how transient problems are solved with finite elements and with boundary elements we consider the example of a vibrating rod, see Fig. 9.1.

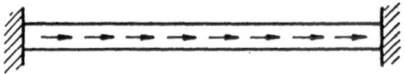

Figure 9.1 Fixed rod

9.1.1 Finite elements

The differential equation of the vibrating rod

$$-EAu''(x,t) + \mu\ddot{u}(x,t) = p(x,t)$$

is supplemented by boundary conditions

$$u(0,t) = u(l,t) = 0$$

and by initial conditions

$$u(x,0) = u_0(x),$$

$$\dot{u}(x,0) = \dot{u}_0(x) \tag{9.1}$$

In the finite element method we choose for the longitudinal displacement the approximation

$$u_h(x,t) = u_i(t)\varphi_i(x) \qquad \varphi_i(0) = \varphi_i(l) = 0$$

and determine the nodal displacements $u_i(t)$ in such a way that the residual is orthogonal to the n basis functions $\varphi_i(x)$,

$$\int_0^l \{-EAu_h''(x,t) + \mu\ddot{u}_h(x,t) - p(x,t)\}\varphi_i(x)\,dx = 0\,,$$

$$i = 1, 2, \ldots, n$$

Integration by parts transforms these n integrals into a system of n differential equations

$$M\ddot{u} + K u = f(t)$$

with constant coefficients

$$M_{ij} = \int_0^l \mu\varphi_i\varphi_j\,dx\,, \qquad K_{ij} = \int_0^l EA\varphi_i'\varphi_j'\,dx\,,$$

The vector on the right-hand side has the components

$$f_i(t) = \int_0^l p(x,t)\varphi_i(x)\,dx$$

This system and the initial conditions (9.1) formulate an initial value problem for the n unknown nodal displacements $u_i(t)$.

To solve this problem approximately we replace the second derivatives by finite differences

$$M\frac{1}{\Delta t^2}(u(t + \Delta t) - 2u(t) + u(t - \Delta t)) + Ku(t) = f(t)$$

and we so obtain the recursion formula

$$M u_{t+1} = (2M - \Delta t^2 K)\,u_t - M u_{t-1} + \Delta t^2 f(t)$$

The values at the first two time steps, $t = 0$ and $t = \Delta t$, are obtained from the initial conditions (9.1).

9.1.2 Boundary elements

The mathematical expression for the influence of the past on the presence is the *convolution*

$$\theta(x,t) = \Phi * \psi = \int\limits_0^t \Phi(x, t-\tau)\psi(x,\tau)\,d\tau \qquad t \geq 0$$

It therefore plays a prominent part in the treatment of transient problems with integral equation methods.

To a convolution the following rules apply

$$\dot{\theta} = \dot{\Phi} * \psi + \Phi(x,0)\psi\,,$$

$$\theta_{,i} = \Phi_{,i} * \psi + \Phi * \psi_{,i}$$

and, therefore, in particular

$$\ddot{u} * \hat{u} = (u * \hat{u})^{\cdot\cdot} - \dot{u}(x,0)\hat{u}(x,t) - u(x,0)\dot{\hat{u}}(x,t)$$

$$= (u * \hat{u})^{\cdot\cdot} - \dot{u}_0\hat{u} - u_0\dot{\hat{u}} \qquad\qquad (9.2)$$

The derivation of Betti's principle for static problems begins with the work integral

$$\int\limits_0^l -EAu''\hat{u}\,dx$$

In transient dynamics the analogous "work integral" is

$$\int\limits_0^l (-EAu'' + \mu\ddot{u}) * \hat{u}\,dx$$

If we apply integration by parts to this integral and consider Eq.(9.2) then we obtain the first identity of the rod under dynamic loads

$$G(u,\hat{u}) = \int\limits_0^l (-EAu'' + \mu\ddot{u}) * \hat{u}\,dx + [N * \hat{u}]_0^l$$

$$- \int\limits_0^l \mu[\dot{u}(x,t)\hat{u}(x,0) - \dot{u}(x,0)\hat{u}(x,t)]\,dx$$

$$-\int_0^l \left(\frac{N * \hat{N}}{EA} - \mu\dot{u} * \hat{u}\right) dx = 0$$

and therefore also the second identity, Betti's principle,

$$B(\hat{u}, u) = G(\hat{u}, u) - G(u, \hat{u}) = 0$$

The fundamental solution of a rod has now a singularity in space and time. Formally it satisfies the differential equation

$$-EAg_0''(y, x; t, \tau) + \mu\ddot{g}_0(y, x; t, \tau) = \delta_0(y - x)\delta_0(t - \tau)$$

It is the response of an infinite rod to a concentrated force $\hat{P} = 1$ which acts at the time mark t and at the point x. If we formulate the second identity with this function and a regular function u then we obtain an influence function for u and thus automatically the coupling condition between the boundary data of u. Formally this condition differs from the coupling condition of the static case only by the two stars

$$\boldsymbol{H} * \boldsymbol{u} = \boldsymbol{G} * \boldsymbol{f} + \boldsymbol{d}(t)$$

which indicate the convolution. Explicitly these conditions read

$$\int_0^t H_{ij}(t - \tau)u_j(\tau) \, d\tau = \int_0^t G_{ij}(t - \tau)f_j(\tau) \, d\tau + d_i(t), \qquad (9.3)$$

$$i = 1, 2$$

The vector $\boldsymbol{d}(t) = \{d_i(t)\}$ represents the influence of the distributed forces and the initial conditions.

Hence, the influence coefficients, the entries of the matrices $\boldsymbol{H}$ and $\boldsymbol{G}$ as the nodal variables are functions of time. If we now want to determine the unknown force and displacement terms on the boundary then it no longer suffices to invert a matrix but we, now, must solve a system of two *Volterra integral equations*.

Integral equations where the parameter x, or as in this case, the time t, is one of the integration limits are called Volterra integral equations. Hence, to satisfy a coupling condition in time means to be consistent with one's own history, or, in mathematical terms: to satisfy a Volterra integral equation.

Let us assume that we want to plot the behavior of the rod in Fig. 9.1 during the first second. The two end displacements $u_j(t)$ are given, they are zero. The two unknown end actions $f_1(t)$ and $f_2(t)$ are to be determined in such a way that they satisfy, at all time marks $0 < t < 1$, the coupling condition

(9.3). To reduce this infinite system of equations to finitely many we subdivide the time interval into ten time steps Δt and we correspondingly approximate the functions f_j by step functions f_{hj}

$$f_1(t) \simeq f_{h1}(t), \qquad f_2(t) \simeq f_{h2}(t),$$

whose 2×10 nodal values we determine in such a way that the two coupling conditions (9.3) are satisfied at the mid-points of the ten time steps,

$$\int\limits_0^{t_k} G_{ij}(t_k - \tau)f_{hj}(\tau)\,d\tau = -d_i(t^k), \qquad t^k = (k - 0.5)\Delta t, \quad k = 1, 2, \ldots, 10$$

The boundary values thus approximated we then substitute into the influence function for u and we obtain the displacement at any point x at any time mark t within the interval $0 < t < 1$.

9.1.3 Summary

In both methods the nodal variables are functions of time. In the finite element method we determine their values by solving a system of n differential equations by proceeding in time; in the boundary element method by solving two Volterra integral equations. The discretization of these two integral equations leads to a linear system of algebraic equations of size

$$n = \text{number of time steps } \Delta t \times \text{number of nodal values}$$

$$= 10 \times 2 = 20$$

In multidimensional problems the number n is considerably higher. If we subdivide the boundary of a membrane into, say, 20 linear elements then this corresponds to 20 unknown nodal variables and ten time steps would therefore lead to a system of size 200×200.

These introductory notes may suffice to give the reader an idea how one-dimensional problems are solved with boundary element methods. In the following we will concentrate on higher dimensional problems. Our first example is a vibrating membrane. We have already discussed in chapter 8 the case of harmonic oscillations (Helmholtz equation). Now we want to focus on the full equation, the wave equation.

9.2 The wave equation

To the wave equation

$$-c^2 \Delta u(\boldsymbol{x}, t) + \ddot{u}(\boldsymbol{x}, t) = p(\boldsymbol{x}, t)$$

belong the identities

$$G(u, \hat{u}) = \int_\Omega (-c^2 \Delta u + \ddot{u}) * \hat{u} \, d\Omega + \int_\Gamma c^2 \frac{\partial u}{\partial n} * \hat{u} \, ds$$

$$- c^2 \int_\Omega \nabla u * \nabla \hat{u} \, d\Omega - \int_\Omega (u * \hat{u})^{\cdot\cdot} d\Omega$$

$$+ \int_\Omega [\dot{u}_0 \hat{u} + u_0 \hat{\dot{u}}] \, d\Omega = 0$$

and

$$B(\hat{u}, u) = G(\hat{u}, u) - G(u, \hat{u}) = 0$$

Next, we consider a loadcase where a concentrated force $\hat{P}$ acts a point $\boldsymbol{x}$. We assume that the magnitude of the force changes with time according to a given function $f(\tau)$ of which we only require that it has a "quiet past"

$$f(\tau) = 0, \qquad \tau \leq 0, \qquad \dot{f}(0) = 0,$$

The corresponding solution of the wave equation is

$$\hat{u}(\boldsymbol{y}, \boldsymbol{x}, \tau) = \frac{1}{c^2 4\pi r} f(\tau - \frac{r}{c})$$

Let us assume that $u = u(\boldsymbol{y}, \tau)$ is a smooth solution of the differential equation

$$-c^2 \Delta u + \ddot{u} = p(\boldsymbol{y}, \tau)$$

It then follows

$$\lim_{\varepsilon \to 0} B(\hat{u}, u)_{\Omega_\varepsilon} = c(\boldsymbol{x}) f(\tau) * u(\boldsymbol{x}, \tau) + \int_\Gamma c^2 \frac{\partial \hat{u}}{\partial \nu} * u \, ds_{\boldsymbol{y}}$$

$$- \int_\Omega [\hat{u}(\boldsymbol{y}, \boldsymbol{x}, t) \dot{u}(\boldsymbol{y}, 0) + \hat{\dot{u}}(\boldsymbol{y}, \boldsymbol{x}, t) u(\boldsymbol{y}, 0)] \, d\Omega_{\boldsymbol{y}}$$

$$- \int_\Gamma \hat{u}(\boldsymbol{y}, \boldsymbol{x}, \tau) * c^2 \frac{\partial u}{\partial \nu}(\boldsymbol{y}, \tau) \, ds_{\boldsymbol{y}}$$

$$- \int_\Omega \hat{u}(\boldsymbol{x}, \boldsymbol{y}, \tau) * p(\boldsymbol{y}, \tau) \, d\Omega_{\boldsymbol{y}} = 0$$

The function $c(\boldsymbol{x})$ is identical with the characteristic function of the Laplacian operator.

If the function $f(\tau)$ converges uniformly to the Dirac-function $\delta(t-\tau)$, so that, in the limit its action on an arbitrary smooth function q becomes

$$f * q = \int\limits_0^t f(t-\tau)q(\tau)\,d\tau = q(t)$$

and therefore as well

$$f(\tau - \frac{r}{c}) * q(\tau) = \int\limits_0^t f(t-\tau-\frac{r}{c})q(t)\,d\tau = q(t-\frac{r}{c}),$$

then follows

$$c(\boldsymbol{x})u(\boldsymbol{x},t) = \int\limits_\Gamma [\hat{u}(\boldsymbol{y},\boldsymbol{x},\tau) * c^2\frac{\partial u}{\partial \nu}(\boldsymbol{y},\tau) - c^2\frac{\partial \hat{u}}{\partial \nu} * u(\boldsymbol{y},\tau)]\,ds\boldsymbol{y}$$

$$+ \int\limits_\Omega [\hat{u}(\boldsymbol{y},\boldsymbol{x},t)\dot{u}(\boldsymbol{y},0) + \dot{\hat{u}}(\boldsymbol{y},\boldsymbol{x},t)u(\boldsymbol{y},0)]\,d\Omega\boldsymbol{y}$$

$$+ \int\limits_\Omega \hat{u}(\boldsymbol{y},\boldsymbol{x},\tau) * p(\boldsymbol{y},\tau)\,d\Omega\boldsymbol{y} \qquad (9.4)$$

We now have

$$\frac{\partial \hat{u}}{\partial \nu} = \frac{1}{4\pi c^2}[-\frac{r_\nu}{r^2}\delta(t-\frac{r}{c}-\tau) - \frac{r_\nu}{cr}\dot{\delta}(t-\frac{r}{c}-\tau)]$$

and because of

$$\int\limits_0^t \dot{\delta}(t-\tau)q(\tau)\,d\tau = \dot{q}(t)$$

therefore as well

$$c^2\frac{\partial \hat{u}}{\partial \nu} * u = -\frac{r_\nu}{4\pi r^2}[u(\boldsymbol{y},t-\frac{r}{c}) + \frac{r}{c}\dot{u}(\boldsymbol{y},t-\frac{r}{c})]$$

Furthermore we have

$$\hat{u}(\boldsymbol{y},\boldsymbol{x},\tau) * c^2 \frac{\partial u}{\partial \nu}(\boldsymbol{y},\tau) = \frac{1}{4\pi r} \frac{\partial u}{\partial \nu}\left(\boldsymbol{y},t-\frac{r}{c}\right)$$

Collecting our results we obtain

$$c(\boldsymbol{x})u(\boldsymbol{x},t) = \int_\Gamma \frac{r_\nu}{4\pi r^2}\left[u\left(\boldsymbol{y},t-\frac{r}{c}\right)+\frac{r}{c}\dot{u}\left(\boldsymbol{y},t-\frac{r}{c}\right)\right]ds\boldsymbol{y}$$

$$+\int_\Gamma \frac{1}{4\pi r}\frac{\partial u}{\partial \nu}\left(\boldsymbol{y},t-\frac{r}{c}\right)ds\boldsymbol{y} + \int_\Omega \frac{1}{4\pi c^2 r}\delta\left(t-\frac{r}{c}\right)\dot{u}(\boldsymbol{y},0)\,d\Omega\boldsymbol{y}$$

$$+\int_\Omega \frac{1}{4\pi c^2 r}\dot{\delta}\left(t-\frac{r}{c}\right)u(\boldsymbol{y},0)\,d\Omega\boldsymbol{y}$$

$$+\int_\Omega \hat{u}(\boldsymbol{y},\boldsymbol{x},\tau) * p(\boldsymbol{y},\tau)\,d\Omega\boldsymbol{y} \qquad (9.5)$$

The integrals of the initial data are equivalent to

$$\int_\Omega \frac{1}{4\pi c^2 r}\delta\left(t-\frac{r}{c}\right)\dot{u}(\boldsymbol{y},0)\,d\Omega\boldsymbol{y} = t\,M_{\boldsymbol{x};ct}[\dot{u}(\boldsymbol{y},0)]$$

and

$$\int_\Omega \frac{1}{4\pi c^2 r}\dot{\delta}\left(t-\frac{r}{c}\right)u(\boldsymbol{y},0)\,d\Omega\boldsymbol{y} = \frac{\partial}{\partial t}(t\,M_{\boldsymbol{x};ct}[u(\boldsymbol{y},0)])\,,$$

where

$$M_{\boldsymbol{x};ct}[u] = \frac{1}{4\pi}\int_0^{2\pi}\int_0^\pi u(\boldsymbol{x}+ct\nabla_{\boldsymbol{y}}r,0)\sin\vartheta\,d\vartheta\,d\varphi\,,\qquad r=|\boldsymbol{y}-\boldsymbol{x}|\,,$$

is the average value of $u(\boldsymbol{y},0)$ on a sphere with radius ct and centre at $\boldsymbol{x}$, see [94, p. 375].

Equation (9.5) is the influence function for the three-dimensional wave equation $u(\boldsymbol{x},t)$. The "boundary" where we formulate the coupling condition consists of the surface Γ and the time axis $t \geq 0$. In two-dimensional problems, if we consider for example the vibrations of a membrane, the boundary can be imagined as a hose which is fixed to the boundary Γ of the membrane and which extends up to the time mark T, the endpoint of the time interval, see Fig. 9.2. Remark: The coupling condition (9.5) has the the same form as that of the rod

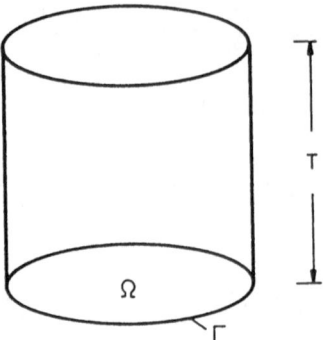

Figure 9.2 The boundary on which we formulate the coupling conditions

$$H * u = G * f + d(t)$$

The transition from Eq.(9.4) to Eq.(9.5) only obscured the underlying convolution.

9.3 The heat equation

The differential equation

$$-\Delta u(\boldsymbol{x}, t) + \dot{u}(\boldsymbol{x}, t) = p(\boldsymbol{x}, t) \tag{9.6}$$

governs the temperature distribution in a homogeneous medium (thermal diffusity $= 1$).

The associated identities are

$$G(u, \hat{u}) = \int_{\Omega} (-\Delta u + \dot{u}) * \hat{u} d\Omega + \int_{\Gamma} \frac{\partial u}{\partial n} * u ds$$

$$- \int_{\Omega} [\nabla u * \nabla \hat{u} + (u * \hat{u})^{\cdot}] d\Omega + \int_{\Omega} u(\boldsymbol{x}, 0)\hat{u}(\boldsymbol{x}, t) d\Omega_y = 0$$

and

$$B(u, \hat{u}) = G(u, \hat{u}) - G(\hat{u}, u) = 0$$

If we place a heat-source at a point $\boldsymbol{x}$ whose magnitude varies in time according to a smooth function $f(t), t > 0$ then the temperature $\hat{u}$ at a point $\boldsymbol{y}$ of the n-dimensional infinite medium at time mark t will be

$$\hat{u}(\boldsymbol{x}, \boldsymbol{y}, t) = \int\limits_0^t \frac{f(\tau)}{[4\pi(t-\tau)^{n/2}]} e^{-r^2/(4(t-\tau))} \tag{9.7}$$

If the function $f(t)$ tends to the Dirac-function $\delta(t-\tau)$ then the solution (9.7) becomes the fundamental solution

$$g_0(\boldsymbol{y}, \boldsymbol{x}) = \frac{1}{[4\pi t]^{n/2}} e^{-r^2/(4t)}$$

If we formulate with Eq.(9.7) and a smooth solution $u(\boldsymbol{x}, t)$ of Eq.(9.6) the limit

$$\lim_{\varepsilon \to 0} B(\hat{u}, u)_{\Omega_\varepsilon} = 0$$

and we let then tend f to the Dirac-function then we obtain the influence function

$$c(\boldsymbol{x}) u(\boldsymbol{x}, t) = \int\limits_\Gamma [\frac{\partial g_0}{\partial n} * u - g_0 * \frac{\partial u}{\partial n}] ds_{\boldsymbol{y}} + \int\limits_\Omega g_0 * p d\Omega_{\boldsymbol{y}}$$

$$+ \int\limits_\Omega [g_0(\boldsymbol{x}, \boldsymbol{y}, 0) u(\boldsymbol{y}, t) - g_0(\boldsymbol{x}, \boldsymbol{y}, t) u(\boldsymbol{y}, 0)] d\Omega_{\boldsymbol{y}}$$

where $c(\boldsymbol{x})$ is the characteristic function of the Laplacian. This equation is the basis of the boundary element method in transient heat conduction.

9.4 Dynamic displacement fields

Associated with the system of differential equations

$$-c_2^2 \Delta \boldsymbol{u} - (c_1^2 - c_2^2)\nabla \mathrm{div}\, \boldsymbol{u} + \ddot{\boldsymbol{u}} = \boldsymbol{p}(\boldsymbol{y}, \tau) \tag{9.8}$$

are the identities

$$G(\boldsymbol{u}, \hat{\boldsymbol{u}}) = \int\limits_\Omega (-c_2^2 \Delta \boldsymbol{u} - (c_1^2 - c_2^2)\nabla \mathrm{div}\, \boldsymbol{u} + \ddot{\boldsymbol{u}}) * \hat{\boldsymbol{u}}\, d\Omega$$

$$+ \int\limits_\Gamma \boldsymbol{\tau}(\boldsymbol{u}) * \hat{\boldsymbol{u}}\, ds - \int\limits_\Omega (\boldsymbol{u} * \hat{\boldsymbol{u}})^{\cdot\cdot} d\Omega$$

$$+ \int\limits_\Omega [\dot{\boldsymbol{u}}_0 \cdot \hat{\boldsymbol{u}}(\boldsymbol{y}, t) + \boldsymbol{u}_0 \cdot \dot{\hat{\boldsymbol{u}}}(\boldsymbol{y}, t)]\, d\Omega - E^*(\boldsymbol{u}, \hat{\boldsymbol{u}}) = 0$$

and

$$B(\hat{u}, u) = G(\hat{u}, u) - G(u, \hat{u}) = 0,$$

where

$$E^*(u, \hat{u}) = \int_\Omega \sigma_{ij} * \hat{\varepsilon}_{ij} \, d\Omega = \int_\Omega S(u) * E(\hat{u}) \, d\Omega = \int_\Omega E(u) * S(\hat{u}) \, d\Omega$$

denotes the convolution of the energy product between the stress and strain tensors $S(u)$ and $E(\hat{u})$.

Let us assume that a concentrated force e_i (unit vector) acts at a point x of the elastic continuum and that the magnitude of this force changes in time according to some given function

$$f(\tau) \in C^2[0, +\infty), \qquad f(\tau) = 0 \quad \text{if } \tau \leq 0, \qquad \dot{f}(0) = 0,$$

In this case the system (9.8) becomes

$$-c_2^2 \Delta u - (c_1^2 - c_2^2) \nabla \operatorname{div} u + \ddot{u} = \delta_0(x - y) e_i f(\tau),$$

and the response of the elastic continuum is the displacement field, see [53, p. 239]

$$g^i(y, x, \tau; f) = \frac{1}{4\pi\rho r} [(3\nabla r \otimes \nabla r - I) \int_{1/c_1}^{1/c_2} \lambda f(\tau - \lambda r) \, d\lambda$$

$$+ \nabla r \otimes \nabla r \left(\frac{1}{c_1^2} f(\tau - \frac{r}{c_1}) - \frac{1}{c_2^2} f(\tau - \frac{r}{c_2}) \right)$$

$$+ \frac{1}{c_2^2} f(\tau - \frac{r}{c_2}) I] e_i, \qquad (9.9)$$

$$\nabla r = \{r_{,i}\}$$

Associated with this field is the stress tensor, see [53, p. 239],

$$S^i(y, x, \tau; f) = \frac{1}{4\pi r^2}[(\nabla r \cdot e_i)f_1 \nabla r \otimes \nabla r - (\nabla r \cdot e_i)f_2 \, I - \nabla r \otimes e_i f_3$$

$$- e_i \otimes \nabla r f_3], \tag{9.10}$$

where[1]

$$f_1(\tau, r) = 5f_0(\tau, r) + 12[f(\tau - \frac{r}{c_2}) - qf(\tau - \frac{r}{c_1})]$$

$$+ \frac{2r}{c_2}[\dot{f}(\tau - \frac{r}{c_2}) - q\dot{f}(\tau - \frac{r}{c_1})],$$

$$f_2(\tau, r) = f_0(\tau, r) + 2f(\tau - \frac{r}{c_2}) + (1 - 4q)f(\tau - \frac{r}{c_1})$$

$$+ \frac{r}{c_1}(1 - 2q)\dot{f}(\tau - \frac{r}{c_1}),$$

$$f_3(\tau, r) = f_0(\tau, r) + 3f(\tau - \frac{r}{c_2}) - 2qf(\tau - \frac{r}{c_1}) + \frac{r}{c_2}\dot{f}(\tau - \frac{r}{c_1}),$$

$$f_0(\tau, r) = -6c_2^2 \int_{1/c_1}^{1/c_2} \lambda f(\tau - \lambda r) \, d\lambda,$$

$$q = c_2^2/c_1^2$$

We denote the components $j = 1, 2, 3$ of the field g^i by

$$u_{ij}(y, x, \tau; f) \tag{9.11}$$

and the components $j = 1, 2, 3$ of the stress tensor $S^i \nu$ by

$$t_{ij}(y, x, \tau; f) \tag{9.12}$$

The solution g^i is called *Stokes state of quiescent past.*

To obtain an influence function for solutions, $u \in C^2(\bar{\Omega}) \times C[0, +\infty)$, of the system (9.8) we proceed as follows:

1) We first formulate the second identity for the pair g^i, u in the punctured domain

$$\Omega_\varepsilon(x) = \Omega - N_\varepsilon(x)$$

and we then let the radius ε tend to zero.

2) In a second step we let the function $f(\tau)$ tend to the Dirac-function $\delta(t - \tau)$.

---

[1] The factor 5 in f_1 is missing in [53].

The first step renders

$$C(\boldsymbol{x})\boldsymbol{u}(\boldsymbol{x},\tau) * f(\tau) + \int_\Gamma [\boldsymbol{T}(\boldsymbol{y},\boldsymbol{x},\tau) * \boldsymbol{u}(\boldsymbol{y},\tau) - \boldsymbol{U}(\boldsymbol{y},\boldsymbol{x},\tau) * \boldsymbol{t}(\boldsymbol{y},\tau)] ds_{\boldsymbol{y}}$$

$$- \int_\Omega \boldsymbol{U}(\boldsymbol{y},\boldsymbol{x},\tau) * \boldsymbol{p}(\boldsymbol{y},\tau) d\Omega_{\boldsymbol{y}}$$

$$- \int_\Omega [\boldsymbol{U}(\boldsymbol{y},\boldsymbol{x},t)\dot{\boldsymbol{u}}_0 + \dot{\boldsymbol{U}}(\boldsymbol{y},\boldsymbol{x},t)\boldsymbol{u}_0] d\Omega_{\boldsymbol{y}} = 0 , \tag{9.13}$$

where $C(\boldsymbol{x})$ is the matrix of the Somigliana identity, see [95 p. 348].

In the second step the function $f(\tau)$ tends to the Dirac-function $\delta(t-\tau)$ so that its action on an arbitrary, smooth function q becomes

$$f * q = \int_0^t f(t-\tau)q(\tau) d\tau = q(t)$$

and we, therefore, have as well

$$f(\tau - \frac{r}{c}) * q(\tau) = \int_0^t f(t-\tau-\frac{r}{c})q(\tau) d\tau = q(t-\frac{r}{c})$$

Hence, Eq.(9.13) becomes

$$C(\boldsymbol{x})\boldsymbol{u}(\boldsymbol{x},t) + \int_\Gamma [\boldsymbol{T}(\boldsymbol{y},\boldsymbol{x},\tau) * \boldsymbol{u}(\boldsymbol{y},\tau) - \boldsymbol{U}(\boldsymbol{y},\boldsymbol{x},\tau) * \boldsymbol{t}(\boldsymbol{y},\tau)] ds_{\boldsymbol{y}}$$

$$- \int_\Omega \boldsymbol{U}(\boldsymbol{y},\boldsymbol{x},\tau) * \boldsymbol{p}(\boldsymbol{y},\tau) d\Omega_{\boldsymbol{y}}$$

$$+ \int_\Omega [\boldsymbol{U}(\boldsymbol{y},\boldsymbol{x},t)\dot{\boldsymbol{u}}_0(\boldsymbol{y},t) - \dot{\boldsymbol{U}}(\boldsymbol{y},\boldsymbol{x},t)\boldsymbol{u}_0(\boldsymbol{y},t)] d\Omega_{\boldsymbol{y}} = 0 , \tag{9.14}$$

where, see [94 p. 405], the convolutions can be expressed as

$$U_{ij} * t_j = u_{ij}(\boldsymbol{y},\boldsymbol{x},t; t_j(\boldsymbol{y},t)) ,$$

$$T_{ij} * u_j = t_{ij}(\boldsymbol{y},\boldsymbol{x},t; u_j(\boldsymbol{y},t)) ,$$

$$U_{ij} * p_j = u_{ij}(\boldsymbol{y},\boldsymbol{x},t; p_j(\boldsymbol{y},t)) ,$$

This means that we simply have to replace the function f in Eqs.(9.11) and (9.12) by $t_j(\boldsymbol{y},t)$, $u_j(\boldsymbol{y},t)$ and $p_j(\boldsymbol{y},t)$, respectively. And, as in the case of the function f (*quiescent past*), we assign the value zero to the functions t_j, u_j and p_j for negative values of time. This results in

$$U_{ij} * t_j = u_{ij}(\boldsymbol{y}, \boldsymbol{x}, t; t_j)$$

$$= \frac{1}{4\pi\rho r} \{(3r_{,i}\, r_{,j} - \delta_{ij}) \int_{1/c_1}^{1/c_2} \lambda t_j(\boldsymbol{x}, t - \lambda r)\, d\lambda$$

$$+ r_{,i}\, r_{,j}\, [\frac{1}{c_1^2} t_j(\boldsymbol{y}, t - \frac{r}{c_1}) - \frac{1}{c_2^2} t_j(\boldsymbol{y} - \frac{r}{c_2})] + \frac{\delta_{ij}}{c_2^2} t_j(\boldsymbol{x}, t - \frac{r}{c_2})\}$$

$$= \frac{1}{4\pi\rho r} \{(3r_{,i}\, r_{,j} - \delta_{ij})[H(t - \frac{r}{c_1}) \int_{r/c_1}^{t} \tau t_j(\boldsymbol{y}, t - \tau)\, d\tau$$

$$- H(t - \frac{r}{c_2}) \int_{r/c_2}^{t} \tau t_j(\boldsymbol{y}, t - \tau)\, d\tau] \frac{1}{r^2}$$

$$+ r_{,i}\, r_{,j}\, [\frac{1}{c_1^2} H(t - \frac{r}{c_1}) t_j(\boldsymbol{y}, t - \frac{r}{c_1}) - \frac{1}{c_2^2} H(t - \frac{r}{c_2}) t_j(\boldsymbol{y}, t - \frac{r}{c_2})]$$

$$+ \frac{\delta_{ij}}{c_2^2} H(t - \frac{r}{c_2}) t_j(\boldsymbol{y}, t - \frac{r}{c_2})\},$$

where

$$H(x) = \begin{cases} 1, & 0 < x, \\ 0, & x < 0, \end{cases}$$

is the *Heaviside function*.

From a physical point of view the appearance of the Heaviside function is a consequence of the finite propagation speed of a disturbance. To reach the points with the distance r the wave needs the time $t_r = r/c_i$ and as long as t is smaller,

$$t - t_r = t - \frac{r}{c_i} < 0,$$

all is quiet at the point, so that only those points $\boldsymbol{y}$ influence the point $\boldsymbol{x}$, whose distance r satisfies the inequality

$$t - \frac{r}{c_i} > 0$$

The domain integral which represents the influence of the initial data can, as in the case of the wave equation, be modified further, see [94, p. 407].

9.5 Numerical treatment

We consider a rigid, massless foundation that is acted upon by an impulsive force which it transmits to the elastic half space. We are investigating the behavior of the half-space in the first second. Because volume forces are absent and because the initial data are zero as well, the coupling conditions between the boundary values of the half-space simplify to

$$H * u = G * f$$

where we denoted the vector of the force terms by f, to distinguish it from the time t.

To keep things simple let us assume that we approximate the boundary data in space and time by piecewise constant functions. We subdivide the time interval into ten time steps Δt and the surface of the half-space into 30 rectangular elements. We do not cover the whole surface with boundary elements but only the immediate vicinity of the foundation and the foundation itself. This is admissible because the boundary data fade off rapidly.

Hence, the vector u of the element displacements consists of ten vectors

$$u^1, u^2, \ldots, u^{10}$$

which represent the displacement of the surface at the centre points of the elements at time marks $t^i = i\Delta t$. Each vector u^i consists of 30×3 components $u_j^i(x^k)$, the three displacements at the 30 nodes x^k.

Similarly the vector f of forces is structured. But because only the soil under the foundation is acted upon by forces and this only at the time step $t^1 = 1\Delta t$ of all the vectors f^i, only the vector f^1 is not zero and only in those components which refer to points under the foundation. Hence, the discrete coupling conditions have a form as in Fig. 9.3. The cascade-like distribution of

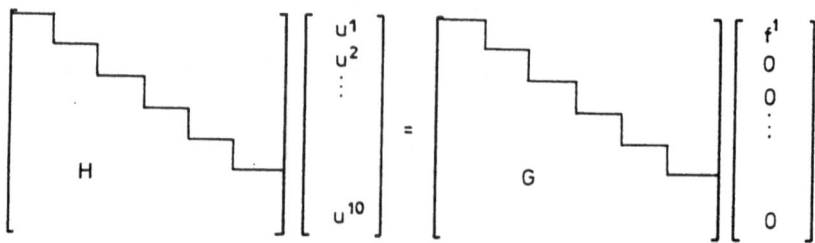

Figure 9.3 The cascade-like shape of the equations

non-zero entries reflects the finite speed of propogation of waves. We assumed that after five time steps the disturbance has reached all elements.

This system of equations is solved as before and then the displacements and stresses in the interior are obtained with the influence function.

9.6 Fourier- and Laplace transforms

We can eliminate from Eq.(9.8) the time variable either by a *Fourier transformation*

$$\tilde{u}(\boldsymbol{x}, \omega) = \frac{1}{2\pi} \int\limits_{-\infty}^{+\infty} u(\boldsymbol{x}, t) e^{i\omega t} \, dt$$

or a *Laplace transformation*

$$\tilde{u}(\boldsymbol{x}, s) = \int\limits_{0}^{\infty} u(\boldsymbol{x}, t) e^{st} \, dt$$

In this respect we speak of the *transition from the time domain into the frequency domain.*

The Fourier transformation leads to the following system of differential equations
$$-c_2^2 \Delta \tilde{\boldsymbol{u}} - (c_1^2 - c_2^2) \nabla \text{div} \, \tilde{\boldsymbol{u}} - \omega^2 \tilde{\boldsymbol{u}} = \tilde{p}(\boldsymbol{x}, \omega)$$

and the Laplace transformation to the system

$$-c_2^2 \Delta \tilde{\boldsymbol{u}} - (c_1^2 - c_2^2) \nabla \text{div} \, \tilde{\boldsymbol{u}} + s^2 \tilde{\boldsymbol{u}} = \tilde{p} + s \, u_0 + \dot{u}_0 \, , \qquad (9.15)$$

which is identical to the first system if we substitute for $s = i\,\omega$. The fundamental solutions are the fundamental solutions of the harmonic problem, see chapter 8, and we, therefore, can solve the boundary value problem of the Laplace transform:

Find a function $\tilde{\boldsymbol{u}}(\boldsymbol{x}, s)$ which satisfies Eq.(9.15) and the boundary conditions
$$\tilde{u}_i = \bar{u}_i(\boldsymbol{x}, s) \qquad \text{on } \Gamma_1 \, , \qquad \tilde{t}_i = \bar{t}_i(\boldsymbol{x}, s) \qquad \text{on } \Gamma_2$$

by the boundary element method.

The only problem is that we must solve the boundary value problem for a whole range of complex-valued parameters s and then apply a (numerical) inverse transformation from the frequency domain back into the time domain

$$u(\boldsymbol{x},t) = \frac{1}{2\pi i} \int\limits_{\beta-i\infty}^{\beta+i\infty} \tilde{u}(\boldsymbol{x},s) e^{st}\, ds$$

to obtain the original solution.

Manolis, [96], solved a plane, dynamical problem by all three methods

    a) Volterra integral equations   "time solution",

    b) Fourier transformation   "Fourier solution",

    c) Laplace transformation   "Laplace solution",

and summarizes his results as follows:

i) For comparable accuracy, the Laplace solution is more economical than either the Fourier solution or time solution by a factor of 1.7 and 5.0, respectively. However, the time solution gives a better picture for very early times up to 1 full transit time. Integral transform solutions tend to reach a plateau corresponding to the solution of the static equivalent of the problem, while the time solution tends to diverge after a very large number of time steps.

ii) The problem with the time solution is that the time step must be kept to a small fraction of the total time interval of interest, which necessitates the use of a relatively large number of times steps. Unfortunately, the solution algorithm becomes more involved at later time steps because the information from all earlier time steps must be used. The basic problem with integral transform solutions is that the solution procedure is performed in complex arithmetic, which requires at least twice the number of operations than a comparable procedure done in real arithmetic.

iii) Some further advantages of the integral transformation is that the steady-state becomes a special case of the Fourier domain solution not requiring a numerical inversion from the transformed domain and that viscoelastic material behavior can be recovered from the Laplace domain solution by a simple change in the elastic constants λ and μ according to the correspondence principle.

9.7 Dynamic stiffness matrices

In the static case the discrete coupling conditions are algebraic equations

$$\boldsymbol{H}\boldsymbol{u} = \boldsymbol{G}\boldsymbol{f}$$

while they are Volterra integral equations

$$\boldsymbol{H}*\boldsymbol{u} = \boldsymbol{G}*\boldsymbol{f} \tag{9.16}$$

in transient analysis. The nodal displacements $u_i = u_i(t)$ and forces $f_i = f_i(t)$ are functions of time.

Theoretically we could, by multiplying Eq.(9.16) with the inverse operator G^{-1} and with the Grams' matrix F, also derive a dynamical stiffness matrix

$$FG^{-1}H * u = Ff,$$

but practically this is impossible because we do not know the inverse operator G^{-1}.

But we know that to Eq.(9.16) corresponds in the frequency domain the equation

$$\tilde{H}\tilde{u} = \tilde{G}\tilde{f},$$

because the *convolution in the time domain* corresponds to the *product in the frequency domain*

$$\mathcal{L}^{-1}\{\tilde{H}\tilde{u}\} = H * u = G * f = \mathcal{L}^{-1}\{\tilde{G}\tilde{f}\},$$

(Laplace transform),

i.e. in the frequency domain we are back to our standard matrix algebra and therefore can formulate in the frequency domain dynamic stiffness matrices. The transition to the frequency domain has the further advantage that in the case of harmonic oscillations the results can be applied directly, without any inverse transformation. Alarcon et al., [97], and Ottenstreuer, [98], formulated in the frequency domain dynamic stiffness matrices for rigid, massless foundations on the elastic halfspace. In the following we quote from [98].

In the frequency domain the coupling condition between the Betti data of the halfspace reads, see chapter 8,

$$\frac{1}{2}u(x) + \int_\Gamma T(y, x)u(y)\,ds_y = \int_\Gamma U(y, x)t(y)\,ds_y$$

(We drop the tilde ~ here and in the following.)

Because the normal derivative r_ν of the distance r between two points on the surface of the halfspace is zero, the matrix T simplifies to

$$\begin{bmatrix} 0 & 0 & * \\ 0 & 0 & * \\ * & * & 0 \end{bmatrix}$$

For a first approximation we can also delete the remaining off-diagonal terms if we neglect the influence of the shear stresses,

$$T_{13} = T_{31} = T_{23} = T_{32} = 0 \tag{9.17}$$

With respect to the calculation of the influence matrices U and T this means that the horizontal and vertical quantities are uncoupled and, therefore, the discrete coupling conditions become

$$\frac{1}{2}u = G\,t$$

If we denote with the index i the elements under the foundation and with the index a the elements on the free surface

$$\frac{1}{2}\begin{bmatrix} u^i \\ u^a \end{bmatrix} = \begin{bmatrix} G^{ii} & G^{ia} \\ G^{ai} & G^{aa} \end{bmatrix} \begin{bmatrix} t^i \\ t^a \end{bmatrix},$$

then we obtain, because the traction vectors t^a are zero,

$$\frac{1}{2}u^i = G^{ii}t^i$$

Hence, there are no contributions from regions outside the foundation and we therefore must only discretize the foundation itself. According to our assumptions the foundation is rigid and it has only six degrees-of-freedom

$$u = \{u_1, u_2, u_3, \varphi_1, \varphi_2, \varphi_3\}^T$$

Hence, the vector of the element displacements

$$u^i = \{u^1, u^2, u^3, \dots, u^n\}^T, \qquad u^k = \{u_1(x^k), u_2(x^k), u_3(x^k)\},$$

(the vectors inside the curly braces are the displacements in the three directions at the centre point x^k of the element (constant elements)), is coupled by a matrix R with u

$$u^i = R\,u$$

Correspondingly the conjugated terms, the equivalent nodal forces Ft, are multiplied with the transposed matrix R^T so that we obtain

$$R^T F G^{-1} R u = R^T F t = p, \qquad G = \frac{1}{2}G^{ii}$$

Because the basis functions are piecewise constant and non-zero only on that element whose index i corresponds to the index i of the basis function, Gram's matrix F is a diagonal matrix whose entries are the areas of the single elements. The vector p of equivalent nodal forces has the six components

$$K_1 = \sum_i \int_{\Gamma_i} t_1 \, ds \,, \qquad M_1 = \sum_i \int_{\Gamma_i} t_3 r_2 \, ds \,,$$

$$K_2 = \sum_i \int_{\Gamma_i} t_2 \, ds \,, \qquad M_2 = -\sum_i \int_{\Gamma_i} t_3 r_1 \, ds \,,$$

$$K_3 = \sum_i \int_{\Gamma_i} t_3 \, ds \,, \qquad M_3 = \sum_i \int_{\Gamma_i} (t_2 r_1 - t_1 r_2) \, ds \,,$$

(here the terms r_j denote the distance with respect to the axis x_j and summation is done over the elements i),

and the (6×6)-matrix

$$K = R^T F G^{-1} R$$

is the complex-valued dynamic stiffness matrix which, for any given frequency ω, relates the rigid-body motions of the foundation to the resulting forces.

In Fig. 9.4 the vertical flexibilities f_{zz} (the elements of the flexibility matrix K^{-1}) of the solid are plotted versus a dimensionless frequency

$$a_0 = \frac{\omega b}{c_2} \,, \qquad b = \text{length of the square foundation} \,,$$

The real part f_{zz}^R is the flexibility of the soil. It decreases if the frequency increases. The higher the frequency the more tense the reaction of the soil. The size of the imaginary part f_{zz}^I is a measure for the damping properties of the solid, how much energy is dissipated.

Numerical investigation showed that the necessary number of elements depends on the frequency a_0. For small values, up to 3, 16 to 25 elements were sufficient. At higher frequencies the mesh must be refined because the wave length becomes too small when compared with the element length.

If we do not neglect the influence of the shear stresses then the displacements u^a of the free surface also influence the displacements u^i under the foundation,

$$\begin{bmatrix} H \end{bmatrix} \begin{bmatrix} u^i \\ u^a \end{bmatrix} = \begin{bmatrix} G \end{bmatrix} \begin{bmatrix} t^i \\ t^a \end{bmatrix} \,, \tag{9.18}$$

and we therefore must theoretically at least discretize the whole surface of the halfspace. But in practical applications it is sufficient to discretize only the immediate vicinity of the foundation because the influence functions decline very rapidly.

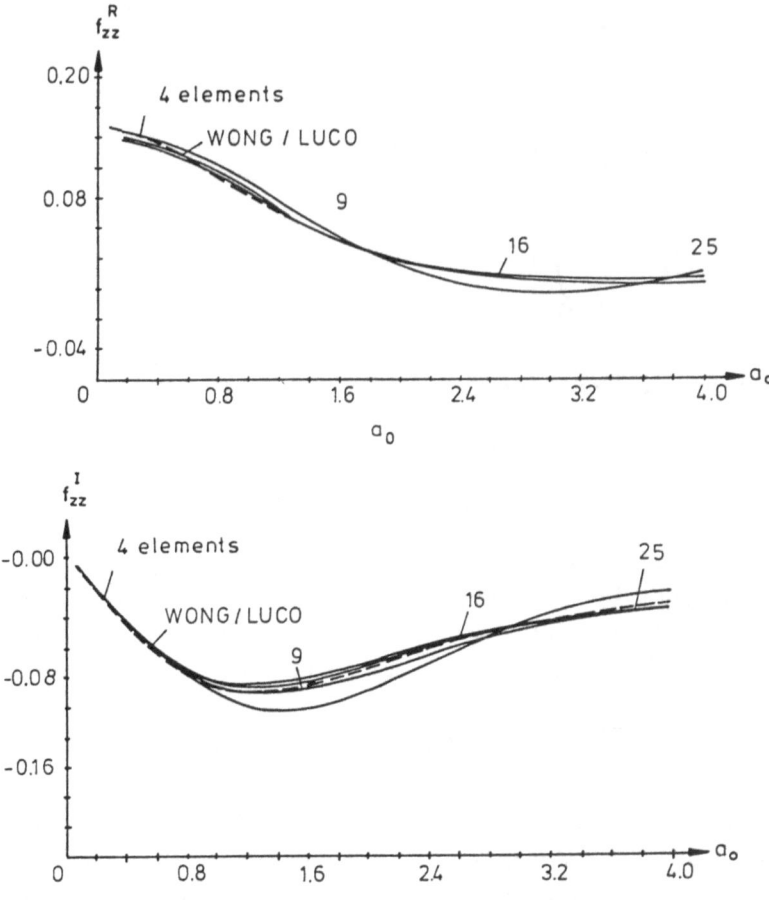

Figure 9.4 The vertical flexibility changes with the frequency a_0

Multiplying Eq.(9.18) with G^{-1} and F, we obtain

$$FG^{-1}H\begin{bmatrix} u^i \\ u^a \end{bmatrix} = F\begin{bmatrix} t^i \\ t^a \end{bmatrix}$$

If we write more concisely $Q = FG^{-1}H$ and if we consider that $t^a = o$ then we obtain

$$\begin{bmatrix} Q^{ii} & Q^{ia} \\ Q^{ai} & Q^{aa} \end{bmatrix}\begin{bmatrix} u^i \\ u^a \end{bmatrix} = F\begin{bmatrix} t^i \\ o \end{bmatrix}$$

and by eliminating u^a

$$F^T(Q^{ii} - Q^{ia}Q^{aa-1}Q^{ai})Fu = F^TRt$$

Hence, the matrix

$$K = F^T(Q^{ii} - Q^{ia}Q^{aa-1}Q^{ai})F$$

is the stiffness matrix if the stresses are coupled.

With such a matrix we can calculate the mutual influence of neighboring foundations.

By switching to complex-valued elastic constants visco elastic effects can be considered as well.

10 Computer programs

As a supplement to this book we offer a package of three programs

> BE-LAPLACE,
> BE-PLATES,
> BE-PLATE-BENDING

which run on the IBM-PC, PS/2 and compatible computers. Hardware requirement are 640 K RAM, a coprocessor 80x87, a hard disk and one of the
following graphics adapters: Hercules card, Color Graphics Adapter, Enhanced
Graphics Adapter or the Olivetti Graphics card.

The programs are easy to use advanced professional tools for the analysis
of potential problems, problems in 2-D elasticity and plate-bending problems.
They feature pop-up menus, help screens and they use graphical representations, see Fig. 10.1. Macro-elements simplify the input, see Fig. 10.2.

10.1 BE-LAPLACE

This program solves the boundary value problem

$$-N\Delta u = p \quad \text{in } \Omega, \quad u = \bar{u} \quad \text{on } \Gamma_1, \quad N\frac{\partial u}{\partial n} = \bar{t} \quad \text{on } \Gamma_2$$

in domains with piecewise straight boundaries and a constant modulus N. The
boundary condition can also be of the type, see Fig. 10.3

$$cu + N\frac{\partial u}{\partial n} = 0 \qquad \text{(Robin)}$$

Elements:

quadratic basis functions for u and t

```
┌─────────────────────────────────────────────────────────────────┐
│                      P R O G R A M - M E N U                      │
├───────────────────────────────────┬─────────────────────────────┤
│          Copyright (c) 1987 by     │ Program BE-LAPLACE: Solution of │
│        Dr.-Ing. Friedel Hartmann   │ potential problems with the BEM │
│    Architectural & Civil Engineering │                             │
│          University of Dortmund    │ Disk: C          Position:  1 │
│                                    │                             │
│   License:                         │ To move the light-bar use cursor │
│   Friedel Hartmann, 46 Dortmund 30 │ keys, choose option with ◄─┘ │
├───────────────────────────────────┼─────────────────────────────┤
│   1. Geometry                      │ 6. Print files              │
│                                    │                             │
│   2. Loading and stress points     │ 7. Graphics                 │
│                                    │                             │
│   3. Calculation                   │ 8. Calculation of the volume │
│                                    │                             │
│   4. Check program limits          │                             │
│                                    │                             │
│   5. View files                    │ 10. Printer configuration   │
├───────────────────────────────────┴─────────────────────────────┤
│ F2  Positions                          F9  Help   F10  Stop      │
└─────────────────────────────────────────────────────────────────┘
```

```
┌───────────────────────────────────┬─────────────────────────────┐
│ BE-LAPLACE            Mask 3       │ Remarks:                    │
├───────────────────────────────────┤                             │
│ Geometry         Position    1     │                             │
│                                    │                             │
│ System of coordinates              │                             │
├───────────────────────────────────┴─────────────────────────────┤
│                                                                  │
│  Which system of coordinates do you use ?                        │
│                                                                  │
│              y ▲              ┌─▷ x                               │
│                └─▷ x        y ▽                                  │
│                                                                  │
│                  type 1          type 2                          │
│                                                                  │
│                        your system :  1                          │
│                                                                  │
├──────────────────────────────────────────────────────────────────┤
│ 1  Next   2  Prev.   3  Erase   4  Copy ↓  5  Beg. ◄─  9  Help  10  Stop │
└──────────────────────────────────────────────────────────────────┘
```

Figure 10.1 a-b Main menu and sample mask of the program BE-LAPLACE

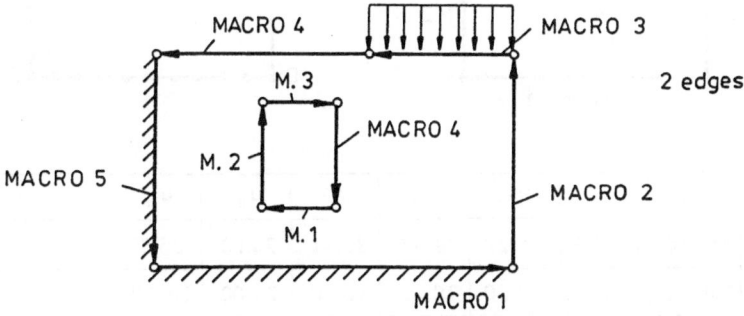

Figure 10.2 Macro-elements

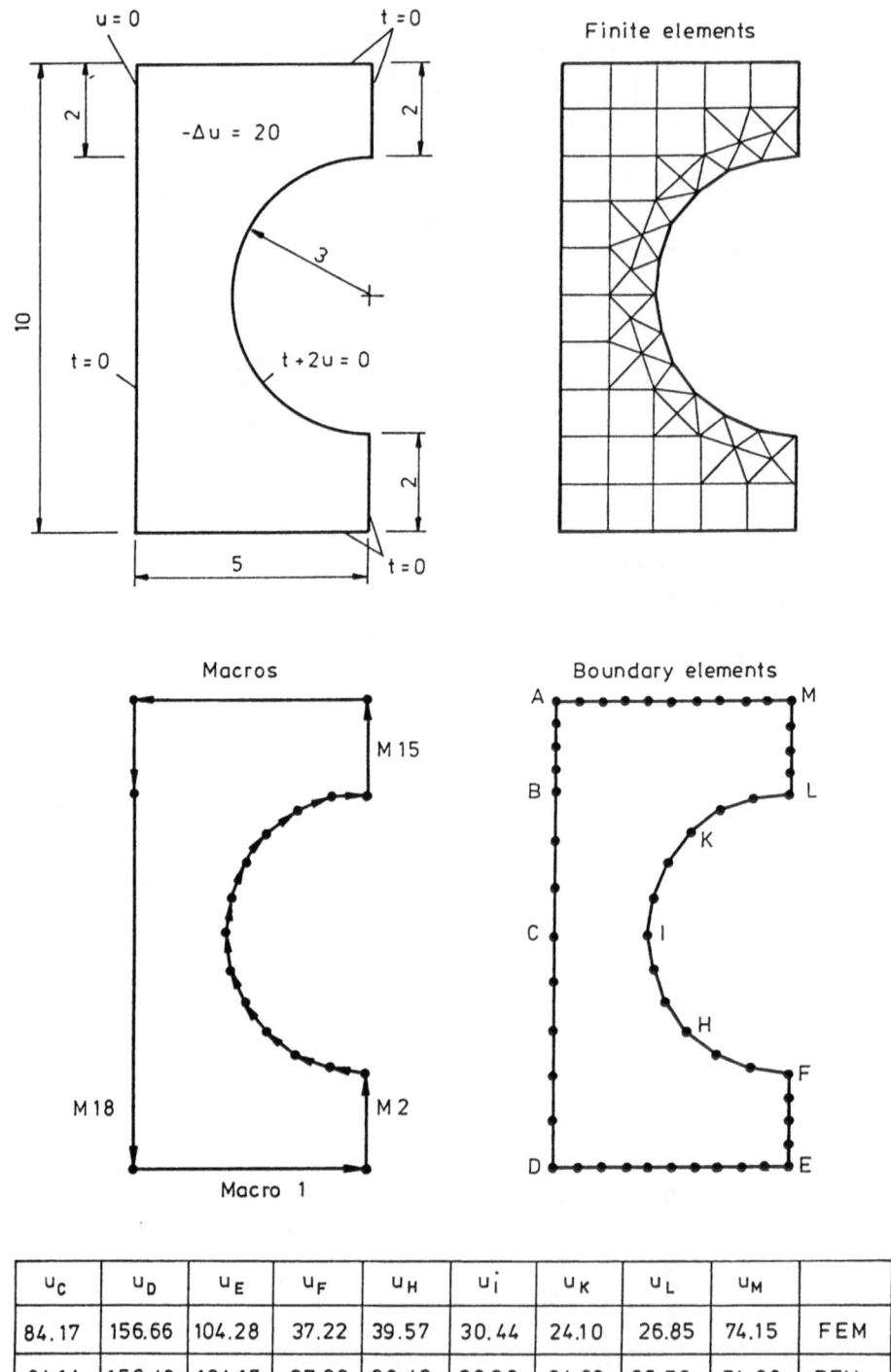

u_C	u_D	u_E	u_F	u_H	$u_{\dot{I}}$	u_K	u_L	u_M	
84.17	156.66	104.28	37.22	39.57	30.44	24.10	26.85	74.15	FEM
84.14	156.49	104.15	37.00	39.42	30.30	24.00	26.70	74.08	BEM

Figure 10.3 Poisson's equation. Comparison between a BE-solution and an FE-solution

Possible right-hand sides:

constant pressure p ,
concentrated forces P ,
line loads

On each part of the edge the deflection u or the traction t can be prescribed.

Program limits:

boundary elements	: 80
boundary elements + elements for interior line loads	: 100

```
┌─────────────────────────────────────────────────────────────────┐
│                      P R O G R A M - M E N U                      │
├───────────────────────────────┬───────────────────────────────────┤
│        Copyright (c) 1987 by   │ Program BE-PLATES:  Solutions of  │
│        Dr.-Ing. Friedel Hartmann│ problems in 2-D elasticity        │
│   Architectural & Civil Engineering│                                │
│        University of Dortmund  │ Disk: C          Position:  1     │
│                                │                                   │
│   License:                     │ To move the light-bar use cursor  │
│   Friedel Hartmann, 46 Dortmund 30│ keys, choose option with ◄─┘   │
├───────────────────────────────┼───────────────────────────────────┤
│ 1. Plate geometry              │ 6. Print files                    │
│                                │                                   │
│ 2. Loading and stress points   │ 7. Graphics                       │
│                                │                                   │
│ 3. Calculation                 │ 8. Superposition of loadcases     │
│                                │                                   │
│ 4. Check program limits        │ 9. Reinforcement                  │
│                                │                                   │
│ 5. View files                  │ 10. Printer configuration         │
├───────────────────────────────┴───────────────────────────────────┤
│ F2   Positions                           F9  Help    F10  Stop     │
└─────────────────────────────────────────────────────────────────┘
```

```
┌─────────────────────────────────────────────────────────────────┐
│ BE-PLATES                 Mask 6 │ Remarks:                        │
├───────────────────────────────────┼───────────────────────────────┤
│ Geometry            Position    1 │ Possible support conditions:    │
│ - Information about macros        │    Rigid         Free           │
│   Edge        1                   │    Elastic       Roller sup.    │
├───────────────────────────────────┴───────────────────────────────┤
│            Number of    Delta  Macro                               │
│ Macro      elements     [ m ]  support c.      stiffness           │
│ ─────────────────────────────────────────────────────────────────  │
│    1 :         4        1.500  Rigid                               │
│    2 :         6        1.333  Free                                │
│    3 :         3        1.000  Free                                │
│    4 :         3        1.000  Free                                │
│    5 :         6        1.333  Free                                │
│                                                                    │
│                                                                    │
│                                                                    │
├────────────────────────────────────────────────────────────────────┤
│ 1  Next   2  Prev.   3  Erase   4  Copy ↓  5  Beg.◄─  9  Help  10  Stop │
└─────────────────────────────────────────────────────────────────┘
```

Figure 10.4 a-b Main menu and sample mask of the program BE-PLATES

openings	:	2
interior line loads	:	12
concentrated forces	:	4

10.2 BE-PLATES

This program calculates the displacements and stresses in elastic plates, see Fig. 10.4.

Elements:

quadratic basis functions for u_i and t_i

Possible support conditions:

free
fixed
roller support
elastic support (parallel to the x- or y-axis)

Possible loading conditions, see Figs. 10.5 and 10.6:

distributed forces
concentrated forces in the interior

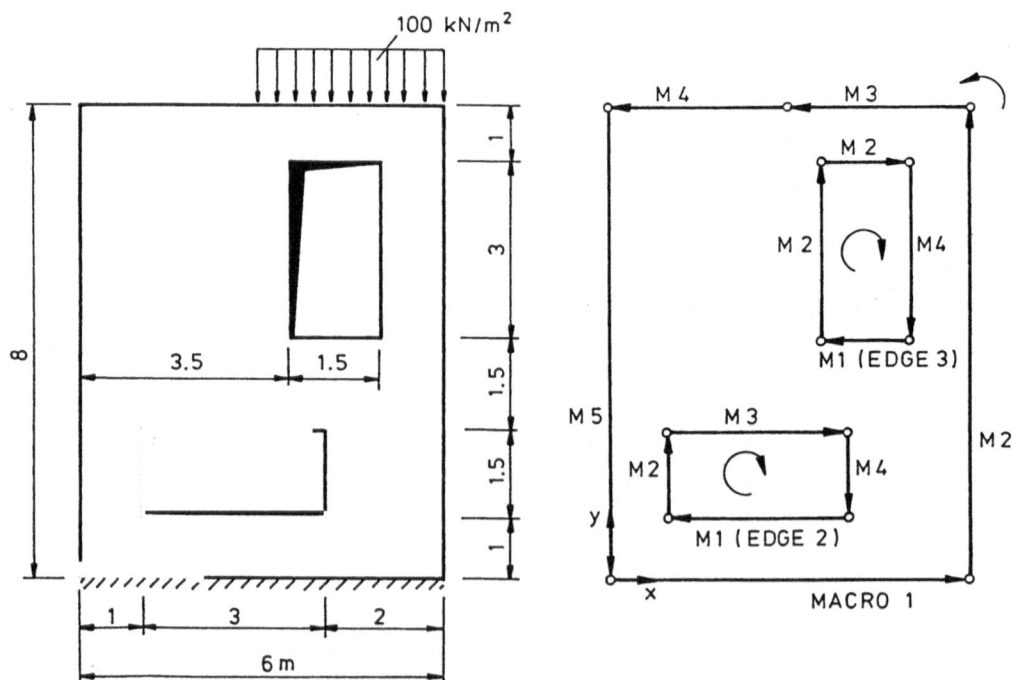

Figure 10.5 The BE-model of an elastic plate with cutouts

369

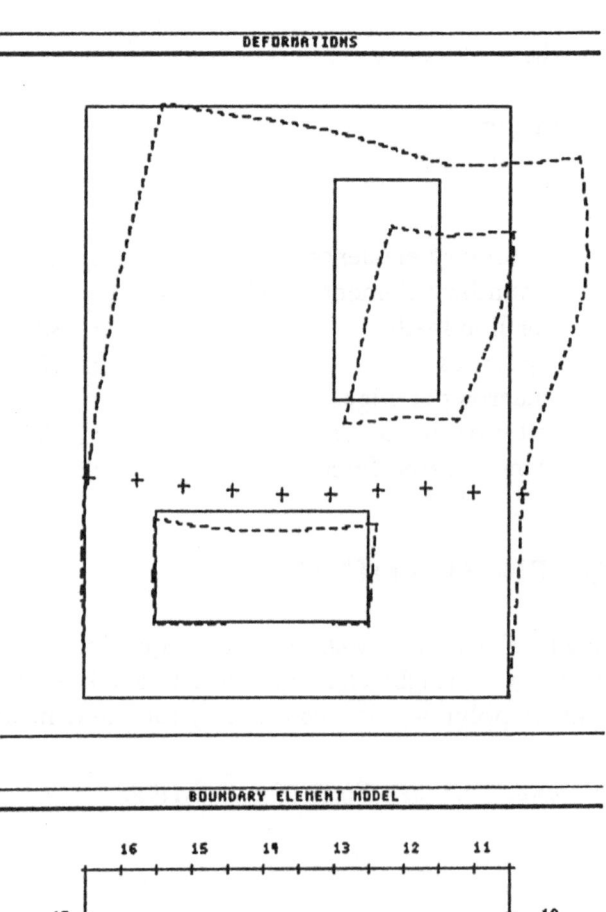

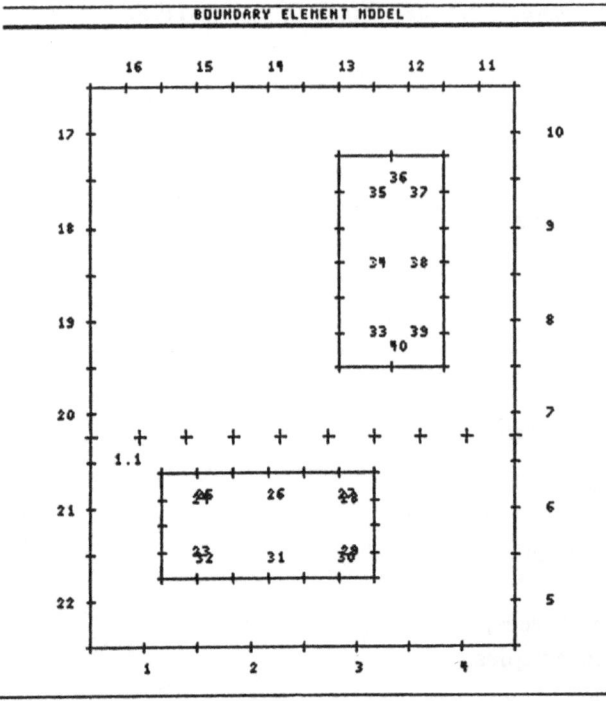

Figure 10.6 The deformed plate and the subdivision of the plate into boundary elements

line loads in the interior
edge loads
edge displacements

Program limits

boundary elements	:	50
boundary elements + elements for line loads	:	80
openings	:	5
macros per edge	:	30
interior line loads	:	12
concentrated forces	:	4

10.3 BE-PLATE-BENDING

This program solves the Kirchhoff plate equation, see Fig. 10.7. The boundary elements are straight elements with linear basis functions and C^1-functions (Hermitian polynomials) respectively for the deflection w.

Possible support conditions, see Figs. 10.8 and 10.9:

hinged,
free,
clamped.

Intermediate supports:

walls,
piers,
springs.

Possible loading conditions:

uniform loads,
linear loads,
subdomain loads,
line loads,
concentrated forces,
concentrated couples,
edge forces,
edge couples

```
┌─────────────────────────────────────────────────────────────────────┐
│                      P R O G R A M - M E N U                          │
├───────────────────────────────────┬─────────────────────────────────┤
│      Copyright (c) 1987 by         │ Program BE-PLATE-BENDING: Analysis│
│     Dr.-Ing. Friedel Hartmann      │ of plates by boundary elements  │
│  Architectural & Civil Engineering │                                 │
│       University of Dortmund        │ Disk: C            Position:  1 │
│                                    │                                 │
│  License:                          │ To move the light-bar use cursor│
│  Friedel Hartmann, 46 Dortmund 30  │ keys, choose option with ◄─┘    │
├───────────────────────────────────┼─────────────────────────────────┤
│  1. Plate geometry                 │ 6. Print files                  │
│                                    │                                 │
│  2. Loading and stress points      │ 7. Graphics                     │
│                                    │                                 │
│  3. Calculation                    │ 8. Superposition of loadcases   │
│                                    │                                 │
│  4. Check program limits           │ 9. Reinforcement                │
│                                    │                                 │
│  5. View files                     │ 10. Printer configuration       │
├───────────────────────────────────┴─────────────────────────────────┤
│  F2  Positions                              F9  Help   F10  Stop      │
└─────────────────────────────────────────────────────────────────────┘
```

```
┌─────────────────────────────────────┬───────────────────────────────┐
│  BE-PLATE-BENDING          Mask  1  │ Remarks:                      │
├─────────────────────────────────────┤                               │
│  Loading           Position   1     │                               │
│  - Loadcase Nr, -description        │                               │
│  - Loadcase parameters              │                               │
├─────────────────────────────────────┴───────────────────────────────┤
│   Project :  Clamped plate                                            │
│                                                                       │
│   Units   :   Length:  m                                              │
│               Forces:  kN                                             │
├──────────────────────────────┬───────────────────────────────────────┤
│  Loadcase-number:        1   │  Magnitude of the                     │
│                              │  - uniform load      :     10.000     │
│  Loadcase description:       │  - linear load       :     0.000000   │
│  Uniform load                │                                       │
│                              │  Number of                            │
│                              │  - point loads       :  0 [max.   100]│
│                              │  - loaded subdomains  :  0 [max.    6] │
│                              │  - line loads        :  0 [max    12]  │
│                              │  - edge loads        :  0 [max    20]  │
├──────────────────────────────┴───────────────────────────────────────┤
│  1 Next   2 Prev.   3 Erase   4 Copy ↓  5 Beg. ◄─ 9 Help  10 Stop     │
└─────────────────────────────────────────────────────────────────────┘
```

Figure 10.7 a-b Main menu and sample mask of the program BE-PLATE-BENDING

Program limits

elements	:	160 (sum)
openings	:	4
subdomains	:	4
line loads	:	6
intermediate supports (walls)	:	6
piers	:	20
springs		
concentrated forces, - couples	:	100 (sum)

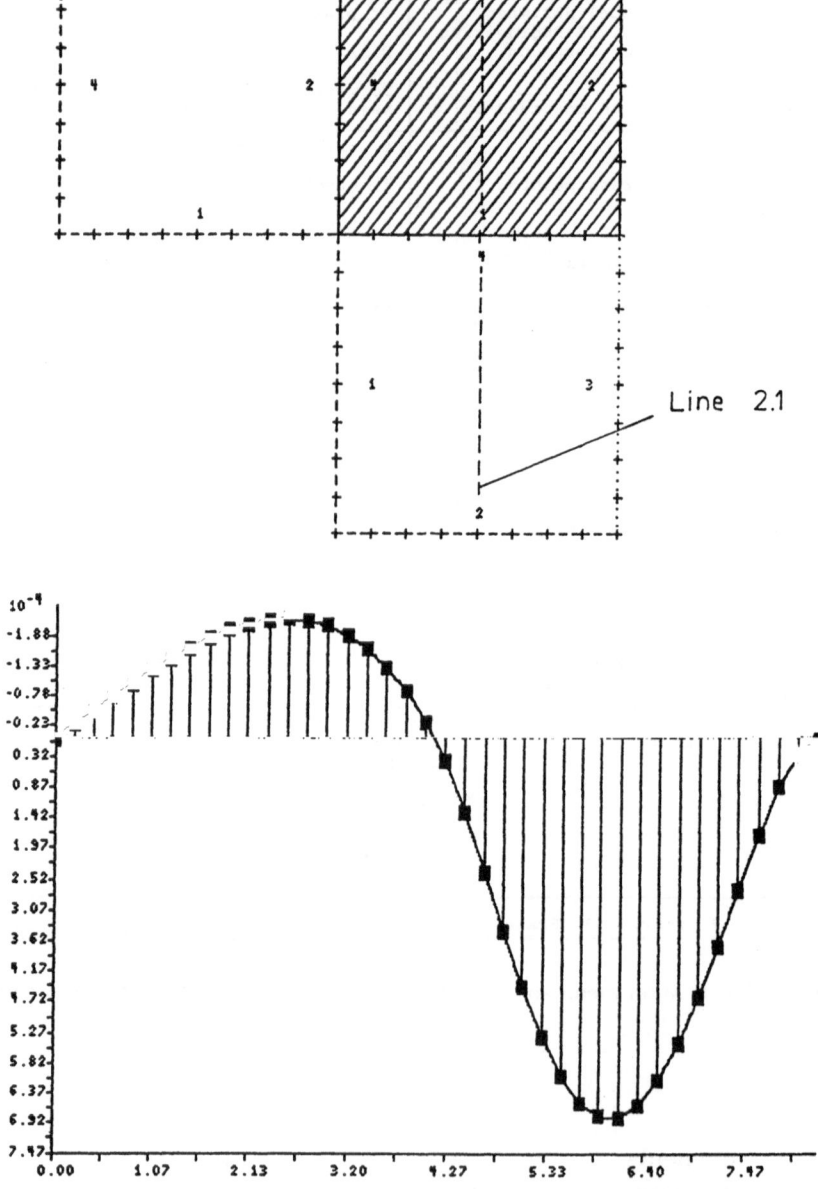

Figure 10.8 a-b L-shaped continuous floor plate: **a** subdomain load; **b** deflection along line 2.1

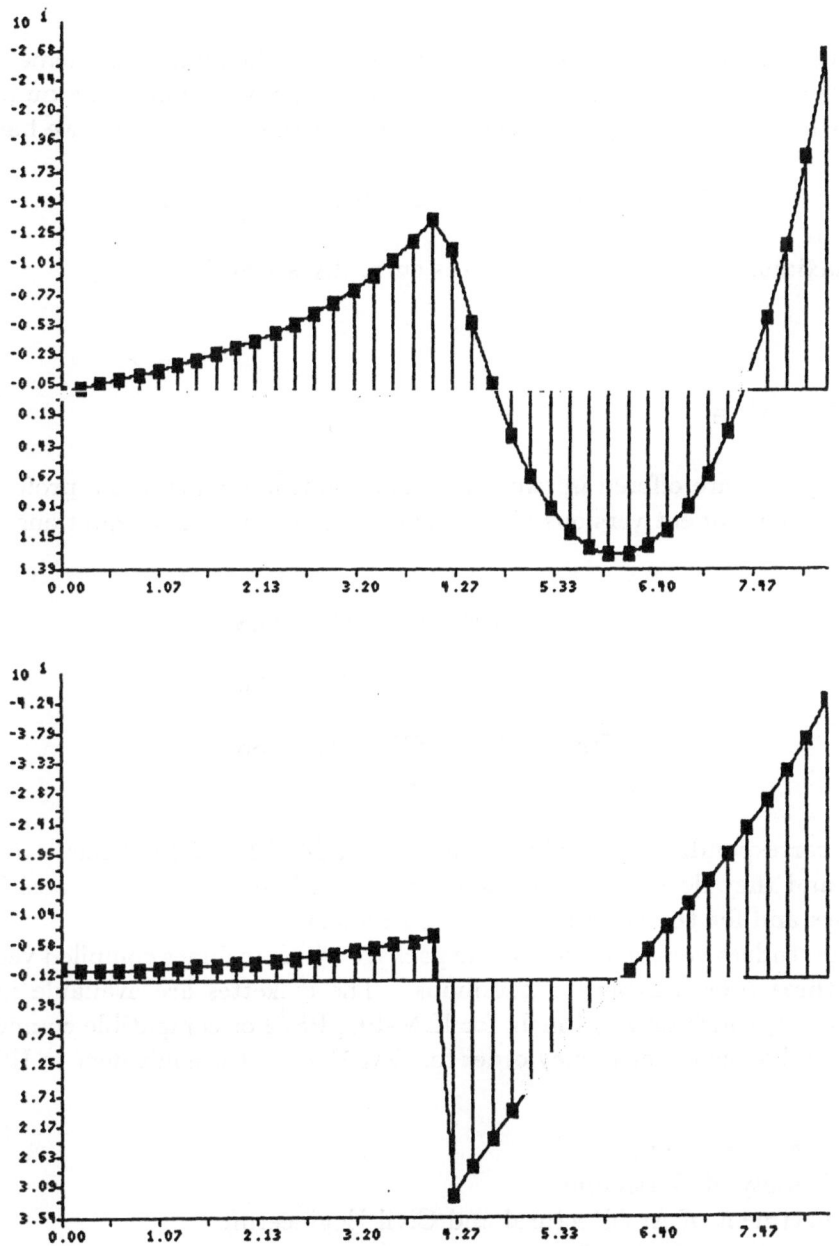

Figure 10.9 L-shaped floor plate: **a** distribution of the bending moment m_y along line 2.1; **b** distribution of the shear force q_y along the same line.

Size of the system matrix

The size of the matrix depends essentially on the number of elements. To each element belongs one collocation point. At smooth points the number of unknowns is 2 and at corner points 3 so that in the simplest case we have

$$\text{size} = \text{elements} \times 2 + \text{number of corner points}$$

The maximum number of equations is currently set at 224.

10.4 Service

The programs are offered in two versions, a student version and a professional version. The student versions allow only a limited number of equations

$$\text{BE-LAPLACE} = 48$$

$$\text{BE-PLATES} = 80$$

$$\text{BE-PLATE-BENDING} = 88$$

(which corresponds to a maximum number of 24, 20 and 40 elements respectively) and they do not include the program modules for the superposition of loadcases and for the design of the reinforcement.

The student versions come in one package which includes compiled versions of the three programs and a handbook. The diskettes are available in two formats, 3 1/2 inch or 5 1/4 inch, for IBM-PC, PS/2 or compatible computers. Please send a cheque or money order for DM 150.- or the equivalent in US $ to

Friedel Hartmann,
University of Dortmund,
Department of Architectural and Civil Engineering,
August-Schmidt-Str. 8,
D-4600 Dortmund 50,
W.Germany

and state diskette format required.

For the price and the actual specifications of the professional versions please contact the author.

Appendix A

In this appendix we derive the integral representation for the slope $\partial w/\partial n$ of a Kirchhoff plate by forming the limit

$$\lim_{\varepsilon\to 0} B(g_1[\boldsymbol{x}], w)_{\Omega_\varepsilon} = \lim_{\varepsilon\to 0}\{\int_{\Gamma_\varepsilon}[V_\nu(g_1(\boldsymbol{y},\boldsymbol{x}))w(\boldsymbol{y}) - M_\nu(g_1(\boldsymbol{y},\boldsymbol{x}))\frac{\partial w}{\partial \nu}(\boldsymbol{y})$$

$$- g_1(\boldsymbol{y},\boldsymbol{x})V_\nu(w)(\boldsymbol{y}) + \frac{\partial}{\partial \nu}g_1(\boldsymbol{y},\boldsymbol{x})M_\nu(w)(\boldsymbol{y})]\,ds_{\boldsymbol{y}} - \int_{\Omega_\varepsilon} g_1(\boldsymbol{y},\boldsymbol{x})p(\boldsymbol{y})\,d\Omega_{\boldsymbol{y}}$$

$$+ \sum_c[(w(\boldsymbol{y}^c) - w(\boldsymbol{x}))F(g_1)(\boldsymbol{y}^c,\boldsymbol{x}) - g_1(\boldsymbol{y}^c,\boldsymbol{x})F(w)(\boldsymbol{y}^c)]\} = 0\,.$$

Let us assume that the source point $\boldsymbol{x}$ is a corner point, see Fig. A1. This is the only interesting case. The two corner points at the intersection of Γ_{N_ε} with the boundary Γ we label $\boldsymbol{y}_1$ and $\boldsymbol{y}_2$. To keep things simple we assume that the boundary Γ is straight near the source point $\boldsymbol{x}$ so that the series of points

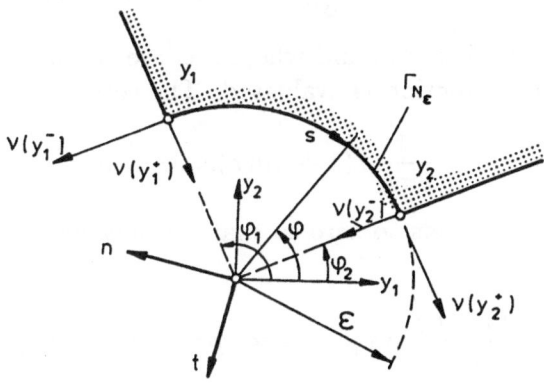

Figure A1 Situation near the source point

$y_1 = (\varepsilon, \varphi_1)$ and $y_2 = (\varepsilon, \varphi_2)$, $\varepsilon \to 0$, will approach x along the two tangents that pass through x.

Let us first investigate the sum over the corner points

$$\sum_c [(w(y^c) - w(x))F(g_1)(y^c, x) - g_1(y^c, x)F(w)(y^c)]$$

$$= \sum_c [[(w(y^c) - w(x))M_{\nu\tau}(g_1(y^c, x)) - g_1(y^c, x)M_{\nu\tau}(w)(y^c)]]$$

where for clarity we re-introduced the double brackets $[[\cdots]]$

$$F(w)(x) = M_{nt}(w)(x^+) - M_{nt}(w)(x^-) = [[M_{nt}(w)]](x)$$

Only two corner terms

$$\sum_c [[\cdots]] = [[\cdots]](y_1) + [[\cdots]](y_2) + \cdots,$$

in our arrangement the first two, depend on ε. Because of

$$M_{\nu\tau}(g_1(y, x)) = -\frac{(1-\nu)}{4\pi r}(r_\tau^2 - r_\nu^2)r_t$$

$$= -\frac{(1-\nu)}{4\pi r}[\sin\varphi\, n_1 - \cos\varphi\, n_2]\cos 2\beta$$

the terms in the double brackets can be expressed as

$$(w(y^c) - w(x))M_{\nu\tau}(g_1(y^c, x)) - g_1(y^c, x)M_{\nu\tau}(w)(y^c) =$$

$$-\frac{(1-\nu)}{4\pi r}[\sin\varphi\, n_1 - \cos\varphi\, n_2]\cos 2\beta(w(y^c) - w(x))$$

$$-\frac{1}{8\pi K}r(1 + 2\ln r)r_n M_{\nu\tau}(w)$$

where $\beta = \alpha - \varphi$, see Fig. 6.9, and where r, φ are the polar coordinates of the point y at which the function is evaluated. The second term

$$\frac{1}{8\pi K}r(1 + 2\ln r)r_n M_{\nu\tau}(w)$$

tends to zero if $r = \varepsilon$ tends to zero so that we only have to consider the limit of the first term,

$$f(y) = -\frac{(1-\nu)}{4\pi r}[\sin\varphi\, n_1 - \cos\varphi\, n_2]\cos 2\beta\,(w(y) - w(x)) \qquad (1)$$

that is

$$\lim_{\varepsilon \to 0} \{ [[\ldots]](\boldsymbol{y}_1) + [[\ldots]](\boldsymbol{y}_2) \} = \lim_{\varepsilon \to 0} \{ f(\boldsymbol{y}_1^+) - f(\boldsymbol{y}_1^-) + f(\boldsymbol{y}_2^+) - f(\boldsymbol{y}_2^-) \} \qquad (2)$$

If we substitute

$$w(\boldsymbol{y}) - w(\boldsymbol{x}) = w_{,1}(\boldsymbol{x})\, \varepsilon \cos \varphi + w_{,2}(\boldsymbol{x})\, \varepsilon \sin \varphi + O(\varepsilon^2)$$

into (1) and neglect the term $O(\varepsilon^2)$ whose contribution to the limit is zero then the terms in (2) become

$$\{ f(\boldsymbol{y}_1^+) - f(\boldsymbol{y}_1^-) + f(\boldsymbol{y}_2^+) - f(\boldsymbol{y}_2^-) \}$$

$$= -\frac{(1-\nu)}{4\pi\varepsilon} \{ [w_{,1}\, \varepsilon \cos \varphi_1 + w_{,2}\, \varepsilon \sin \varphi_1][\sin \varphi_1\, n_1 - \cos \varphi_1\, n_2]$$

$$\times [\cos 2\beta_1^+ - \cos 2\beta_1^-] + [w_{,1}\, \varepsilon \cos \varphi_2 + w_{,2}\, \varepsilon \sin \varphi_2]$$

$$\times [\sin \varphi_2\, n_1 - \cos \varphi_2\, n_2][\cos 2\beta_2^+ - \cos 2\beta_2^-] \}$$

According to Fig. A1 we have

$$\beta_1^- = \frac{\pi}{2} \qquad \beta_1^+ = \pi \qquad \beta_2^- = \pi \qquad \beta_2^+ = \frac{3}{2}\pi$$

so that

$$[\cos 2\beta_1^+ - \cos 2\beta_1^-] = 2 \qquad [\cos 2\beta_2^+ - \cos 2\beta_2^-] = -2$$

After some intermediate steps we so obtain the expression

$$\{ f(\boldsymbol{y}_1^+) - f(\boldsymbol{y}_1^-) + f(\boldsymbol{y}_2^+) - f(\boldsymbol{y}_2^-) \}$$

$$= -\frac{1}{2\pi}(1-\nu)[w_{,1}\,([0.5 \sin 2\varphi]_{\varphi_2}^{\varphi_1}\, n_1 - [\cos^2 \varphi]_{\varphi_2}^{\varphi_1}\, n_2)$$

$$+ w_{,2}\,([\sin^2 \varphi]_{\varphi_2}^{\varphi_1}\, n_1 - [0.5 \sin 2\varphi]_{\varphi_2}^{\varphi_1}\, n_2)]$$

which does not depend on ε so that it is also the limit of (2).

Next we calculate the limits of the integrals over Γ_{N_ε}. Because of

$$g_1 = O(\varepsilon) \qquad \frac{\partial g_1}{\partial \nu} = O(\ln \varepsilon) \qquad ds = O(\varepsilon)$$

and

$$M_\nu = O(1) \qquad V_\nu(w) = O(1)$$

it follows that

$$\lim_{\varepsilon \to 0} \int_{\Gamma_{N_\varepsilon}} [\frac{\partial g_1}{\partial \nu} M_\nu(w) - g_1 V_\nu(w)] ds_y = 0$$

so that we must only consider the limit of the hyper-singular and the singular kernel

$$A = \int_{\Gamma_{N_\epsilon}} [V_\nu(g_1)(w(\boldsymbol{y}) - w(\boldsymbol{x})) - M_\nu(g_1)\frac{\partial w}{\partial \nu}]ds\boldsymbol{y}$$

At points $\boldsymbol{y}$ on Γ_{N_ϵ} holds

$$\frac{\partial w}{\partial \nu}(\boldsymbol{y}) = -w_{,1}(\boldsymbol{x})\cos\varphi - w_{,2}(\boldsymbol{x})\sin\varphi + O(\varepsilon)$$

If we substitute this into the expression A and neglect the term $O(\varepsilon)$ which is zero in the limit then we obtain

$$A = \frac{1}{4\pi\varepsilon^2}\int_{\varphi_2}^{\varphi_1}\{(3-\nu)(\cos\varphi\, n_1 + \sin\varphi\, n_2)[w_{,1}\,\varepsilon\cos\varphi + w_{,2}\,\varepsilon\sin\varphi]$$

$$- \varepsilon(1+\nu)(\cos\varphi\, n_1 + \sin\varphi\, n_2)[-w_{,1}\cos\varphi - w_{,2}\sin\varphi]\}\varepsilon\, d\varphi$$

and after some intermediate steps — note that

$$\int_{\varphi_2}^{\varphi_1}\cos\varphi\,\sin\varphi\, d\varphi = [0.5\sin^2\varphi]_{\varphi_2}^{\varphi_1} = -[0.5\cos^2\varphi]_{\varphi_2}^{\varphi_1}$$

— the result

$$A = \frac{1}{2\pi}[w_{,1}([0.5\sin 2\varphi + \varphi]_{\varphi_2}^{\varphi_1}\,n_1 - [\cos^2\varphi]_{\varphi_2}^{\varphi_1}\,n_2)$$

$$- w_{,2}([\sin^2\varphi]_{\varphi_2}^{\varphi_1}\,n_1 - [0.5\sin 2\varphi + \varphi]_{\varphi_2}^{\varphi_1}\,n_2])]$$

which does not depend on ε so that it is also the limit.

It is not too difficult to see that all these results are also obtained if the boundary is curved near $\boldsymbol{x}$.

The convergence of the domain integral

$$\lim_{\varepsilon \to 0}\int_{\Omega_\epsilon}g_1[\boldsymbol{x}]pd\Omega\boldsymbol{y} = \int_\Omega g_1[\boldsymbol{x}]pd\Omega\boldsymbol{y}$$

and the two boundary integrals (Γ_ϵ' is the original boundary minus that part which is closer to $\boldsymbol{x}$ than ε)

$$\lim_{\varepsilon \to 0}\int_{\Gamma_\epsilon'}g_1[\boldsymbol{x}]V_\nu ds\boldsymbol{y} = \int_\Gamma g_1[\boldsymbol{x}]V_\nu ds\boldsymbol{y}$$

$$\lim_{\varepsilon \to 0} \int_{\Gamma'_\varepsilon} M_\nu(g_1[\boldsymbol{x}]) \frac{\partial w}{\partial \nu} ds\boldsymbol{y} = \int_{\Gamma} M_\nu(g_1[\boldsymbol{x}]) \frac{\partial w}{\partial \nu} ds\boldsymbol{y}$$

is straightforward. The only critical limit is the integral

$$\lim_{\varepsilon \to 0} \int_{\Gamma'_\varepsilon} [V_\nu(g_1[\boldsymbol{x}])(w(\boldsymbol{y}) - w(\boldsymbol{x})) - M_\nu(g_1[\boldsymbol{x}]) \frac{\partial w}{\partial \nu}] ds\boldsymbol{y}$$

but if we expand the deflection w into a Taylor series then we can show that this Cauchy principal value exists. (Note that this is one of the points where we need the smoothness of w and note also that neither the hyper-singular nor the singular integral alone does have a Cauchy principal value. Only the two together converge).

If we collect our results then we obtain just (6.7).

Appendix B

by Peter Schoepp

Double integrals

For Galerkin's method we need to evaluate double-integrals as

$$\int_\Gamma \int_\Gamma g_0(y,x)\varphi(y)ds_y\varphi(x)ds_x = \sum_m \sum_n \int_{\Gamma_m} \int_{\Gamma_n} g_0(y,x)\varphi(y)ds_y\varphi(x)ds_x$$

While the "off-diagonal" terms, $m \neq n$, can be approximated numerically we need to integrate the diagonal terms, $m = n$ analytically. Let us choose the index $e = m = n$ for this particular element. On this element Γ_e the basis functions φ are represented by three local basis functions (if we use quadratic elements)

$$\varphi = \varphi_1^e + \varphi_2^e + \varphi_3^e$$

so that the double-integral becomes

$$\int_{\Gamma_e} \int_{\Gamma_e} g_0(y,x)\varphi(y)ds_y\varphi(x)ds_x =$$

$$\sum_{i=1}^3 \sum_{j=1}^3 \int_{\Gamma_e} \int_{\Gamma_e} g_0(y,x)\varphi_i^e(y)ds_y\varphi_j^e(x)ds_x$$

To calculate this double-integral we first express for each index i the inner integral as a function of x

$$f(x) = \int_{\Gamma_e} g_0(y,x)\varphi_i^e(y)ds_y$$

and we then integrate the product of this function for each index j with the weighting function φ_j^e

$$\int_{\Gamma_e} f(x)\varphi_j^e(x)ds_x$$

In the following we represent these inner and outer integrals for the case of straight boundary elements and quadratic basis functions

$$\varphi_1^e = (1-\xi)(1-2\xi), \quad \varphi_2^e = 4\xi(1-\xi), \quad \varphi_3^e = 2\xi(\xi-0.5)$$

The first table, Table B1, lists simply the double integrals of these basis functions. You need these results for the integral-free terms.

Because all the integrations are done on the master element $[0,\xi]$ we give the inner integrals in terms of the coordinate $0 \le \xi \le 1$ of the master element. This is no loss of generality because the mapping between a point x on Γ_e and its image ξ on the master element is unique.

Laplace equation

$$g_0(y,x) = \frac{-1}{2\pi N}\ln r$$

$$\int_{\Gamma_e} g_0(y,x)\varphi_1^e(y)ds_y = \frac{-1}{2\pi N}\frac{l}{36}\{[36\xi - 54\xi^2 + 24\xi^3]\ln \xi$$

$$+ [6 - 36\xi + 54\xi^2 - 24\xi^3]\ln(1-\xi) - 17 + 6\ln l + 42\xi - 24\xi^2\}$$

$$\int_{\Gamma_e} g_0(y,x)\varphi_2^e(y)ds_y = \frac{-1}{2\pi N}\frac{l}{36}\{[72\xi^2 - 48\xi^3]\ln \xi$$

$$+ [24 - 72\xi^2 + 48\xi^3]\ln(1-\xi) - 20 + 24\ln l - 48\xi + 48\xi^2\}$$

$$\int_{\Gamma_e} g_0(y,x)\varphi_3^e(y)ds_y = \frac{-1}{2\pi N}\frac{l}{36}\{[-18\xi^2 + 24\xi^3]\ln \xi$$

$$+ [6 + 18\xi^2 - 24\xi^3]\ln(1-\xi) + 1 + 6\ln l + 6\xi - 24\xi^2\}$$

$$l = \text{length of element } \Gamma_e$$

The outer integrals are listed in Table B2.

Table B1 Double integrals of the quadratic basis functions

$\varphi_i \quad \psi_j$	$\int_{r_e} \frac{1}{2} \varphi_i \psi_j \, ds_x$
	$\dfrac{1}{15}$
	$\dfrac{1}{30}$
	$-\dfrac{1}{60}$
	$\dfrac{1}{30}$
	$\dfrac{41}{15}$
	$\dfrac{1}{30}$
	$-\dfrac{1}{60}$
	$\dfrac{1}{30}$
	$\dfrac{1}{15}$
	$\dfrac{1}{6}$
	$\dfrac{1}{12}$
	$\dfrac{1}{12}$
	$\dfrac{1}{6}$
	$\dfrac{1}{2}$

Table B2 Double integrals of the Laplace kernel g_0. The kernel g_1 is zero on Γ_e

ψ_i $\quad$ ψ_j	$\displaystyle\iint\limits_{\Gamma_e\,\Gamma_e} g_0\,\psi_i\,ds_y\,\psi_j\,ds_x$
(shape) (shape)	$-\dfrac{1}{2\pi N}\cdot\dfrac{l^2}{144}\,[\,4\ln l - 15\,]$
(shape) (shape)	$-\dfrac{1}{2\pi N}\cdot\dfrac{l^2}{144}\,[\,16\ln l - 20\,]$
(shape) (shape)	$-\dfrac{1}{2\pi N}\cdot\dfrac{l^2}{144}\,[\,4\ln l + 3\,]$
(shape) (shape)	$-\dfrac{1}{2\pi N}\cdot\dfrac{l^2}{144}\,[\,16\ln l - 20\,]$
(shape) (shape)	$-\dfrac{1}{2\pi N}\cdot\dfrac{l^2}{144}\,[\,64\ln l - 112\,]$
(shape) (shape)	$-\dfrac{1}{2\pi N}\cdot\dfrac{l^2}{144}\,[\,16\ln l - 20\,]$
(shape) (shape)	$-\dfrac{1}{2\pi N}\cdot\dfrac{l^2}{144}\,[\,4\ln l + 3\,]$
(shape) (shape)	$-\dfrac{1}{2\pi N}\cdot\dfrac{l^2}{144}\,[\,16\ln l - 20\,]$
(shape) (shape)	$-\dfrac{1}{2\pi N}\cdot\dfrac{l^2}{144}\,[\,4\ln l - 15\,]$
(shape) (shape)	$-\dfrac{1}{2\pi N}\cdot\dfrac{l^2}{16}\,[\,4\ln l - 7\,]$
(shape) (shape)	$-\dfrac{1}{2\pi N}\cdot\dfrac{l^2}{16}\,[\,4\ln l - 5\,]$
(shape) (shape)	$-\dfrac{1}{2\pi N}\cdot\dfrac{l^2}{16}\,[\,4\ln l - 5\,]$
(shape) (shape)	$-\dfrac{1}{2\pi N}\cdot\dfrac{l^2}{16}\,[\,4\ln l - 7\,]$
(shape) (shape)	$-\dfrac{1}{2\pi N}\cdot\dfrac{l^2}{2}\,[\,2\ln l - 3\,]$

Plane elasticity

The fundamental solutions

$$U_{ij}(y,x) = \frac{1}{8\pi\mu(1-\nu)}[(3-4\nu)\ln\frac{1}{r}\delta_{ij} + r_{,i}\,r_{,j}]$$

$$T_{ij}(y,x) = -\frac{1}{4\pi(1-\nu)r}[\frac{\partial r}{\partial\nu}((1-2\nu)\delta_{ij} + 2r_{,i}\,r_{,j})$$

$$-(1-2\nu)\{r_{,i}\,\nu_j(y) - r_{,j}\,\nu_i(y)\}]$$

The inner integrals

$$\int_{\Gamma_e} U_{11}(y,x)\varphi_1^e(y,x)dsy = \frac{-(3-4\nu)}{8\pi\mu(1-\nu)}\frac{l}{36}\{[36\xi - 54\xi^2 + 24\xi^3]\ln\xi$$

$$+[6 - 36\xi + 54\xi^2 - 24\xi^3]\ln(1-\xi)$$

$$-17 + 6\ln l + 42\xi - 24\xi^2 + \frac{l}{6}\cos\alpha_e\}$$

$$\int_{\Gamma_e} U_{11}(y,x)\varphi_2^e(y,x)dsy = \frac{-(3-4\nu)}{8\pi\mu(1-\nu)}\frac{l}{36}\{[72\xi^2 - 48\xi^3]\ln\xi$$

$$+[24 - 72\xi^2 + 48\xi^3]\ln(1-\xi)$$

$$-20 + 24\ln l - 48\xi + 48\xi^2 + \frac{l}{3}\cos\alpha_e\}$$

$$\int_{\Gamma_e} U_{11}(y,x)\varphi_3^e(y,x)dsy = \frac{-(3-4\nu)}{8\pi\mu(1-\nu)}\frac{l}{36}\{[-18\xi^2 + 24\xi^3]\ln\xi$$

$$+[6 + 18\xi^2 - 24\xi^3]\ln(1-\xi)$$

$$1 + 6\ln l + 6\xi - 24\xi^2 + \frac{l}{6}\cos\alpha_e\}$$

For the kernel U_{22} replace $\cos\alpha_e$ by $\sin\alpha_e$ where α_e is the angle between the horizontal axis and the element, see section 4.4.

The kernels U_{12} and U_{21} are constant regular terms so that the calculation of their inner integrals is straightforward. The outer integrals of the four kernels are listed in Table B3 and B4.

Table B3 Double integrals of the displacement kernels U_{11} and U_{22} of plane elasticity

ψ_i ψ_j	$\displaystyle\iint_{\Gamma_e \Gamma_e} U_{11} \, \psi_i \, ds_y \, \psi_j \, ds_x$ $\quad U_{11}: \ K = \cos\alpha_e$ $\quad U_{22}: \ K = \sin\alpha_e$
(shape) (shape)	$\dfrac{1}{8\pi\mu(1-\nu)}\left\{ -(3-4\nu)\dfrac{l^2}{144}[4\ln l - 15] + \dfrac{l^2}{36}\cdot K^2 \right\}$
(shape) (shape)	$\dfrac{1}{8\pi\mu(1-\nu)}\left\{ -(3-4\nu)\dfrac{l^2}{144}[16\ln l - 20] + \dfrac{l^2}{9}\cdot K^2 \right\}$
(shape) (shape)	$\dfrac{1}{8\pi\mu(1-\nu)}\left\{ -(3-4\nu)\dfrac{l^2}{144}[4\ln l + 3] + \dfrac{l^2}{36}\cdot K^2 \right\}$
(shape) (shape)	$\dfrac{1}{8\pi\mu(1-\nu)}\left\{ -(3-4\nu)\dfrac{l^2}{144}[16\ln l - 20] + \dfrac{l^2}{9}\cdot K^2 \right\}$
(shape) (shape)	$\dfrac{1}{8\pi\mu(1-\nu)}\left\{ -(3-4\nu)\dfrac{l^2}{144}[64\ln l - 112] + \dfrac{4l^2}{9}\cdot K^2 \right\}$
(shape) (shape)	$\dfrac{1}{8\pi\mu(1-\nu)}\left\{ -(3-4\nu)\dfrac{l^2}{144}[16\ln l - 20] + \dfrac{l^2}{9}\cdot K^2 \right\}$
(shape) (shape)	$\dfrac{1}{8\pi\mu(1-\nu)}\left\{ -(3-4\nu)\dfrac{l^2}{144}[4\ln l + 3] + \dfrac{l^2}{36}\cdot K^2 \right\}$
(shape) (shape)	$\dfrac{1}{8\pi\mu(1-\nu)}\left\{ -(3-4\nu)\dfrac{l^2}{144}[16\ln l - 20] + \dfrac{l^2}{9}\cdot K^2 \right\}$
(shape) (shape)	$\dfrac{1}{8\pi\mu(1-\nu)}\left\{ -(3-4\nu)\dfrac{l^2}{144}[4\ln l - 15] + \dfrac{l^2}{36}\cdot K^2 \right\}$
(shape) (shape)	$\dfrac{1}{8\pi\mu(1-\nu)}\left\{ -(3-4\nu)\dfrac{l^2}{16}[4\ln l - 7] + \dfrac{l^2}{4}\cdot K^2 \right\}$
(shape) (shape)	$\dfrac{1}{8\pi\mu(1-\nu)}\left\{ -(3-4\nu)\dfrac{l^2}{16}[4\ln l - 5] + \dfrac{l^2}{4}\cdot K^2 \right\}$
(shape) (shape)	$\dfrac{1}{8\pi\mu(1-\nu)}\left\{ -(3-4\nu)\dfrac{l^2}{16}[4\ln l - 5] + \dfrac{l^2}{4}\cdot K^2 \right\}$
(shape) (shape)	$\dfrac{1}{8\pi\mu(1-\nu)}\left\{ -(3-4\nu)\dfrac{l^2}{16}[4\ln l - 7] + \dfrac{l^2}{4}\cdot K^2 \right\}$
(shape) (shape)	$\dfrac{1}{8\pi\mu(1-\nu)}\left\{ -(3-4\nu)\dfrac{l^2}{2}[2\ln l - 3] + l^2\cdot K^2 \right\}$

Table B4 Double integrals of the displacement kernels U_{12} and U_{21} of plane elasticity

$\psi_i \qquad \psi_j$	$\displaystyle\iint\limits_{\Gamma_e \ \Gamma_e} U_{12 \atop (21)} \ \psi_i \ ds_y \ \psi_j \ ds_x$
	$\dfrac{1}{8\pi\mu\,(1-\nu)} \ \dfrac{l^2}{36} \ \cos\alpha_e \sin\alpha_e$
	$\dfrac{1}{8\pi\mu\,(1-\nu)} \ \dfrac{l^2}{9} \ \cos\alpha_e \sin\alpha_e$
	$\dfrac{1}{8\pi\mu\,(1-\nu)} \ \dfrac{l^2}{36} \ \cos\alpha_e \sin\alpha_e$
	$\dfrac{1}{8\pi\mu\,(1-\nu)} \ \dfrac{l^2}{9} \ \cos\alpha_e \sin\alpha_e$
	$\dfrac{1}{8\pi\mu\,(1-\nu)} \ \dfrac{4l^2}{9} \ \cos\alpha_e \sin\alpha_e$
	$\dfrac{1}{8\pi\mu\,(1-\nu)} \ \dfrac{l^2}{9} \ \cos\alpha_e \sin\alpha_e$
	$\dfrac{1}{8\pi\mu\,(1-\nu)} \ \dfrac{l^2}{36} \ \cos\alpha_e \sin\alpha_e$
	$\dfrac{1}{8\pi\mu\,(1-\nu)} \ \dfrac{l^2}{9} \ \cos\alpha_e \sin\alpha_e$
	$\dfrac{1}{8\pi\mu\,(1-\nu)} \ \dfrac{l^2}{36} \ \cos\alpha_e \sin\alpha_e$
	$\dfrac{1}{8\pi\mu\,(1-\nu)} \ \dfrac{l^2}{4} \ \cos\alpha_e \sin\alpha_e$
	$\dfrac{1}{8\pi\mu\,(1-\nu)} \ \dfrac{l^2}{4} \ \cos\alpha_e \sin\alpha_e$
	$\dfrac{1}{8\pi\mu\,(1-\nu)} \ \dfrac{l^2}{4} \ \cos\alpha_e \sin\alpha_e$
	$\dfrac{1}{8\pi\mu\,(1-\nu)} \ \dfrac{l^2}{4} \ \cos\alpha_e \sin\alpha_e$
	$\dfrac{1}{8\pi\mu\,(1-\nu)} \ l^2 \ \cos\alpha_e \sin\alpha_e$

The inner integrals of the traction kernel T_{12} are

$$\int_{\Gamma_e} T_{12}(\boldsymbol{y},\boldsymbol{x})\varphi_1^e(\boldsymbol{y})ds_{\boldsymbol{y}} = \frac{-1-2\nu}{4\pi(1-\nu)}\{(1-3\xi+2\xi^2)[\ln(1-\xi)-\ln\xi]-2+2\xi\}$$

$$\int_{\Gamma_e} T_{12}(\boldsymbol{y},\boldsymbol{x})\varphi_2^e(\boldsymbol{y})ds_{\boldsymbol{y}} = \frac{-1-2\nu}{4\pi(1-\nu)}\{(4\xi-42\xi^2)[\ln(1-\xi)-\ln\xi]+2-4\xi\}$$

$$\int_{\Gamma_e} T_{12}(\boldsymbol{y},\boldsymbol{x})\varphi_3^e(\boldsymbol{y})ds_{\boldsymbol{y}} = \frac{-1-2\nu}{4\pi(1-\nu)}\{(-\xi+2\xi^2)[\ln(1-\xi)-\ln\xi]+2\xi\}$$

For the kernel T_{21} simply switch the sign. The kernels T_{11} and T_{22} are zero on Γ_e because the normal derivative r_ν of r is zero. The outer integrals are listed in Table B5.

Table B6 contains the double integrals of the kernel functions that result when the domain integrals are transformed into boundary integrals.

Plate bending

The double integrals for plate bending are given in Tables B7 through B11. For the expressions of the inner integrals see [99].

Table B5 Double integrals of the traction kernels T_{12} and T_{21} of plane elasticity

φ_i $\quad$ ψ_j	$\displaystyle\iint_{\Gamma_e \Gamma_e} T_{12 \atop (21)} \varphi_i\, ds_y\, \psi_j\, ds_x$ $\qquad$ $T_{12} : K = 1$ $\quad$ $T_{21} : K = -1$
	0
	$\dfrac{1 - 2\nu}{4\pi\,(1 - \nu\cdot)} \cdot \dfrac{1}{2} \cdot K$
	0
	$-\dfrac{1 - 2\nu}{4\pi\,(1 - \nu)} \cdot \dfrac{1}{2} \cdot K$
	0
	$\dfrac{1 - 2\nu}{4\pi\,(1 - \nu)} \cdot \dfrac{1}{2} \cdot K$
	0
	$-\dfrac{1 - 2\nu}{4\pi\,(1 - \nu)} \cdot \dfrac{1}{2} \cdot K$
	0
	0
	$\dfrac{1 - 2\nu}{4\pi\,(1 - \nu)} \cdot \dfrac{1}{2} \cdot K$
	$\dfrac{1 - 2\nu}{4\pi\,(1 - \nu)} \cdot \dfrac{1}{2} \cdot K$
	0
	0

Table B6 Double integrals of the equivalent boundary kernels for the distributed load

ψ_j	$\dfrac{1}{16\pi\mu(1-\nu)}\,(p_1\cos\alpha_e+p_2\sin\alpha_e)\cdot v_{\underset{(2)}{1}}\displaystyle\iint\limits_{\Gamma_e\,\Gamma_e} r\,(2\ln r+1)\,ds_y\,\psi_j\,ds_x$
	$\dfrac{1}{16\pi\mu(1-\nu)}\,(p_1\cos\alpha_e+p_2\sin\alpha_e)\cdot v_{\underset{(2)}{1}}\cdot\dfrac{l^3}{72}\,[\,6\,(2\ln l+1)-7\,]$
	0
	$\dfrac{1}{16\pi\mu(1-\nu)}\,(p_1\cos\alpha_e+p_2\sin\alpha_e)\cdot v_{\underset{(2)}{1}}\cdot\dfrac{l^3}{72}\,[-6\,(2\ln l+1)+7\,]$
	$\dfrac{1}{16\pi\mu(1-\nu)}\,(p_1\cos\alpha_e+p_2\sin\alpha_e)\cdot v_{\underset{(2)}{1}}\cdot\dfrac{l^3}{72}\,[\,6\,(2\ln l+1)-7\,]$
	$\dfrac{1}{16\pi\mu(1-\nu)}\,(p_1\cos\alpha_e+p_2\sin\alpha_e)\cdot v_{\underset{(2)}{1}}\cdot\dfrac{l^3}{72}\,[-6\,(2\ln l+1)+7\,]$
	0

Table B7 Double integrals of the basis functions for plate bending

$\varphi_i^{(w_i)}$	φ_j	$\int_{r_e} \frac{1}{2} \varphi_i^{(w_i)} \varphi_j \, ds_x$
		$\dfrac{7}{40}\, l$
		$\dfrac{1}{40}\, l^2$
		$\dfrac{3}{40}\, l$
		$-\dfrac{1}{60}\, l^2$
		$\dfrac{3}{40}\, l$
		$\dfrac{1}{60}\, l^2$
		$\dfrac{7}{40}\, l$
		$-\dfrac{1}{40}\, l^2$
		$\dfrac{1}{6}\, l$
		$\dfrac{1}{12}\, l$
		$\dfrac{1}{12}\, l$
		$\dfrac{1}{6}\, l$

Table B8 Double integrals of the deflection kernels g_0 and g_1 and their normal derivatives (plate bending)

$\varphi_i \quad \varphi_j$	$\displaystyle\iint_{\Gamma_e \Gamma_e} g_0\,\varphi_i\,ds_y\,\varphi_j\,ds_x$	$\displaystyle\iint_{\Gamma_e \Gamma_e} g_1\,\varphi_i\,ds_y\,\varphi_j\,ds_x$
	$\dfrac{1}{8\pi K}\cdot\dfrac{l^4}{144}\,[\,4\ln l-3\,]$	0
	$\dfrac{1}{8\pi K}\cdot\dfrac{l^4}{144}\,[\,8\ln l-4\,]$	$\dfrac{r_n}{8\pi K}\cdot\dfrac{l^3}{144}\,[\,-24\ln l+2\,]$
	$\dfrac{1}{8\pi K}\cdot\dfrac{l^4}{144}\,[\,8\ln l-4\,]$	$\dfrac{r_n}{8\pi K}\cdot\dfrac{l^3}{144}\,[\,24\ln l-2\,]$
	$\dfrac{1}{8\pi K}\cdot\dfrac{l^4}{144}\,[\,4\ln l-3\,]$	0

$\varphi_i \quad \varphi_j$	$\displaystyle -\iint_{\Gamma_e \Gamma_e}\frac{\partial}{\partial v}\,g_0\,\varphi_i\,ds_y\,\varphi_j\,ds_x$	$\displaystyle -\iint_{\Gamma_e \Gamma_e}\frac{\partial}{\partial v}\,g_1\,\varphi_i\,ds_y\,\varphi_j\,ds_x$
	0	$-\dfrac{r_t\,r_\tau}{8\pi K}\cdot\dfrac{l^2}{8}\,[\,4\ln l-5\,]$
	0	$-\dfrac{r_t\,r_\tau}{8\pi K}\cdot\dfrac{l^2}{8}\,[\,4\ln l-3\,]$
	0	$-\dfrac{r_t\,r_\tau}{8\pi K}\cdot\dfrac{l^2}{8}\,[\,4\ln l-3\,]$
	0	$-\dfrac{r_t\,r_\tau}{8\pi K}\cdot\dfrac{l^2}{8}\,[\,4\ln l-5\,]$

Table B9 Double integrals of the bending moment kernels

φ_i φ_j	$-\iint\limits_{\Gamma_e \Gamma_e} M_v (g_0) \; \varphi_i \; ds_y \; \varphi_j \; ds_x$	$-\iint\limits_{\Gamma_e \Gamma_e} M_v (g_1) \; \varphi_i \; ds_y \; \varphi_j \; ds_x$
	$\dfrac{1}{8\pi} \cdot \dfrac{l^2}{8} \; [\, 4(1+v) \, \ln l \, - 5 - v \,]$	0
	$\dfrac{1}{8\pi} \cdot \dfrac{l^2}{8} \; [\, 4(1+v) \, \ln l \, - 3 + v \,]$	$-\dfrac{(1+v) \, r_n \cdot l}{8\pi}$
	$\dfrac{1}{8\pi} \cdot \dfrac{l^2}{8} \; [\, 4(1+v) \, \ln l \, - 3 + v \,]$	$\dfrac{(1+v) \, r_n \cdot l}{8\pi}$
	$\dfrac{1}{8\pi} \cdot \dfrac{l^2}{8} \; [\, 4(1+v) \, \ln l \, - 5 - v \,]$	0

Table B10 Double integrals of the Kirchhoff shear kernels

$w_i \qquad \varphi_j$	$\displaystyle\iint_{\Gamma_e \Gamma_e} V_\nu(g_0)\, w_i\, ds_y\, \varphi_j\, ds_x$	$\displaystyle\iint_{\Gamma_e \Gamma_e} V_\nu(g_1)\, w_i\, ds_y\, \varphi_j\, ds_x$
◺ ◿	0	$\dfrac{(1+\nu)\, r_\tau\, r_t}{4\pi} \cdot \dfrac{7}{12}$
◺ ◹	0	$-\dfrac{(1+\nu)\, r_\tau\, r_t}{4\pi} \cdot \dfrac{7}{12}$
◿ ◺	0	$\dfrac{(1+\nu)\, r_\tau\, r_t}{4\pi} \cdot \dfrac{7}{12}$
◿ ◹	0	$-\dfrac{(1+\nu)\, r_\tau\, r_t}{4\pi} \cdot \dfrac{7}{12}$

$w_i \qquad \varphi_j$	$\displaystyle\iint_{\Gamma_e \Gamma_e} V_\nu(g_0)\, w_i\, ds_y\, \varphi_j\, ds_x$	$\displaystyle\iint_{\Gamma_e \Gamma_e} V_\nu(g_1)\, w_i\, ds_y\, \varphi_j\, ds_x$
⌒ ◺	0	$\dfrac{(1+\nu)\, r_\tau\, r_t\, l}{4\pi} \cdot \dfrac{1}{24}$
⌒ ◹	0	$-\dfrac{(1+\nu)\, r_\tau\, r_t \cdot l}{4\pi} \cdot \dfrac{1}{24}$
⌒ ◺	0	$\dfrac{(1+\nu)\, r_\tau\, r_t \cdot l}{4\pi} \cdot \dfrac{1}{24}$
⌒ ◹	0	$-\dfrac{(1+\nu)\, r_\tau\, r_t \cdot l}{4\pi} \cdot \dfrac{1}{24}$

Table B11 Double integrals of the equivalent boundary kernels for the domain load

φ_j	$\dfrac{1}{64\,\pi\,K}\displaystyle\int_{r_e}\int_{r_e} r^3\left(2\ln r - \frac{1}{2}\right) r_v\; ds_y \;\cdot\; \varphi_j\; ds_x$
	0
	0

φ_j	$\dfrac{1}{64\,\pi K}\displaystyle\int_{r_e}\int_{r_e}\left\{ r^2\left[\left(6\ln r + \frac{1}{2}\right) r_n\, r_v + \left(2\ln r - \frac{1}{2}\right) r_\tau\, r_t\right]\right\} ds_y\; \varphi_j\; ds_x$
	$\dfrac{1}{64\,\pi K}\cdot\dfrac{r_\tau\, r_t\; l^4}{72}\left[\,3\,(4\ln l - 1) - 7\,\right]$
	$\dfrac{1}{64\,\pi K}\cdot\dfrac{r_\tau\, r_t\; l^4}{72}\cdot\left[\,3\,(4\ln l - 1) - 7\,\right]$

Literature

1 Mikhlin, S.G.: Multdimensional Singular Integrals and Integral Equations. London: Pergamon Press 1965
2 Hartmann F.: The Mathematical Foundation of Structural Mechancis. Berlin Heidelberg New York Tokyo: Springer-Verlag 1985
3 Wendland, W.: On some mathematical aspects of boundary element methods for elliptic problems. In: The Mathematics of Finite Elements and Applications V, Mafelap 1984. J.R. Whiteman (Ed.). London: Academic Press 1985, 193-227
4 Nedelec, J.C.: Integral equations with non integrable kernels. Integral Equations Operator Theory 5 (1982) 562-572
5 Gunther, Potential Theory and its Applications to Basic Problems of Mathematical Physics. New York: Ungar 1967
6 Aziz, A.K. (Ed.): The Mathematical Foundations of the Finite Element Method with Applications to Partial Differential Equations.. New York London: Academic Press 1972
7 Arnold, D.N.; Wendland, W.L: Collocation versus Galerkin procedures for boundary integral methods. In: Boundary Element Methods in Engineering, Proc. 4th Int. Seminar Southampton 1982, C.A. Brebbia (Ed.). Berlin Heidelberg New York: Springer-Verlag 1982, 18-33
8 Ruotsalainen, K.; Saranen, J.: Some boundary element methods using Dirac's distributions as trial functions. SIAM J. Numer. Anal. 24 (1987) 816-827
9 Wendland, W.L.: On Galerkin collocation methods for integral equations of elliptic boundary value problems. Int. Schriftenreihe Numer. Math. 53 (1980) 244-275
10 Wendland, W.L., Strongly elliptic boundary integral equations. In: The State of the Art in Numerical Analysis. A. Iserless, M. Powell (eds.) IMA, Oxford Univ. Press (1987)
11 Prössdorf, S.; Rathsfeld, A.: A spline collocation method for singular integral equations with piecewise continuous coefficients. Integral Equ. and Operator Theory 7 (1984) 536-560
12 Prössdorf, S.; Rathsfeld, A.: On quadrature methods and spline approximation of singular integrals equations. In: Boundary Elements IX (eds. C.A. Brebbia, W.L. Wendland, G. Kuhn) Southampton: Computational Mechanics Publications, Berlin: Springer-Verlag 1987, 193-212
13 Schmidt, G.: On spline collocation for singular integral equations. Math. Nachr. 111 (1983) 177-189
14 Schmidt, G.: The convergence of Galerkin and collocation methods with splines for pseudodifferential equations on closed curves. Zeitschr. Anal. und Anwendungen 3 (1984) 371-384
15 Schmidt, G.: On spline collocation methods for boundary integral equations in the plane. Math. Methods in the Appl. Sci. 7 (1985) 74-89
16 Arnold, D.N.; Wendland, W.: The convergence of spline collocation for strongly elliptic equations on curves. Numer. Math. 47 (1985) 317-341
17 Saranen, J.; Wendland, W.L.: On the asymptotic convergence of collocation methods with spline functions of even degree. Math. Comp. 45 (1985) 91-108

18 Schmidt, G.: On ε-collocation for pseudodifferential equations on a closed curve. Math. Nachr. 126 (1986) 183-196

19 Wendland, W.L.: Asymptotic accuracy and convergence for point collocation methods. In: Topics in Boundary Element Research 2, Chap. 9. (C.A. Brebbia Ed.). Berlin: Springer-Verlag (1985)

20 Niessner, H.; Ribaut, M.: Condition of boundary integral equations arising from flow computations. J. Computational and Appl. Math. 12 & 13 (1985) 491-503

21 Hartmann, F.: Elastic potentials on piecewise smooth surfaces. J. Elasticity 12 (1982) 31-50

22 Hartmann, F.: The physical nature of elastic layers. J. Elasticity 12 (1982) 19-29

23 Han, P.S.; Olson, M.D.: An adaptive boundary element method. Int. J. Numer. Methods Eng. 24 (1987) 1187-1202

24 Migeot, J.L.: Scale effect in the BEM solution of 2D potential problems. In: Boundary Elements VII (eds. C.A. Brebbia, G. Maier) Proc. 7th Int. Conf., Villa Olmo, Lake Como, Italy, September 1985. Berlin Heidelberg New York Tokyo: Springer-Verlag 1985, I-47 - I-62

25 Kuhn, G.; Löbel, G.; Potrc, I.: Kritisches Lösungsverhalten der Randelementmethode mit logarithmischen Kern. GAMM Meeting 1986, Dortmund

26 Trefftz, E.: Ein Gegenstück zum Ritzschen Verfahren. 2. Int. Kongress f. Techn. Mechanik, Zürich 1926, S. 131-137

27 Rektorys, K.: Variational Methods in Mathematics, Science and Engineering, 2nd edition. Dordrecht Boston London: D. Reidel Publishing Company

28 Hörmander, L.: On the theory of general partial differential operators. Acta Mathematica 94 (1955) 161-248

29 Kitahara, M.: Boundary Integral Equation Methods in Eigenvalue Problems of Elasto-dynamics and Thin Plates. New York: Elsevier 1985

30 Ortner, V.N.: Regularisierte Faltung von Distributionen, Teil 1: Zur Berechnung von Fundamentallösungen. ZAMP 31 (1980) 133-154

31 Ortner, V.N., Regularisierte Faltung von Distributionen, Teil 2: Eine Tabelle von Fundamentallösungen. ZAMP 31 (1980) 155-173

32 Ortner, V.N., Construction of fundamental solutions. In: Topics in Boundary Element Research Vol. 3 C.A. Brebbia (Ed.) Berlin Heidelberg New York: Springer-Verlag 1987

33 Tosaka, N. New integral equation formulations for continuum mechanics. In: Boundary Elements IX (eds. C.A. Brebbia, W.L. Wendland, G. Kuhn) Vol. 1, 131-142, Computational Mechanics Publications Southampton Boston, Springer-Verlag Berlin Heidelberg New York London Paris Tokyo 1987

34 Antes, H.: On boundary integral equations for circular cylindrical shells. In: Boundary Element Methods, Proc. 3rd Int. Seminar Irvine, C.A. Brebbia (Ed.). Berlin Heidelberg New York: Springer-Verlag 1981, 224-238

35 Newton, D.A., Tottenham, H.: Boundary value problems in thin shallow shells of arbitrary plane form, J.Eng. Math. 2 (1968) 211-224

36 Simmonds, J.G., Bradley, M.R.: The fundamental solution for a shallow shell with an arbitrary quadratic midsurface. J. Appl. Mech. 43 Trans. ASME (1976) 286-290

37 Tepavitcharov, A.D.: Fundamental solutions and boundary integral equations in the bending theory of shallow spherical shells. In: Boundary Element Methods, Proc. 7th Int. Conf. Lake Como 1985, C.A. Brebbia (Ed.). Berlin Heidelberg New York: Springer-Verlag, 4-53 - 4-62

38 Hansen, E.B.: Stress concentration in a stretched cylindrical shell with two elliptical holes. J. Appl. Mech. 45 Trans. ASME (1978) 839-844

39 Hein, J.C.: Ein gemischtes Randwertproblem der Kreiszylinderschale mit beliebigen Ausschnitten. Math. Meth. Appl. Sci. 4 (1982) 354-381

40 Matsui, T.; Matsuoka, O.: The fundamental solution in the theory of shallow shells. Int. J. Solids Struct. 14 (1978) 971-981

41 Hersh, R.; Griego, R.J.: Brownian motion and potential theory. Sci. Am. 220 (1969) 66-77

42 Schulze, B.W.; Wildenhain, G.: Zur Potentialtheorie für stark elliptische Systeme mit konstanten Koeffizienten. Math. Nachrichten 62 (1974) 189-215

43 Maz'ja, V.G.; Plamenevskij, B.A.: The first boundary value problem for classical equations of mathematical physics in domains with piecewise smooth boundaries II. Z. Analysis und ihre Anwendungen 2 (1983) 523-551

44 Fichera, G.: Il teorema del massimo modulo per l'equazione dell' elastostatica tridimensionale. Arch. Rational Mech. Anal. 7 (1961) 373-387

45 Adler, G.: Majoration des tensions dans un corps élastique à l'aide des déplacements superficiels. Arch. Rational Mech. Anal. 16 (1964) 345-372

46 Neureiter, W.; Kuhn, G.: Boundary-Element-Methode mit Substrukturtechnik. ZAMM 61 (1981) T 112 - T114

47 Lamp, U.; Schleicher, T.; Stephan, E.; Wendland, W.: Theoretical and experimental asymptotic convergence of the boundary integral method for a plane mixed boundary value problem. In: Boundary Element Methods in Engineering, Proc. 4th Int. Seminar Southampton 1982, C.A. Brebbia (Ed.). Berlin Heidelberg New York: Springer-Verlag 1982, 3-17

48 Atkinson, C.; Xanthis, L.S.; Bernal, M.J.M: Boundary integral equation crack-tip analysis and applications to elastic media with spatially varying elastic properties. Comp. Meth. Appl. Mech. Eng. 29 (1981) 35-49

49 Atkinson, C.: Fracture Mechanics Stress Analysis. In: Progress in Boundary Element Methods 2, C.A. Brebbia (Ed.), London Plymouth: Pentech Press 1983

50 Jin, H.; Tullberg, O.: More on boundary elements for three-dimensional potential problems. In: Boundary Element Methods, Proc. 7th Int. Conf. Lake Como 1985, C.A. Brebbia (Ed.). Berlin Heidelberg New York: Springer-Verlag, 2-13 - 2-24

51 Hartmann, F.: The Somigliana identity on piecewise smooth surfaces. J. Elasticity 11 (1981) 403-423

52 Danson, A boundary element formulation of problems in linear isotropic elasticity with body forces. In: Boundary Element Methods, Proc. 3rd Int. Sem. Irvine, CA, C.A. Brebbia (Ed). Berlin Heidelberg New York: Springer-Verlag 1981

53 Gurtin, E. M.: The linear theory of elasticity. In: Encyclopedia of Physics (S. Flügge Ed.) Vol. VIa/2 Solid Mechanics II (C. Truesdell Ed.). Berlin Heidelberg New York: Springer-Verlag 1972

54 Brebbia, C.A.; Telles, J.C.F.; Wrobel, L.C.: Boundary Element Techniques. Berlin Heidelberg New York Tokyo: Springer-Verlag 1984

55 Stippes, M.; Rizzo, F.J.: A note on the body force integral of classical elastostatics. ZAMP 28 (1977) 339-341

56 Mitra, A.K., Ingber, M.S.: Resolving difficulties in the BIEM caused by geometric corners and discontinuous boundary conditions. In: Boundary Elements IX Vol. 1 (C.A. Brebbia, W.L. Wendland, G. Kuhn, eds.) Berlin Heidelberg New York: Springer-Verlag 1987, 519 - 532

57 Kröner, H.: Ein Verfahren zum Auffinden elliptischer Löcher in elastisch isotropen Scheiben mit Hilfe einer Randelementmethode. Fortschrittsberichte VDI Reihe 1: Konstruktionstechnik/Maschinenelemente Nr. 128. Düsseldorf: VDI Verlag 1985

58 Cruse, T.A.: Two- and three-dimensional problems of fracture mechanics. In: Developments in Boundary Element Methods 1, (P.K. Banerjee, R. Butterfield eds.) London: Applied Science Publishers Ltd 1979

59 Blandford, G.E.; Ingraffea, A.R.; Liggett, J.A.: Two-dimensional stress intensity factor computations using the boundary element method. Int. J. Numer. Methods Eng. 17 (1981) 387-404

60 Li, H.-B.; Han, G.-H.; Mang, A.H.: A new method for evaluating singular integrals in stress analysis of solids by the direct BEM. Int. J. Numer. Method Eng. 21 (1986) 2071-2098

61 Möhrmann, W.: DBETSY Industrial application of the BEM. In: Boundary Elements IX Vol. 1 (C.A. Brebbia, W.L. Wendland, G. Kuhn, eds.) Berlin Heidelberg New York: Springer-Verlag 1987, 593 - 607

62 Cody, W.J.: Chebyshev Approximations for the Complete Elliptic Integrals K and E, Math. of Computations 19 (1965) 105-112

63 Márkus, G., Theorie und Berechnung rotationssymmetrischer Bauwerke, Düsseldorf: Werner-Verlag 1978

64 Ruotsalainen, K.; Wendland, W.: On the boundary element methods for some nonlinear boundary value problems. To appear.

65 Bialecki, R.; Nowak, A.J.: Boundary value problems for nonlinear material and nonlinear boundary conditions. Applied Math. Modelling 5 (1981) 417- 421

66 Kikuta, M.; Togoh, H.; Tanaka, M.: Boundary element analysis of nonlinear transient heat conduction problems. Comp. Methods Appl. Mech. Eng. 62 (1987) 321-329

67 Mukherjee, S.: Boundary element methods in creep and fracture. London, New York: Applied Science Publishers 1982

68 Tan, C.L.; Lee, K.H.: Elastic-plastic stress analysis of a cracked thick-walled cylinder. In: Boundary Integral Equation Methods in Stress Analysis. London: Mechanical Engineering Publications 1983, 50-57

69 Hartmann, F.: Integral representations of plate bending solutions. Unpublished manuscript; excerpts in appendix A.

70 Zotemantel, R.: Berechnung von Platten nach der Methode der Randelemente. Dissertation University of Dortmund 1985

71 Hartmann, F.: A note on the domain force integral of Kirchhoff plates. Engineering Analysis 2 (1985) 111-112

72 Blum, Numerical treatment of corner and crack singularities. In: CISM Lecture Notes Udine, Wien: Springer-Verlag 1986

73 Melzer, H.; Rannacher, R.: Spannungskonzentrationen in Eckpunkten der Kirchhoffschen Platte. Bauingenieur 55 (1980) 181-184

74 Rüsch, H.; Hergenröder, A.: Einflußfelder der Momente schiefwinkliger Platten. 3. Auflage. Düsseldorf: Werner-Verlag 1969, Blatt 98

75 Puttonen, J.; Varpasuo, P.: Boundary element analysis of a plate on elastic foundations. Int. J. Numer. Meth. Eng. 23 (1986) 287-303

76 Costa, J.A.; Brebbia, C.A.: Elastic buckling of plates using the boundary element method. In: Boundary Element Methods, Proc. 7th Int. Conf. Lake Como 1985, C.A. Brebbia (Ed.). Berlin Heidelberg New York: Springer-Verlag, 4-29 - 4-42

77 Zienkiewicz, O.C.; Kelly, D.W.; Bettess, P.: Marriage à la mode - the best of both worlds (finite elements and boundary integrals). In: Energy Methods in Finite Element Analysis. R. Glowinski, E.Y. Rodin, O.C. Zienkiewicz (Eds.). Chichester New York: John Wiley & Sons 1979

78 Johnson, C.; Nédélec, J.C.: On the coupling of boundary integral and finite element methods. Mathematics of Computation 35 (1980) 1063-1079

79 Wendland,W.: On asymptotic error estimates for combined BEM and FEM. In: CISM Lecture Nodes Udine (1986) Wien: Springer-Verlag

80 Costablel, M.: A symmetric method for the coupling of finite elements and boundary elements. MAFELAP VI (ed. J.R. Whiteman) London: Academic Press 1987

81 Polizzotto, C.: A consistent formulation of the BEM within elastoplasticity. Proc. Symposium on Advanced Boundary Element Methods, San Antonio (eds. T.A. Cruse et al.)

82 Beer, G.: Finite element, boundary element and coupled analysis of unbounded problems in elastostatics. Int. J. Numer. Meth. Eng. 19 (1983) 567-580

83 Li, H.-B.; Han, G.-M.; Mang, H.A.; Torzicky, P.: A new method for the coupling of finite element and boundary element discretized subdomains of elastic bodies. Comp. Meth. Appl. Mech. Eng. 54 (1986) 161-185

84 Tullberg, O.; Bolteus, S.: A critical study of different boundary element matrices. In: Boundary Element Methods in Engineering, Proc. 4th Int. Seminar Southampton 1982, C.A. Brebbia (Ed.). Berlin Heidelberg New York: Springer-Verlag 1982, 621 - 635

85 Tullberg, O.: BEMDYN - A boundary element program for two-dimensional elastodynamics. In: Boundary Elements, Proc. 5th Int. Conf. Hiroshima 1983. Berlin Heidelberg New York: Springer-Verlag 1983, 835-845

86 Wong, G.K.K.; Hutchinson, J.R.: An improved boundary element method for plate vibrations. In: Boundary Element Methods, Proc. 3rd Int. Seminar Irvine, C.A. Brebbia (Ed.). Berlin Heidelberg New York: Springer-Verlag 1985, 272-290

87 Akyol, T.P.: Ein Beitrag zur Berechnung der von rotationssymmetrischen Maschinenstrukturen abgestrahlten Schalleistung mit Hilfe einer Randelement-Methode. Dissertation Universität Dortmund 1984

88 Burton, A.J.; Miller, G. F.: The application of integral equation methods to the numerical solution of some exterior boundary-value problems. Proc. Roy. Soc. London A323 (1971)

89 Kress, R.: Minimizing the condition number of boundary integral operators in acoustic and electromagnetic scattering. NAM-Bericht Nr. 5, Inst. f. num. u. angew. Math. Universität Göttingen 1983

90 Nardini, D.; Brebbia, C.A.: A new approach to free vibration analysis using boundary elements. In: Boundary Element Methods in Engineering, Proc. 4th Int. Conference Boundary Element Methods, Southampton University, 1982, Brebbia, C.A. (Ed.). Berlin Heidelberg New York: Springer-Verlag 1982

91 Wrobel, L.C.; Telles, J.C.F.; Brebbia, C.A.: A dual reciprocity boundary element formulation for axisymmetric diffusion problems. In: Boundary Elements VIII, Proceedings of the 8th Int. Conf., Tokyo Japan September 1986, Tanaka, M., Brebbia, C.A. (Eds.). Berlin Heidelberg New York Tokyo: Springer-Verlag 1986

92 Latz, K., Dynamische Analyse elastischer Scheiben mit Hilfe der Randelement-Methode, Entwurf am Institut für Angewandte Mechanik, TU Braunschweig, 1987

93 Nardini, D.; Brebbia, C.A.: A new approach to free vibration analysis using boundary elements. In: Proc. 4th Int. Conf. BEM Southampton 1982: Berlin: Springer-Verlag 1982

94 Eringen, C.A.; Şuhubi, E.S.: Elastodynamics, Vol. II. New York London: Academic Press 1974

95 Hartmann, F.: Methode der Randelemente, Boundary Elements in der Mechanik auf dem PC. (German edition of this book). Berlin Heidelberg New York London Paris Tokyo: Springer-Verlag 1986

96 Manolis, G.D.: A comparative study of three boundary element method approaches to problems in elastodynamics. Int. J. Numer. Meth. Eng. 19 (1983) 73-91

97 Alarcon, E.; Dominguez; del Cano, F.: Dynamic stiffnesses of foundations. In: New developments in Boundary Element Methods. CML Publications: Southampton 1980, 264-280

98 Ottenstreuer, M.: Das Verfahren der Randelemente — Ein Beitrag zur Darstellung der Wechselwirkung zwischen Bauwerk und Baugrund. Dissertation Ruhr-University Bochum 1981

99 Schoepp, P.: Die Anwendung des Galerkin-Verfahrens auf die Methode der Randelemente. Dissertation University of Dortmund 1988

100 Gipson, G.S.: Boundary Element Fundamentals - Basic Concepts and Recent Developments in the Poisson Equation. Topics in Engineering Series Vol. 2. Southampton: Computational Mechanics Publications 1987

101 Flügge, W.: Handbook of Engineering Mechanics. New York Toronto London: McGraw-Hill Book Company 1962

102 Brebbia, C.A.; Dominguez, J.: Boundary Elements An Introductory Course. Southampton Boston New York St. Louis San Francisco Computational Mechanics Publications McGraw-Hill Book Company 1988

103 Crouch, S.L.; Starfield, A.M.: Boundary Element Methods in Solid Mechanics. London Boston Sydney: George Allen & Unwin 1983

Bibliography

1906

Fredholm, J.: Solution d'un problème fondamental de la Théorie de l'élasticité. Arkiv för Matematik Astronomi och Fysk 2, Vol. 28, 3-8

1907

Lauricella, G.: Sull' integrazione delle equazioni dei corpi elastici isotropi. Rendiconto Academia dei Lincei, Vol. XV, ser. 5, fas. 8, 426-432

Lauricella, G.: Alcune Applicazioni della teoria delle equazioni funzionali alle fisica matematica. Nuovo Cimento, 55, Vol. 3.

1929

Kellogg, O.D.: Foundations of Potential Theory. Berlin: Springer-Verlag 1929 (reprint 1967)
Nemenyi, P: Eine neue Singularitätenmethode für die Elastizitätstheorie. ZAMM 9 (1929) 480-490

1931

Weinel, E.: Die Integralgleichungen des ebenen Spannungszustandes und der Plattentheorie. ZAMM 11 (1931) 349-360

1936

Sobolev, S.: Méthode nouvelle à résoudre le problème de Cauchy pour les équations linéaires hyperboliques normales. Mat. Sb. 1 (1936) 39-72

1949

Massonet, C.: Résolution graphomécanique des problèmes généraux de l'élasticité plane. Bull. Centre Et. Rech. Essais Sc. Génie Civil. Vol. 4, 169-180

1953

Bergmann, S.; Schiffer, M.: Kernel Functions and Elliptic Differential Equations in Mathematical Physics. New York London: Academic Press 1953
Muskhelishivili, N.I.: Singular Integral Equations. Groningen: P. Noordhoff N.V. 1953

1956

Massonet, C.: Solution générale du problème aux tensions de l'élasticité tridimensionnelle. 9e Congrès Int. Méc. Appl. Bruxelles, Vol. 8, 168-180

1960

Babich. V.M.: Fundamental solutions of the dynamical equations of elasticity for nonhomogeneous media. PMM 25 (1960) 38-45

1961

Fichera, G.: Linear elliptic equations of higher order in two independent variables and singular integral equations. In: Proc. Conf. Partial Differential Equations on Continuum Mechanics, R.E. Langer (ed.) University Press (Madison-Wisconsin)

1963

Jaswon, M.A.: Integral equation methods in potential theory-I. Proc. Roy. Soc. 273 A (1963) 237-246
Nordgren, R.P.: On the method of Green's function in the thermoelastic theory of shallow shells. Int. J. Engng. Sci. 1 (1963) 279-308

1964

Hess, J.L.; Smith, A.M.O.: Calculations of nonlifting potential flow about arbitrary three-dimensional bodies. J. Ship. Res. 8(2) (1964) 22-44

Jahanshahi, A.: Some notes on singular solutions and the Green's functions in the theory of plates and shells. J. Appl. Mech. Trans. ASME 31 (1964) 441-446

Symm, G.T.: Integral equation methods in elasticity and potential theory. Ph.D. thesis London University 1964

1965

England, H.: On stress singularities in linear elasticity. Int. J. Engngr. Sci. 9 (1965) 571-585

Günter, N.M.: Potential Theory and its Applications to Basic Problems of Mathematical Physics. New York: Ungar Pub. 1965

Kupradze, V.D.: Potential Methods in the Theory of Elasticity. Jerusalem: Israel program for scientific translations 1965

Massonet, Ch.: Numerical use of integral procedures. In: Stress Analysis, Zienkiewicz, O.C.; Holister, S.G. (Ed.). London New York: John Wiley 1965

1966

Hess, J.L.; Smith, A.M.O.: Calculations of potential flow about arbitrary bodies. Progress in Aeronautical Sciences 8 (1966) 1-138

Seeley, R.T.: Singular integrals and boundary value problem. Amer. J. Math. 88 (1966) 781-809

Stroud, A.H.; Secrest, D.: Gaussian Quadratur Formulas. Englewood Cliffs: Prentice-Hall 1966

1967

Rim, K.; Henry, A.S.: An integral equation method in plane elasticity. NASA Report No. CR-779-1967

Rizzo, F.J.: An integral equation approach to boundary value problems of classical elastostatics. Quart. Appl. Math. 25 (1967) 83-95

Jaswon, M.A.; Maiti, M.; Symm, G.T.: Numerical biharmonic analysis and some applications. Int. J. Solids Struct. 3 (1967) 309-312

1968

Cruse, T.A.; Rizzo, F.J.: A direct formulation and numerical solution of the general transient elastodynamic problem I. J. Math. Analysis and Appl. 22 (1968) 244-259

Cruse, T.A.: A direct formulation and numerical solution of the general transient elastodynamic problem II. J. Math. Analysis and Appl. 22 (1968) 341-355

Jaswon, M.A.; Maiti, M.: An integral equation formulation of plate bending problems, J. Engng. Math. 2 (1968) 83-93

Oliveira, E.R.A.: Plane stress analysis by a general integral method. J. Eng. Mech. Div. Proc. ASCE 94 (1968) 79-101

Segedin, C.M.; Brickell, D.G.A.: Integral equation method for a corner plate. J. Struct. Div. Proc. ASCE 94 (1968) 41-52

1969

Cruse, T.A.: Numerical solutions in three dimensional elastostatics. Int. J. Solids Structures 5 (1969) 1295-1274

Forbes, D.J.; Robinson, A.R.: Numerical analysis of elastic plates and shallow shells by an integral equation method. University of Illinois, Structural Research Series Report 345 (1969)

1971

Bergmann, S.: Integral Operators in the Theory of Linear Partial Differential Equations, (3rd ed.). Berlin Heidelberg New York: Springer-Verlag 1971

Christiansen, S.: Numerical solution of an integral equation with a logarithmic kernel. BIT 11 (1971) 267-287

Stroud, A.H.: Approximate Calculation of Multiple Integrals. Englewood Cliffs: Prentice-Hall 1971

Swedlow, J.L.; Cruse, T.A.: Formulation of boundary integral equations for three-dimensional elasto-plastic flow. Int. J. Solids and Structs. 7 (1971) 144-151

1972

Guerrero, I.; Turteltaub, M.J.: The elastic sphere under arbitrary concentrated surface loads. J. Elasticity 2 (1972) 21-33

Hajdin, N.; Krajcinovic, D.: Integral equation method for solution of boundary value problems of structural mechanics, part I, ordinary differential equations. Int. J. Numer. Methods Eng. 4 (1972) 509-522

Hajdin, N.; Krajcinovic, D.: Integral equation method for solution of boundary value problems of structural mechanics, part II, elliptic partial differential equations. Int. J. Numer. Methods Eng. 4 (1972) 523-539

Watson, J.O.: The analysis of thick shells with holes by integral representation of displacement. Ph.D. Thesis, University of Southampton 1972

1973

Cruse, T.A.: Application of the boundary integral equation method to three-dimensional stress analysis. Computers & Structures 3 (1973) 509-527

Hsiao, G.; MacCamy, R.C.: Solution of boundary value problems by integral equations of the first kind. SIAM Rev. 15 (1973) 687-705

1974

Christiansen, S.: On Green's third identity as a basis for derivation of integral equations. ZAMM 54 (1974) T 185 - T 186

Cruse, T.A.: An improved boundary-integral equation method for three dimensional elastic stress analysis. Computers & Structures 4 (1974) 741-754

Hess, J.L.: The problem of three-dimensional lifting potential flow and its solution by means of surface singularity distributions. Computer Meth. in Appl. Mech. Engng. 4 (1974) 283- 319

Maiti, M.; Chakrabarty, S.K.: Integral equation solutions for simply supported polygonal plates. Int. J. Engng. Sci. 12 (1974) 93-806

McDonald, B.H.; Friedmann, M.; Wexler, A.: Variational solution of integral equations. IEEE Transactions on microwave theory and techniques (MTT-22) 3 (1974) 237-248

1975

Christiansen, S.: Integral equations without a unique solution can be made useful for solving some plane harmonic problems. J. Inst. Math. Appl. 16 (1975) 143-159

Christiansen, S.; Hansen, E.: A direct integral equation method for computing the hoop stress at holes in plane isotropic sheets. J. Elasticity 5 (1975) 1-14

Cruse, T.A.: Two-dimensional BIE fracture mechanics analysis, Appl. Math. Modelling 2 (1975) 287

Davis, P.J.; Rabinowitz, P.: Methods of Numerical Integration. New York: Academic Press 1975

Hess, J.L.; Review of integral-equation techniques for solving potential-flow problems with emphasis on the surface-source method. Comut. Meth. Appl. Mech. Engrg. 5 (1975) 145-196

Heise, U.: The calculation of Cauchy principal values in integral equations for bvp of the plane and three-dimensional theory of elasticity. J. Elasticity 5 (1975) 99-110

Kutt, H.R.: The numerical evaluation of principal value integrals by finite part integration. Numer. Math. 24 (1975) 205-210

Kutt, H.R.: Quadrature formulae for finite part integrals. Report WISK 178, The National Research Institute for Mathematical Science, Pretoria (1975)

Lachat, J.C.: Further development of the boundary integral techniques for elasto-statics. Ph.D. thesis, Southampton University

Snyder, M.D.; Cruse, T.A.: Boundary integral analysis of cracked anisotropic plates. Int. J. Frac. 11 (1975) 315

Vogel, S. M; Rizzo, F.J.: An integral equation formulation of three-dimensional anisotropic elastostatic boundary value problems. J. Elasticity 3 (1975) 203-216

Zabreyko, P.P.: Integral Equations - A Reference Text. Leyden: Noordhoff International Publishing 1975

1976

Banerjee, P.K.: Integral equation methods for analysis of piecewise non-homogeneous three-dimensional elastic solids of arbitrary shape. Int. J. Mech. Sci 18 (1976) 293-303

Fichera, G.; Ricci, P.: The single layer potential approach of boundary-value problems for elliptic equations. In: Lecture Notes in Mathematics 561, pp. 39-50, Berlin: Springer-Verlag 1976

Hansen, E.B.: Numerical solution of integro-differential and singular integral equations for plate bending problems. J. Elasticity 6 (1976) 39-56

Ivanov, V.V.; The Theory of Approximate Methods and Their Application to the Numerical Solution of Singular Integral Equations. Leyden; Noordhoff International Publishing 1976

Lachat, J.C.; Watson, J.O.: Effective numerical treatment of boundary integral equations: a formulation for three-dimensional elastostatics. Int. J. Numer. Methods. Eng. 10 (1976) 991-1005

Maiti, M.; Bela Das; Palit, S.S.: Somigliana's method applied to plane problems of elastic half-spaces. J. Elasticity 6 (1976) 429-439

Nédélec, J.C., Curved finite element methods for the solution of singular integral equations on surfaces in R^3, Comp. Meth. Appl. Mech. Eng. 8 (1976) 61-80

Schweiger, W.; Mayr, M.: On the solution of a certain fluid-structure-coupling problem using the boundary-integral equation method. Mech. Res. Comm 3 (1976) 495-500

Simmonds, J.G.; Bradley, M.R.: The fundamental solution for a shallow shell with an arbitrary quadratic midsurface. J. Appl. Mech. Trans. ASME 43 (1976) 286-290

Wang, S.T.; Blandford, G.E.: Comparison of boundary integral equation and FE methods. J. Eng. Mech. Div. Proc. Am. Soc. Civil Eng. 102 (1976) 1941-1947

de Wolff, S.; de Mey, G.: Numerical solution of integral equations for potential problems by a variational principle. Inform. Proc. Lett. 4 (1976) 136-139

1977

Cruse, T.A.; Snow, D.W.; Wilson, R.B.: Numerical solutions in axisymmetric elasticity. Computers & Structures 7 (1977) 445-451
Filippi, P.: Potentiels de couche pour les ondes mécaniques scalaires. Rév. CETHEDEC 51 (1977) 121-175
Filippi, P.: Layer potentials and acoustic diffraction. J. Sound Vibration 54 (1977) 473-500
Hsiao, G.C.; Wendland, W.L.: A finite element method for some integral equations of the first kind. J. Math. Anal. Appl. 58 (1977) 449-481
Jaswon, M.A.; Symm,G.T.: Integral Equation Methods in Potential Theory and Elastostatics. London New York San Francisco: Academic Press 1977
Katsikadelis, J.T.; Massalas, C.V.; Tzivanidis, G.J.: An integral equation solution of the plane problem of the theory of elasticity. Mech. Res. Com. 4 (1977) 199-208
Melnikov, Y.A.: Some application of the Green's function method in mechanics. Int. J. Solids Structures 13 (1977) 1045-1058
Prössdorf, S.; Silbermann, B.: Projektionsverfahren und die näherungsweise Lösung singulärer Gleichungen. Leipzig: Teubner 1977
Shippy, D.J.; Rizzo, F.J.: An advanced boundary integral equation method for three-dimensional thermoelasticity. Int. J. Numer. Methods Eng. 11 (1977) 1753-1768
Zienkiewicz, O.C.; Kelly, D.W.; Bettess, P.: The coupling of the finite element and boundary solution procedures. Int. J. Numer. Methods Eng. 11 (1977) 355-375

1978

Altiero, N.; Sikarskie, D.: A boundary integral method applied to plates of arbitrary plane form. Computers & Structures 9 (1978) 163-168
Brebbia, C.A.: The Boundary Element Method for Engineers. London: Pentech Press 1978
Brebbia, C.A. (Ed.): Recent Advances in Boundary Element Methods, Proc. 1st Int. Conf. Boundary Element Methods, Southampton University, 1978. London: Pentech Press 1978
Christiansen, S.: A review of some integral equations for solving the Saint-Venant torsion problem. J. Elasticity 8 (1978) 1-20
Clements, D.L.; Rizzo, F.J.: A method for the numerical solution of boundary value problems governed by second order elliptic systems. J. Inst. Maths. Applies. 22 (1978) 197
Giroire, J.; Nedelec, J.C.: Numerical solution of an exterior Neumann problem using a double layer potential. Math. Comp. 32 (1978) 973-990
Heise, U.: The spectra of some integral operators for plane elastostatic boundary value problems. J. Elasticity 8 (1978) 47-49
Heise, U.; Müller, C.H.; Numerical properties of integral equations in which the given boundary values and the sought solution are defined on different curves. Computers & Structures 8 (1978) 199-205
Richter, G.R.: Numerical solution of integral equations of the first kind with nonsmooth kernels. SIAM J. Numer. Anal. 15 (1978) 511-522

1979

Banerjee, P.K.; Butterfield, R., (Ed.): Developments in Boundary Element Methods - 1. London: Applied Science Publishers Ltd. 1979
Friedmann, M.J.: A finite element method for the solution of a potential theory integral equation. Math. Meth. in the Appl. Sci. 1 (1979) 581-587
Krenk, S.: Stress concentration around holes in anisotropic sheets. Appl. Math. Modelling 3 (1979) 137-142
Kupradze, V.D.,(Ed.): Three-dimensional Problems of the Mathematical Theory of Elasticity and Thermoelasticity. Amsterdam: North-Holland Publishing Co. 1979
Lukasiewicz, S.: Local Loads in Plates and Shells. Alphen aan den Rijn: Sitjhoff and Noordhoff and PWN Polish Scientific Publishers 1979
Stakgold, I.: Green's functions and boundary value problems. New York: John Wiley & Sons 1979
Stanisic, M.M.: On the response of thin elastic plates by means of Green's functions. Ing. Archiv 48 (1979) 279-288
Stern, M.: A general boundary integral formulation for the numerical solution of plate bending problems. Int. J. Solids Structures 15 (1979) 769-782
Wendland, W.L.; Stephan, E.; Hsiao, G.C.: On the integral equation method for the plane mixed boundary value problem of the Laplacian. Math. Methods Appl. Sci. 1 (1979) 265-321

1980

Albrecht, J.; Collatz, L.: Numerical Treatment of Integral Equations, ISMN 53. Basel Boston Stuttgart: Birkhäuser Verlag 1980
Altiero, N.J.; Gavazza, S.C.: On a unified boundary-integral equation method. J. Elasticity 10 (1980) 1-9
Banerjee, P.K.; Cathie, D.N.: A direct formulation and numerical implementation of the boundary element method for two-dimensional problems in elasto-plasticity. Int. J. Mech. Sci. 22 (1980) 233-245

Biollay, Y.; First boundary value problem in elasticity: bounds for the displacements and Saint-Venant's principle. ZAMP 31 (1980) 556-567

Brebbia, C.A., (Ed.): New Developments in Boundary Element Methods, Proc. 2nd Int.Conference Boundary Element Methods, Southampton University, 1980. Southampton: CML Publications 1980

Clements, D.L.; A boundary integral equation method for the numerical solution of a second order elliptic equation with variable coefficients. J. Austral. Math. Soc. 22 (Series B) (1980) 218

Ha Duong T.: A finite element method for the double layer potential solutions of the Neumann exterior problem. Math. Methods Appl. Sci. 2 (1980) 191-208

Heise, U.: Systematic compilation of integral equations of the Rizzo type and of Kupradze's functional equations for boundary value problems of plane elastostatics. J. Elasticity 10 (1980) 23-56

Heise, U.; Müller, C.: Numerical calculation of eigenvalues of integral operators for plane elastostatic boundary value problems. Computer Methods in Appl. Mech. Eng. 21 (1980) 17-43

Heise, U.: Integral equations for the mixed boundary value problem in plane elastostatics. Appl. Math. Modelling 4 (1980) 63-66

Mattioli, F.: Numerical instabilities of the integral approach to the interior boundary-value problem for the two-dimensional Helmholtz equation. Int. J. Num. Meth. Eng. 15 (1980) 1303-1313

Lyness, J.N.: Quadrature error functional expansions for the simplex when the integrand function has singularities at the vertices. Math. Comp. 34 (1980) 213-225

Mayr, M.; Drexler, W.; Kuhn, G.: A semianalytical boundary integral approach for axisymmetric elastic bodies with arbitrary boundary conditions. Int. J. Solid Structures 16 (1980) 863-871

Michlin, S.G.; Prößdorf, S.: Singuläre Integraloperatoren. Berlin: Akademie Verlag 1980

1981

Banerjee, P.K.; Butterfield, R.: Boundary Element Methods in Engineering Science. London: McGraw Hill (UK) 1981

Bettess, P.: Operation counts for boundary integral and finite element methods. Int. J. Numer. Methods Eng. 17 (1981) 306-308

Bezine, G.: A boundary integral equation method for plate flexure with conditions inside the domain. Int. J. Numer. Methods Eng. 17 (1981) 1647-1657

Bialecki, R.; Nowak, A.J.: Boundary value problems in heat conduction with nonlinear material and nonlinear boundary conditions. Appl. Math. Modelling 5 (1981) 417-421

Brebbia, C.A., (Ed.): Boundary Element Methods, Proc. of the 3rd Int.Seminar, Irvine, California, July 1981. Berlin Heidelberg New York: Springer-Verlag und CML Publications 1981

Brebbia, C.A., (Ed.): Progress in Boundary Element Methods, Vol. 1. London, Plymouth: Pentech Press 1981

Christiansen, S.: Condition number of matrices derived from two classes of integral equations. Math. Methods. Appl. Sci. 3 (1981) 364-392

Clements, D.L.; Jones, O.A.C.: The boundary integral equation method for the solution of a class of problems in anisotropic elasticity. J. Aust. Math. Soc. 22 (Series B) (1981) 394

Fabrikant, V.; Hoa, S.V.; Sankar, T.S.: On the approximate solution of singular integral equations. Comp. Methods Appl. Mech. Eng. 29 (1981) 19-33

Fusco, F.B. Jr.: A unified formulation of the finite and boundary element methods using energy methods. Appl. Math. Modelling 5 (1981) 263-268

Heise, U.: Removal of the zero eigenvalues of integral operators in elastostatic boundary value problems. Acta Mechanica 41 (1981) 41-46

Heise, U.: Comparison of round-off errors in integral equation formulations of elastostatical boundary value problems. Comp. Meth. Appl. Mech. Eng. 28 (1981) 145-177

Hsiao, G.C.; Wendland, W.L.: Super approximation for boundary integral methods. In: Advances in Computer Methods for Partial Differential Equations IV, R. Vichnevetsky & R.S. Steplemand (Eds.), IMACS Syp. Rutgers Univ. Dept. Comp. Sc. New Brunswick, N.J., 200-205, 1981

Ioakimidis, N.I.: On the weighted Galerkin method of numerical solution of Cauchy type singular integral equations, SIAM J. Numer. Anal. 18 (1981) 1120-1127

Irschik, H.; Ziegler, F.: Application of the Green's function method to thin elastic polygonal plates. Acta Mechanica 39 (1981) 155-169

Kramer, M.A.; Calo, J.M.: An improved computational method for sensitivity analysis: Green's function method with 'AIM', Appl. Math. Modelling 5 (1981) 432-441

Liggett, J.A.; Salmon, J.R.: Cubic spline boundary elements. Int. J. Numer. Methods Eng. 17 (1981) 543-556

Mukherjee, S.; Morjaria, M.: A boundary element formulation for planar time-dependent inelastic deformation of plates with cutouts. Int. J. Solids Structures 17 (1981) 115-126

Okabe, M.: A boundary integral approach in the geoelectrical cavity prospecting. Computer Meth. in Appl. Mech. Eng. 29 (1981) 297-311

Prössdorf, S.; Schmidt, G.: A finite element collocation method for singular integral equations. Math. Nachr. 100 (1981) 33-66

Telles, J.C.F.; Brebbia, C.A.: Boundary element solution for half-plane problems. Int. J. Solid Structures 17 (1981) 1149-1158

1982

Banerjee, P.K.; Shaw, R.P., (Eds.): Developments in Boundary Element Methods - 2. London, New York: Applied Science Publishers 1982

Bird, H.W.K; Sheperd, R.: Wave interaction with large submerged structures. Proc. ASCE 108 (1982) 146-161

Brebbia, C.A., (Ed.): Boundary Element Methods in Engineering, Proc. 4th Int. Conference Boundary Element Methods, Southampton University, 1982. Berlin Heidelberg New York: Springer-Verlag 1982

Butkovskiy, A.G.: Greens Functions and Transfer Functions Handbook. New York: John Wiley 1982

Christiansen, S.: On two methods for elimination of non-unique solutions of an integral equation with logarithmic kernel. Applicable Anal. 13 (1982) 1-18

Crotty, J.M.: A block equation solver for large unsymmetric matrices arising in the boundary integral equation method. Int. J. Numer. Methods Eng. 18 (1982) 997-1017. See also: Letter to the editor by Dr. Stabrowski. 21 (1985) 967-970 and Letter to the editor by Crotty 23 (1986) 725-730

Fenyö, S.; Stolle, H.W.: Theorie und Praxis der linearen Integralgleichungen, Bd. 1-4. Basel, Boston, Sutttgart: Birkhäuser Verlag 1982

Fischer, T.M.: An integral equation procedure for the exterior three-dimensional slow viscous flow. Integral Equations and Operator Theory 5 (1982) 490-505

Gospodinov, G.; Ljutskanov, A.: The boundary element method applied to plates. Appl. Math. Modelling 6 (1982) 237-244

Groenenboom, P.H.L.: The application of boundary elements to steady and unsteady potential fluid flow problems in two and three dimensions. Appl. Math. Modelling 6 (1982) 35-40

Howell, G.C.; Doyle, W.S.: An assessment of the boundary integral equation method for in-plane elastostatic problems. Appl. Math. Modelling 6 (1982) 245-256

Ioakimidis, N.I.: Application of finite-part integrals to the singular integral equations of crack problems in plane and three-dimensional elasticity. Acta Mechanica 45 (1982) 31-47

Johnson, W.C.; Lee, J.K: An integral equation approach to the inclusion problem of elastoplasticity. Trans. ASME 49 (1982) 312-318

Kamiya, N.; Sawaki, Y.: Integral equation formulation for nonlinear bending of plates - formulation by weighted residual method. ZAMM 62 (1982) 651-655

Kamiya, N.; Sawaki, Y.; Nakamura, Y.; Fukui, A.: An approximate finite deflection analysis of a heated elastic plate by the boundary element method. Appl. Math. Modelling 6 (1982) 23-27

Liu, P.L-F.; Abbaspour, M.: An integral equation method for the diffraction of oblique waves by an infinite cylinder. Int. J. Numer. Methods. Eng. 18 (1982) 1497-1504

Mansur, W.J.; Brebbia, C.A.: Numerical implementation of the boundary element method for two dimensional transient scalar wave propagation problems. Appl. Math. Modelling 6 (1982) 299-306

Mukherjee, S.; Morjaria, M.: Comparison of boundary element and finite element methods in the inelastic torsion of prismatic shafts. Int. J. Numer. Methods Eng. 18 (1982) 1576-1588

Mustoe, G.W.; Volait, F.; Zienkiewicz, O.C.: A symmetric direct boundary integral equation method for two-dimensional elastostatics. Res. Mechanica 4 (1982) 57-82

Nedelec, J.C.: Integral equations with non-integrable kernels. Integral Equations and Operator Theory 5 (1982) 562-572

Novati, G.; Brebbia, C.: Boundary element formulation for geometrically nonlinear elastostatics. Appl. Math. Modelling 6 (1982) 136-138

Rangogni, R.; Reali, M.: The coupling of the finite difference method and the boundary element method. Appl. Math. Modelling 6 (1982) 233-236

Redekop, D.: Fundamental solutions for the collocation method in planar elastostatics. Appl. Math. Modelling 6 (1982) 390-393

Utuku, M.; Carey, G.F.: Boundary penalty techniques. Comput. Methods Appl. Mech. Engng 30 (1982) 103-118

van der Weeën, F.: Application of the boundary integral equation method to Reissner's plate model. Int. J. Numer. Methods Eng. 18 (1982) 1-10

Watson, J.O.: Hermitian cubic boundary elements for plane problems of fracture mechanics. Res. Mechanica 4 (1982) 23-43

1983

Arnold, D.N.; Wendland, W.L.: On the asymptotic convergence of collocation methods. Math. Comp. 41 (1983) 349-381

Arvay, K.: Calculation of structures with application of a set of integral equations. ZAMM 63 (1983) T 339 - T 340

Atkinson, K.; Graham, I.; Sloan, I.: Piecewise continuous collocation for integral equations. SIAM J. Numer. Anal. 20 (1983) 172-186

Barber, J.R.: The solution of elasticity problems for the half-space by the method of Green and Collins. Appl. Sci. Res. 40 (1983) 135-157

Bettess, J.A.: Economical solution technique for boundary integral matrices. Int. J. Numer. Methods Eng. 19 (1983) 1073-1077

Brebbia, C.A.,(Ed.): Progress in Boundary Element Methods, Vol. 2. London, Plymouth: Pentech Press 1983

Brebbia, C.A.; Futagami, T.; Tanaka, M.: Boundary Elements, Proceedings of the 5th Int. Conf., Hiroshima, Japan, November 1983. Berlin Heidelberg New York Tokyo: Springer-Verlag 1983

Christiansen, S.: Numerical investigation of an integral equation of Hsiao and MacCamy. ZAMM 63 (1983) T 341 - T 343

Costabel, M.; Stephan, E.: The normal derivative of the double layer potential on polygons and Galerkin approximation. Applicable Anal. 16 (1983) 205-228

Hodous, M.F.; Katnik, R.B.; Bozek, D.G.; Kline, K.A.: Vector processing applied to boundary element algorithms on the CDC CYBER-205. In: Vector and Parallel Computing in Scientific Applications. Paris: Pluralis 1983

Howell, G.C.; Dolye, W.S.: The plane stress/strain analysis of non-homogeneous continua by the boundary integral equation method. Computers & Structures 17 (1983) 603-610

Kuhn, G.; Möhrmann, W.: Boundary element method in elastostatics: theory and applications. Appl. Math. Modelling 7 (1983) 97-105

Lazarenko, M.V.; Tarakanov, V.I.: On the fundamental solutions of the equations of cylindrical shells. In: Mechanics of continuous media. Tomsk: 1983, pp. 35-47 (in Russian)

Nardini, D.; Brebbia, C.A.: A new approach to free vibration analysis using boundary elements. Appl. Math. Modelling 7 (1983) 157 - 162

Noblesse, F.: Integral identities of potential theory of radiation and diffraction of regular water waves by a body. J. Engg. Math. 17 (1983) 1-13

Phan-Thien, N.: On the image system for the Kelvin-state. J. Elasticity 13 (1983) 231-235

Redekop, D.; Thompson, J.C.: Use of fundamental solutions in the collocation method in axisymmetric elastostatics. Computers & Structures 17 (1983) 485-490

Rudolphi, T.J.: An implementation of the boundary element method for zoned media with stress discontinuities. Int. J. Numer. Methods. Eng. 19 (1983) 1-15

Telles, J.C.F.: The Boundary Element Method Applied to Inelastic Problems. Lecture Notes in Engineering, Vol. 1. Berlin Heidelberg New York Tokyo: Springer-Verlag 1983

Tsamasphyros, G.; Theocaris. P.S.; Stassinakis, C.A.: A numerical solution of singular integral equations without using special collocation points. Int. J. Numer. Methods Eng. 19 (1983) 421-430

Venturini, W.S.: Boundary Element Method in Geomechanics. Lecture Notes in Engineering, Vol. 4. Berlin Heidelberg New York Tokyo: Springer-Verlag 1983

van der Weeën, F.: Mixed mode fracture analysis of rectilinear anisotropic plates using singular boundary elements. Computers & Structures 17 (1983) 469-474

1984

Alliney, S.; Tralli, A.: A weak formulation of boundary integral equations, with applications to elasticity problems. Appl. Math. Modelling 8 (1984) 75-80

Banerjee, P.K.; Mukherjee, S.: Developments in Boundary Element Methods - 3. London and New York: Elsevier Applied Science Publishers 1984

Brebbia, C.A. (ed.): Boundary Elements VI. Southampton: Computational Mechanics Publications 1984

Brebbia, C.A.: Boundary Element Techniques in Computer-Aided Engineering. Dordrecht Boston Lancaster: Martinus Nijhoff Publishers 1984

Brebbia, C.A.(Ed.): Topics in Boundary Element Research Vol. 1, Berlin Heidelberg New York Tokyo: Springer-Verlag 1984

Brebbia, C.A.; Telles, J.C.F.; Wrobel, L.C.: Boundary Element Techniques. Theory and Applications in Engineering. Berlin: Springer-Verlag 1984

Burgess, G.; Mahajerin, E.: A comparison of the boundary element and superposition methods. Computers & Structures 19 (1984) 697-705

Costabel, M.: Starke Elliptizität von Randintegraloperatoren erster Art. Habilitationsschrift, TH Darmstadt 1984

Gakwaya, A.; Dhatt, G.; Cardou, A.: An implementation of stress discontinuity in the boundary element method and application to gear teeth. Appl. Math. Modelling 8 (1984) 319-327

Hasegawa, M.; Nakai, S.; Fukuwa, N.; Tamura, T.: Dynamic analysis of a structure embedded in a multilayered medium by the boundary element method. Tokyo: Shimizu Tech. Res. Bull., No.4, 1-7, 1984

Herrera, I.: Boundary Methods. An Algebraic Approach. London: Pitman 1984

Hromadka II, T.V.: The Complex Variable Boundary Element Method. Lecture Notes in Engineering, Vol. 9. Berlin Heidelberg New York Tokyo: Springer-Verlag 1984

Hsiao, G.C.; Kopp, P.; Wendland, W.L.: Some applications of a Galerkin-Collocation method for boundary integral equations of the first kind. Math. Methods Appl. Sci. 6 (1984) 280-325

Ingham, D.B.; Kelmanson, M.A.: Boundary Integral Equation Analysis of Singular, Potential and Biharmonic Problems. Lecture Notes in Engineering, Vol. 7. Berlin Heidelberg New York Tokyo: Springer-Verlag 1984

Irschik, H.: A boundary-integral equation method for bending of orthotropic plates. Int. J. Solids Structures 20 (1984) 245-255

Kawase, H.; Nakai, S.: Dynamic structure-soil-structure interaction analysis by boundary element method. Tokyo: Shimizu Tech. Res. Bull, No.3, 19-25, 1984

Kitahara, M.; Niwa, Y.; Hirose, S.; Yamazaki, M.: Coupling of numerical Green's matrix and boundary integral equations for the elastodynamic analysis of inhomogeneous bodies on elastic halfspace. Appl. Math. Modelling 8 (1984) 397-407

Kobayashi, S.; Hishimura, N.; Kawakami, T.: Simple layer potential method for domains having external corners. Appl. Math. Modelling 8 (1984) 62-66

Mackerle, J.; Andersson, T.: Boundary element software in engineering. Advances in Eng. Software 6 (1984) 66-102

Martinez, J.; Dominguez, J.: On the use of quarter-point boundary elements for stress intensity factor computations. Int. J. Numer. Methods Eng. 20 (1984) 1941-1950

McCartney, L.N.: A new boundary element technique for solving plane problems of linear elasticity: improved theory and an application to fracture mechanics. Appl. Math. Modelling 8 (1984) 243-250

Meric, R.A., Boundary element methods for optimization of distributed parameter systems, Int. J. Numer. Methods Eng. 20 (1984) 1291-1306

Mukehrjee, S., Morajaria, M., On the efficiency and accuracy of the boundary element method and the finite element method. Int. J. Numer. Methods Eng. 20 (1984) 515-522

Quinghua Du; Zhenhan Yao; Guoshu Song: Solution of some plate bending problems using the boundary element method. Appl. Math. Modelling 8 (1984) 15-22

Radaj, D.; Möhrmann,W.; Schilberth,G.: Economy and convergence of notch stress analysis using boundary and finite element methods. Int. J. Numer. Methods Eng. 20 (1984) 565-572

Rank, E.; A-posteriori error estimates and adaptive refinement for some boundary integral element methods. In: Proceedings of the ARFEC-Conference Lisbon, Portugal, 19 - 22 June 1984

Shi, Z.C.: On the convergence rate of the boundary penalty method. Int. J. Numer. Methods Eng. 20 (1984) 2017-2032

Sládek, J.; Sládek, V.: Boundary integral equation method in thermoelasticity: part II crack analysis. Appl. Math. Modelling 8 (1984) 27-36

Sládek, J.; Sládek, V.: Boundary integral equation method in thermoelasticity: part III uncoupled thermoelasticity. Appl. Math. Modelling 8 (1984) 413-418

Stephan, E.: Boundary Integral Equations for Mixed Boundary Value Problems, Screen and Transmission Problems in R^3. Habilitationsschrift, TH Darmstadt 1984

Stephan, E.; Wendland, W.: An augmented Galerkin procedure of the boundary integral method applied to two-dimensional screen and crack problems. Applicable Analysis 18 (1984) 183-220

Tepavitcharov, A.; Gospòdinov, G.: The boundary integral equation method applied to shallow membrane shell of positive Gaussian curvature. Appl. Math. Modelling 8 (1984) 179-187

Theocaris, P.S.; Tsamasphyros, G.; Theotokoglou, E.E.: A combination of the finite element and singular-integral equation methods for the solution of the generally cracked body, Int. J. Numer. Methods Eng. 20 (1984) 2065-2075

Venturini, W.S.; Brebbia, C.A.: Boundary element formulation for nonlinear applications in geomechanics. Appl. Math. Modelling 8 (1984) 251-260

Wu, J.C.: Fundamental solutions and numerical methods for flow problems. Int. J. Numer. Methods Fluids 4 (1984) 185-201

Zietsman, J.F.W.: The coupled finite element and boundary integral analysis of ocean wave loading: a versatile tool. Computer Meth. Appl. Mech. Eng. 44 (1984) 153-176

1985

Adeyeye, J.O.; Bernal, M.J.; Pitman, K.E.: An improved boundary integral equation method for Helmholtz problems. Int. J. Numer. Methods Eng. 21 (1985) 779-787

Alarcon, E.; Reverter, A.; Molina, J.: Hierarchical boundary elements. Computers & Structures 20 (1985) 151-156

Aliabadi, M.H.; Hall, W.S.: Taylor expansions for singular kernels in the boundary element method. Int. J. Numer. Methods. Eng. 21 (1985) 2221-2236

Athanasiadis, G.: Direct and indirect boundary element methods for solving the heat conduction problem. Computer Methods in Applied Mechanics and Engineering 49 (1985) 37-54

Bamberger, A.; Ha Duong, T.: Diffraction d'une onde acoustique par une paroi absorbante: Nouvelles équations intégrales. Rapport Int. 121, Centre Math. Appl. Ecole Polytechnique, Palaiseau, France.

Bezine, G.; Cimetiere, A.; Gelbert, J.P.: Unilateral buckling of thin elastic plates by the boundary integral equation method, Int. J. Numer. Methods Eng. 21 (1985) 2189-2199

Brebbia, C.A. (ed.). Topics in Boundary Element Research. Time-dependent and Vibration Problems. Berlin: Springer-Verlag 1985

Brebbia, C.A.; Maier, G. (Eds.): Boundary Elements VII, Proceedings of the 7th Int. Conf., Villa Olmo, Lake Como, Italy, September 1985. Berlin Heidelberg New York Tokyo: Springer-Verlag 1985

Brebbia, C.A.; Noye, B.J.: BETECH 85, Proceedings of the 1st boundary element technology conference, South Australia Inst. of Technology, Adelaide, Australia, Nov. 1985. Berlin Heidelberg New York Tokyo: Springer-Verlag 1985

Burczynski, T.: The boundary element method for stochastic potential problems. Appl. Math. Modelling 9 (1985) 189-194

Burczynski, T.; Adamczyk, T.: The boundary element formulation for multiparameter structural shape optimization. Appl. Math. Modelling 9 (1985) 195-200

Chandra, A.; Mukherjee, S.: A boundary element formulation for sheet metal forming. Appl. Math. Modelling 9 (1985) 175-182

Clements, D.L.; Haselgrove, M.: The boundary integral equation method for the solution of problems involving elastic slabs. Int. J. Numer. Methods Eng. 21 (1985) 663-670

Costabel, M.; Stephan, E.: A direct boundary integral equation method for transmission problems. J. Math. Anal. Appl. 106 (1985) 367-413

Cruse, T.A.: Recent advances in boundary element analysis methods. Comp. Methods Appl. Mech. Eng. 62 (1985) 227-244

Delves, L.M.; Mohamed, J.L.: Computational Methods for Integral Equations. Cambridge London New York New Rochelle Melbourne Sydney: Cambridge University Press 1985

Hartmann, F.; Katz, C.; Protopsaltis, B.: Boundary elements and symmetry. Ing. Archiv 55 (1985) 440-449

Herrera, I.: Unified approach to numerical methods, part 1. Green's formulas for operators in discontinuous fields. Numer. Meth. Partial Diff. Equations 1 (1985) 12-37

Hromadka, T.V.: The Complex Variable Boundary Element Method. Lecture Notes in Engineering, Vol. 9. Berlin Heidelberg New York Tokyo: Springer-Verlag 1985

Hromadka II, T.V.; Yen, C.C.; Guymon, G.L.: The complex variable boundary element method: applications. Int.J.Numer. Methods Eng. 21 (1985) 1013-1025

Hromadka II, T.V.; Pardoen, G.C.: Application of the CVBEM to non-uniform St.Venant torsion. Computer Meth. in Appl. Mech. Eng. 53 (1985) 149-161

Hume III, E.C.; Brown, R.A.; Deen, W.M.: Comparison of boundary and finite element methods for moving-boundary problems governed by a potential. Int. J. Numer. Methods Eng. 21 (1985) 1295-1314

Kawahara, M.; Kashimaya, K.: Boundary type finite element method for surface wave motion based on trigonometric function interpolation. Int. J. Numer. Methods Eng. 21 (1985) 1833-1852

Lamp, U.; Schleicher, K.-T.; Wendland, W.: The fast Fourier transform and the numerical solution of one-dimensional boundary integral equations. Numer. Math. 47 (1985) 15-38

de Mey, G.: The auxiliary boundary element method for time dependent problems. Journal of Computational and Applied Mathematics 12 & 13 (1985) 239-245

Michlin, S.G.: Fehler in numerischen Prozessen. Berlin: Akademie-Verlag 1985

Mitsui, Y.; Ichikawa, Y.; Obara, Y.; Kawamoto, T.: A coupling scheme for boundary and finite elements using a joint element. Int. J. Numer. Methods Eng. 9 (1985) 161-172

Mykhas'kiv, V.V.; Khai, M.V.: Reduction of three-dimensional dynamical elasticity theory problems with arbitrarily located plane slits to integral equations. Prikl. Matem. Mekhan. 49 (1985) 719-724

Niessner, H.; Ribaut, M.: Condition of boundary integral equations arising from flow computations. J. Computational and Appl. Math. 12 & 13 (1985) 491-503

Perucchio, R.; Ingraffea, A.R.: An integrated boundary element analysis system with interactive computer graphics for three-dimensional linear-elastic fracture mechanics. Computers & Structures 20 (1985) 157-171

Rannacher, R.; Wendland, W.L.: On the order of pointwise convergence of some boundary element methods, part 1. Operators of negative and zero order. RAIRO Model. Math. Anal. Numerér 19 (1985) 65-88

Rizzo, F.J.; Shippy, D.J.; Rezayat, M.: A boundary integral equation method for radiation and scattering of elastic waves in three dimensions. Int. J. Numer. Methods. Eng. 21 (1985) 115-129

Sgallari, F.: Primal-dual variational problems by boundary and finite elements. Appl. Math. Modelling 9 (1985) 246-252

Sgallari, F.: A weak formulation of boundary integral equations for time dependent parabolic formulations. Appl. Math. Modelling 9 (1985) 295-301

Stephan, E.; Wendland, W.: An augmented Galerkin procedure for the boundary integral method applied to mixed boundary value problems. Applied Numerical Mathematics 1 (1985) 121-143

Vable, M.: An algorithm based on the boundary element method for problems in engineering mechanics, Int. J. Numer. Methods. Eng. 21 (1985) 1625 - 1640

Wendland, W.L.: Splines versus trigonometric polynomials, h-versus p.version in 2-D boundary integral methods. Dundee Biennial Conference on Numerical Analysis 1985 (D. Griffiths, R. Mitchell eds.)

Ye, T.Q.; Liu, Y.: Finite deflection analysis of elastic plates by the boundary element method. Appl. Math. Modelling 9 (1985) 183-188

Zastrow, U.: On the formulation of the fundamental solutions for orthotropic plane elasticity. Acta Mechanica 57 (1985) 113-122

1986

Alarcon, E.; Reverter, A.: p-adaptive boundary elements. Int. J. Numer. Methods Eng. 23 (1986) 801-829

Ang, W.T.; Clements, D.L.: A boundary element method for determining the effect of holes on the stress distribution around a crack. Int. J. Numer. Methods Eng. 23 (1986) 1727-1737

Antes, H.; Spyrakos, C.C.: Time domain boundary element method approaches in elastodynamics: a comparative study. Computer & Structures

Antes, H.: A boundary element procedure for transient wave propagations in two-dimensional isotropic elastic media. Finite Elements in Analysis and Design 1 (1986) 313-323

Aramaki, G.: Boundary elements for thin layers with high permeability in Biot's consolidation analysis. Appl. Math. Modelling 10 (1986) 82-86

Bakr, A.A., The Boundary Integral Equation Method in Axisymmetric Stress Analysis Problems. Lecture Notes in Engineering, Vol. 14. Berlin Heidelberg New York Tokyo: Springer-Verlag 1986

Boutros, Y.Z.; Anwar, M.N.; Tewfick, A.H.: Application of boundary integral equation method for modelling unsteady nonlinear water waves. Appl. Math. Modelling 10 (1986) 11-15

Castellano, L.; Thoma, U.: Application of the boundary element method in free jet impingement problems. Appl. Math. Modelling 10 (1986) 93-96

Coleman, C.J.: A boundary integral formulation of the Stefan problem. Appl. Math. Modelling 10 (1986) 445-449

Costabel, M.; Wendland, W.L.: Strong ellipticity of boundary integral operators. Journal für die reine und angewandte Mathematik (Crelles Journal) 372 (1986) 34-63

Cruse, T.A.; Polch, E.Z.: Elastoplastic BIE analysis of cracked plates and related problems. Part 1: Formulation, Int. J. Numer. Methods Eng. 23 (1986) 429-437

Cruse, T.A.; Polch, E.Z.: Elastoplastic BIE analysis of cracked plates and related problems. Part 2: Numerical results, Int. J. Numer. Methods Eng. 23 (1986) 439-452

Curran, D.A.S.; Lewis, B.A.; Cross, M.: A boundary element method for the solution of the transient diffusion equation in two dimensions. Appl. Math. Modelling 10 (1986) 107-113

Fabrikant, V.I.: Computer evaluation of singular integrals and their applications. Int. J. Numer. Methods Eng. 23 (1986) 1439-1453

Hartmann, F.; Zotemantel, R.: The direct boundary element method in plate bending. Int. J. Numer. Methods Eng. 23 (1986) 2049-2069

Hebeker, F.K.: Efficient boundary element methods for three-dimensional exterior viscous flows. Numer. Meth. Partial Diff. Equations (1986)

Hsiao, G.C.: On the stability of integral equations of the first kind with logarithmic kernels. Arch. Rational Mech. Analysis 94 (1986) 179-192

Ingber, M.S.; Ambar, K.M.: Grid optimization for the boundary element method. Int. J. Numer. Methods Eng. 23 (1986) 2121-2136

Ingham, D.B.; Kelmanson, M.A.: A note on the comparison between BIE and FD techniques for solving elliptic bvps with boundary singularities. Communications in Applied Num. Methods 2 (1986) 189

Jirousek, J.; Guex, L.: The Hybrid-Trefftz finite element model and its applications to plate bending. Int. J. Numer. Methods. Eng. 23 (1986) 651-694

Katsikadelis, J.T.; Kallivokas, L.F.: Clamped plates on Pasternak-type elastic foundation by the boundary element method. J. Appl. Mech. ASME 53 (1986) 909-917

Karabalis, D.L.; Beskos, D.E.: Dynamic response of 3-D embedded foundations by the boundary element method. Computer Methods Appl. Mech. Eng. 56 (1986) 91-119

Manolis, G.D.; Banerjee, P.K.: Conforming versus non-conforming boundary elements in three-dimensional elastostatics. Int. J. Numer. Methods Eng. 24 (1986) 1885-1904

Korkut, L.; Mikelić, A.: The potential integral for a polynomial distribution over a curved triangular domain. Int. J. Numer. Methods Eng. 23 (1986) 2277-2285

Manolis, G.D.; Beskos, D.E.; Pineros, M.F.; Beam and plate stability by boundary elements. Computers & Structures 22 (1986) 917-923

Minato, A; Tone, T; Miya, A.: Three-dimensional analysis of magnetic field distortion of ferromagnetic beam-plates by the boundary element method. Int. J. Numer. Methods Eng. 23 (1986) 1201-1216

Murai, T.; Kagawa, Y.: Boundary element iterative techniques for determining the interface boundary between two Laplace domains - a basic study of impedance plethysmography as an inverse problem. Int. J. Numer. Methods Eng. 23 (1986) 35-47

Niwa, Y.; Hirose, S.; Kitahara, M.: Elastodynamic analyis of inhomogeneous anisotropic bodies. Int. J. Solids Structures 22 (1986) 1541-1555

Ohga, M.; Shigematsu, T.; Hara, T.: Structural analysis by a combined boundary element-transfer matrix method. Computers & Structures 24 (1986) 385-389

Paris, F.; De Leon, S.: Simply supported plates by the boundary integral equation method. Int. J. Numer. Methods Eng. 23 (1986) 173-191

Paris, F.; De Leon, S.: Boundary element method applied to the analysis of thin plates. Computers & Structures 25 (1986) 225-233

Payne, F.R.; Corduneanu, C.C.: Haji-Sheikh, A.; Huang, T. (eds.). Integral Methods in Science and Engineering. Berlin: Springer-Verlag 1986.

Peng Xiaolin; He Guangqian: Computation of fundamental solutions of the boundary element method for shallow shells. Appl. Math. Modelling 10 (1986) 185-189

Providakis, C.P.; Beskos, D.E.: Dynamic analysis of beams by the boundary element method. Computers & Structures 22 (1986) 957-964

Rabinowitz, P.: Some practical aspects in the numerical evaluation of Cauchy principal value integrals. Intern. J. Computer Math. 20 (1986) 283-298

Rangogni, R.: Numerical solution of the generalized Laplace equation by coupling the boundary element method and the perturbation method. Appl. Math. Modelling 10 (1986) 266-270

Sawada, T.; Imanari, M.: Error estimates of numerical integration in boundary element method analysis. Bulletin of JSME 29 (1986) 4072-4079

Schippers, H.: Multigrid methods for boundary integral equations. Numer. Math. (1986)

Sládek, V.; Sládek, J.; Balas, J.: Boundary integral equation formulation of crack problems. ZAMM 66 (1986) 83-94

Sládek, J.; Sládek, V.: Dynamic stress intensity factors studied by boundary integro-differential equations. Int. J. Numer. Methods Eng. 23 (1986) 919-928

Sládek, J.: Sládek, V.: Improved computation of stresses using the boundary element method. Appl. Math. Modelling 10 (1986) 249-255

Tanaka, Y.; Honma, T.; Kaji, I.: On mixed boundary element solutions of convection-diffusion problems in three dimensions. Appl. Math. Modelling 10 (1986) 170-175

Tanaka, M.; Brebbia, C.A. (eds.): Boundary Elements VIII. Southampton: Computational Mechanics Publications 1986

Wendland, W.L.; Christiansen, S.: On the condition number of the influence matrix belonging to some first kind integral equations with logarithmic kernel. Applicable Analysis 21 (1986) 175-183

Zamani, N.G.; Peters, F.H.: Application of the boundary element method in elctrodeposition problems. Appl. Math. Modelling 10 (1986) 262-265

1987

Amini, S.: An iterative method for the boundary element solution of the exterior acoustic problem. Journal of Computational and Applied Mathematics 20 (1987) 109-117

Belotserkovskii, S.M.; Lifanov, I.K.; Soldatov, M.M.: The method of discrete singularities in plane problems of the theory of elasticity with non-smooth boundaries. PMM U.S.S.R. 51 (1987) 219-266

Beskos, D.E.: Boundary element methods in dynamic analysis. Applied Mechanics Reviews 40 (1987)

Brebbia, C.A. (ed.): Topics in Boundary Element Research. Computational Aspects. Berlin: Springer-Verlag 1987

Brebbia, C.A. (ed.): Topics in Boundary Element Research. Applications in Geomechanics. Berlin: Springer-Verlag 1987

Brebbia, C.A.; Connor, J.J. (eds.): Betech 86. Southampton: Computational Mechanics Publications 1987

Brebbia, C.A.; Wrobel, L.C.; Telles, J.C.F.: Boundary Elements - A Course for Engineers. Southampton: Computational Mechanics Publications 1987

Brebbia, C.A.; Venturini, W.S. (eds.): Boundary Element Techniques - Applications in Fluid Flow and Computational Aspects. Southampton: Computational Mechanics Publications 1987

Brebbia, C.A.; Venturini, W.S. (eds.): Boundary Element Techniques - Applications in Stress Analysis and Heat Transfer. Southampton: Computational Mechanics Publications 1987

Chandra, A.; Mukherjee, S.: A boundary element analysis of metal extrusion processes. J. Appl. Mech. 54 (1987) 335-340

Cheng, A., H.-D.; Predelanu, M.: Transient boundary element formulation for linear poroelasticity. Appl. Math. Modelling 11 (1987) 285-290

Chuang, J.M.; Zamani, N.G.; Hsiung, C.C.: Some computational aspects of BEM simulation of cathodic protection systems. Appl. Math. Modelling 11 (1987) 371-379

Cruse, T.A.: Recent advances in boundary element analysis methods. Computer Meth. in Appl. Mech. Eng. 62 (1987) 227-244

Cruse, T.A. (ed.): Advanced Boundary Element Methods. (IUTAM Symposium), in prep.

Das, S.; Kostrov, B.V.: On the numerical boundary integral equation method for three-dimensional dynamic shear crack problems. Journal of Applied Mechanics 54 (1987) 99-104

Dohner, J.L.; Shoureshi, R.; Bernhard, R.: Transient analysis of three-dimensional wave propagation using the boundary element method. Int. J. Numer. Methods Eng. 24 (1987) 621-634

Dumir, P.C.; Mehta, A.K.: Boundary element solution for elastic orthotropic half-plane problems. Computers & Structures 26 (1987) 431-438

Evans, G.A.; Forbes, R.C.; Hyslop, J.: Solution of non-linear Fredholm integral equations by a variational approach. Intern. J. Computer Math. 22 (1987) 149-159

Fan, T.Y.: A perturbation-boundary integral equation method for transient dynamics of nonlinear elastic materials. Appl. Math. Modelling 11 (1987) 296-300

Fares, N.; Li, V.C.: An indirect boundary element method for 2-D finite/infinite regions with multiple displacement discontinuities. Engineering Fracture Mechanics 26 (1987) 127-141

Fawzi, T.H.; Safar, Y.A.: Boundary methods for the analysis and design of high-voltage insulators. Comp. Methods Appl. Mech. Eng. 60 (1987) 343-369

Gerstle, W.H.; Martha, L.F.; Ingraffea, A.R.: Finite and boundary element modelling of crack propagation in two and three dimensions. Engineering with Computers 2 (1987) 167-183

Gilbert, R.P.; Manganini, R.: The boundary integral method for two-dimensional orthotropic materials. Journal of Elasticity 18 (1987) 61-82

Gosh, N.; Mukherjee, S.: A new boundary element method formulation for three dimensional problems in linear elasticity. Acta Mechanica 67 (1987) 107-119

Gründemann, H.: A general boundary integral approach to elliptical boundary value problems. Engineering Analysis 4 (1987) 165-173

Guiggiani, M.; Casalini, P.: Direct computations of Cauchy principal value integrals in advanced boundary elements. Int. J. Numer. Methods Eng. 24 (1987) 1711-1720

412

Haitjema, H.M.: Evaluating solid angles using contour integrals. Appl. Math. Modelling 11 (1987) 69-71

Hebeker, F.K.: An integral equation of the first kind for a free boundary value problem of the stationary Stoke's equation. Math. Mech. in the Appl. Sci. 9 (1987) 550-575

Heise, U.: Dependence of the round-off error in the solution of boundary integral equations on a geometrical scale factor. Computer Meth. in Appl. Mech. Eng. 62 (1987) 115-126

Heuer, R.; Irschik, H.: A boundary element method for eigenvalue problems of polygonal membranes and plates. Acta Mechanica 66 (1987) 9-20

Ioakimidis, N.I.: Validity of the hypersingular integral equation of crack problems in three-dimensional elasticity along the crack boundaries. Engineering Fracture Mechanics 26 (1987) 783-788

Ioakimidis, N.I.: Application of Betti's reciprocal work theorem to the construction of the hypersingular integral equation of a plane crack in three-dimensional elasticity. J. Elasticity 18 (1987) 165-171

Joe, S.: Discrete Galerkin methods for Fredholm integral equations of the second kind. IMA J. Num. Analyis 7 (1987) 149- 164

Johnson, D.: Plate bending by a boundary point method. Computers & Structures 26 (1987) 673-680

Karageorghis, A.: Numerical solution of a shallow dam problem by a boundary element method. Computer Meth. in Appl. Mech. Eng. 61 (1987) 265-276

Kagiwada, H.H.; Kalaba, R.E.: The b and h functions for integral equations with displacement kernels: a computational method and an application to radiative transfer. Applied Mathematics and Computation 23 (1987) 93-101

Kang Yong Lee; Dong Sung Won; Hyung Jip Choi: Boundary element analysis of stress intensity factors for z-shaped cracks. Engineering in Fracture Mechanics 27 (1987) 75-82

Katsikadelis, J.T.; Kokkinos, F.T.: Static and dynamic analysis of composite shear walls by the boundary element method. Acta Mechanica 68 (1987) 231-250

Kelmanson, M.A.: A direct boundary formulation for the Oseen flow past a two-dimensional cylinder of arbitrary cross-section. Acta Mechanica 68 (1987) 99-119

Kikuta, M.; Togoh, H.: Boundary element analysis of nonlinear transient heat conduction problems. Computer Meth. in Appl. Mech. Eng. 61 (1987) 321-329

Lefeber, D.: Solving Problems with Singularities Using Boundary Elements. Southampton: Computational Mechanics Publications 1987

Lin, H.-T.; Tassoulas, J.L.: Discrete Green functions for layered strata. Int. J. Numer. Methods Eng. 24 (1987) 1645-1658

Loines, J.; Bernal, M.J.M.: An integral equation method for the solution of 3-dimensional, non-linear, magnetostatic problems. Int. J. Numer. Methods Eng. 24 (1987) 1551-1562

Luchi, M.L.; Rizzuti, S.: Boundary elements for three-dimensional elastic crack analysis. Int. J. Numer. Methods Eng. 24 (1987) 2253-2271

Mackerle, J.; Brebbia, C.A.: The boundary element software handbook. Southampton: Computational Mechanics Publications 1987

Mariem, J.B.; Hamdi, M.A.: A new boundary finite element method for fluid-structure interaction problems. Int. J. Numer. Methods Eng. 24 (1987) 1251-1267

Mullen, R.L.; Rencis, J.J.: Iterative methods for solving boundary element equations. Computers & Structures 25 (1987) 713-723

Neta, B.; Nelson, P.: An adaptive method for the numerical solution of Fredholm integral equations of the second kind. I. Regular kernels. Appl. Math. Computation 21 (1987) 171-184

Pangiotopoulos, P.D.: Multivalued boundary integral equation for inequality problems. The convex case. Acta Mechanica 70 (1987) 145-167

Power, H.; Miranda, G.; Second kind integral equation formulation of Stokes' flows past a particle of arbitrary shape. SIAM J.appl. math. 47 (1987) 689-698

Quinghua, D. (Ed.) : Boundary Elements. Proceedings of the international conference, Beijing 1986. New York: Pergamon Press 1987

Rannacher, R.; Wendland, W.L.: On the order of pointwise convergence of some boundary element methods. Part II: Operators of positive order. Math. Modelling and Numer. Analysis, to appear

Schnack, E.: A hybrid BEM model. Int. J. Numer. Methods Eng. 24 (1987) 1015-1025

Sheng, C.F.: Boundary element method by dislocation distribution. Journal of Applied Mechanics 54 (1987) 105-109

Sládek, J.; Sládek, V.: A boundary integral equation method for dynamic crack problems. Engin. Fracture Mech. 27 (1987) 269-277

Sloan, I.H.: A quadrature-based approach to improving the collocation method. Numer. Math. to appear

Stadelmann, R.; Praxisgerechte Plattenberechnung mittels Randelementmethode. Heft 8 (1987) Institut für Baustatik, TU Graz, Austria (plate bending and BEM)

Stabrowski, M.M.: A block equation solver for large unsymmetric linear equation systems with dense coefficient matrices. Int. J. Numer. Methods Eng. 24 (1987) 289-300

Tanaka, M.; Itoh, H.: New crack elements for boundary element analysis of elastostatics considering arbitrary stress singularities. Appl. Math. Modelling 11 (1987) 357-363

Tanaka, M. (ed.): Theory and applications of boundary element methods. Proc. 1st Japan-China Symposium, 1-6 June 1987. New York: Pergamon Press 1987

Tanaka, Y.; Toshishisa, H.; Kaji, I.: Mixed boundary element solution for three-dimensional convection-diffusion problem with a velocity profile. Appl. Math. Modelling 11 (1987) 402-410

Telles, J.C.F.: A self-adaptive co-ordinate transformation for efficient numerical evaluation of general boundary element integrals. Int. J. Numer. Methods Eng. 24 (1987) 959-973

Teong, A.W.; Clements, D.L.: A boundary integral equation method for the solution of a class of crack problems. Journal of Elasticity 17 (1987) 9-21

Tsamasphyros, G.: Methods for combination of finite element and singular integral equation methods. Comp. Methods Appl. Mech. Eng. 60 (1987) 45-46

Tsamasphyros, G.; Androudlidakis, P.: The tanh transformation for the solution of singular integral equations. Int. J. Numer. Methods Eng. 24 (1987) 543-556

Umetani, S.: Adaptive boundary element methods in elastostatics. Southampton: Computational Mechanics Publications 1987

Vable, M.: Making the boundary element method less sensitive to changes or errors in the input data. Int. J. Numer. Methods Eng. 24 (1987) 1533-1540

Wang, X.: Existence and uniqueness theorems for the solution of boundary integral equations for linearized elastostatics. Computers & Structures 25 (1987) 365-369

Wrobel, L.C.; Brebbia, C.A.: The dual reciprocity boundary element formulation for nonlinear diffusion problems. Comp. Methods Appl. Mech. Eng. 65 (1987) 147-164

Zamani, N.G.; Chuang, J.M.; Porter, J.F.: BEM simulation of cathodic protection systems employed in infinite electrolytes. Int. J. Numer. Methods Eng. 24 (1987) 605-620

Zielinski, A.P.; Herrera, I.: Trefftz method: fitting boundary conditions. Int. J. Numer. Methods Eng. 24 (1987) 871-891

1988

Achenbach, J.D.; Kechter, G.E.; Xu, Y.-L.: Off-boundary approach to the boundary element method. Computer Meth. in Appl. Mech. Eng. 701 (1988) 191-201

Ahmad, S.; Banerjee, P.K.: Multi-domain bem for two-dimensional problems of elastodynamics. Int. J. Numer. Methods Eng. 26 (1988) 891-911

Anwar, M.N.; Sherief, H.H.: Boundary integral equation formulation of generalized thermoelasticity in a Laplace- transform domain. Appl. Math. Modelling 12 (1988) 161-166

Azevedo, J.P.S.; Wrobel, L.C.: Non-linear heat conduction in composite bodies: a boundary element formulation. Int. J. Numer. Methods Eng. 26 (1988) 19-38

Banerjee, P.K.: Advanced elastic and inelastic three-dimensional analysis of gas turbine engine structures by bem. Int. J. Numer. Methods Eng. 26 (1988) 393-411

Banerjee, P.K.; Ahmad, S.; Wang, H.C.: A new BEM formulation for the acoustic eigenfrequency analysis. Int. J. Numer. Methods Eng. 26 (1988) 1299-1309

Brebbia, C.A.; Dominguez, J.: Boundary Elements - An Introductory Course. Southampton: Computational Mechanics Publications 1988

Budreck, D.E.; Achenbach, J.E.: Scattering from three-dimensional planar cracks by the boundary integral equation method. J. Appl. Mech. ASME 55 (1988) 405-412

Chen, D.R.; Sheu, M.J.: Investigation of numerical solutions of integral equation methods for multi-element aerofoils. Comp. Methods Appl. Mech. Eng. 68 (1988) 345-364

Choi, J.H.; Kwak, B.M.: Boundary integral equation method for shape optimization of elastic structures. Int. J. Numer. Methods Eng. 26 (1988) 1579-1595

Costa Jr., J.A.: Plate vibrations using B.E.M. Appl. Math. Modelling 12 (1988) 78-85

Delaye, A.: Quadrature formulae for singular functions. Intern. J. Computer Math. 23 (1988) 167-176

Denda, M.: A complex variable Green's function representation of plane inelastic deformation in isotropic solids. Acta Mechanica 72 (1988) 205-221

Dravanski, M; Mossessian, T.K.: On evaluation of the Green functions for harmonic line loads in a viscoelastic half space. Int. J. Numer. Methods Eng. 26 (1988) 823-841

Fabrikant, V.I.: Green's functions for a penny-shaped crack under normal loading. Eng. Fracture Mech. 30 (1988) 87-104

Gu, H.; Yew, C.H.: Finite element solution of a boundary integral equation for mode I embedded three-dimensional fractures. Int. J. Numer. Methods Eng. 26 (1988) 1525-1540

Gupta, R.S.; Banik, N.C.: Constrained integral methods for solving moving boundary problems. Comp. Methods Appl. Mech. Eng. 67 (1988) 211-221

Henry Jr., D.P.; Banerjee, P.K.: A variable stiffness type boundary element formulation for axisymmetric elastoplastic media. Int. J. Numer. Methods Eng. 26 (1988) 1005-1027

Henry Jr., D.P.; Banerjee, P.K.: A new BEM formulation for two- and three-dimensional thermoelasticity using particular integrals. Int. J. Numer. Methods Eng. 26 (1988) 2061-2077

Henry Jr., D.P.; Banerjee, P.K.: A new BEM formulation for two- and three-dimensional elastoplasticity using particular integrals. Int. J. Numer. Methods Eng. 26 (1988) 2079-2096

Ioakimidis, N.I.; Pitta, M.S.; Remarks on the Gaussian quadrature rule for finite-part integrals with a second-order singularity. Computer Meth. in Appl. Mech. Eng. 69 (1988) 325-343

Jost, G.: Integral equations with modified fundamental solution in time-harmonic electromagnetic scattering. IMA J. Appl. Math. 40 (1988) 129-144

414

Karageorghis, A.; Fairweather, G.: The Almansi method of fundamental solutions for solving biharmonic problems. Int. J. Numer. Methods Eng. 26 (1988) 1665-1682

Kleinman, R.E.; Martin, P.A.: On single integral equations for the transmission problem of acoustics. SIAM J. Appl. Math. 48 (1988) 307-325

Lee, K.Y.: Boundary element analysis of stress intensity factors for bimaterial interface cracks. Eng. Fracture Mechanics 29 (1988) 461-472

Lefeber, D.: Solving Problems with Singularities Using Boundary Elements - Topics in Engineering Series. Vol. 4. Southampton: Computational Mechanics Publications 1988

Manolis, G.D.; Beskos, D.E.: Boundary Element Methods in Elastodynamics. London: Unwin Hyman Ltd. 1988

Meric, A.R.: Shape design sensitivity analysis for non-linear anisotropic heat conducting solids and shape optimization by the BEM. Int. J. Numer. Methods Eng. 26 (1988) 109-120

Ohga, M.; Shigematsu, T.: Bending analysis of plates with variable thickness by boundary element-transfer matrix method. Computers & Structures 28 (1988) 635-640

Papia, M.: Analysis of infilled frames using a coupled finite element and boundary element solution scheme. Int. J. Numer. Methods Eng. 26 (1988) 731-742

Polizzotto, C.: An energy approach to the boundary element method. Part I: Elastic solids. Computer Meth. in Appl. Mech. Eng. 69 (1988) 167-184

Polizzotto, C.: An energy approach to the boundary element method. Part II: Elastic-plastic solids. Computer Meth. in Appl. Mech. Eng. 69 (1988) 263-276

Pullan, A.J.: Boundary element solutions of quasilinearised time-dependent infiltration. Appl. Math. Modelling 12 (1988) 9-17

Ruotsalainen, K.; Saranen, J.: A dual method to the collocation method. Math. Meth. in the Appl. Sci. 10 (1988) 439- 445

Sandgren, E.; Wu, S.-J.: Shape optimization using the boundary element method with substructuring. Int. J. Numer. Methods Eng. 26 (1988) 1913-1924

Sturla, F.A; Barber, J.R.: Thermoelastic Green's functions for plane problems in general anisotropy. J. Appl. Mech. ASME 55 (1988) 245-247

Wang, H.-C.; Banerjee, P.K.; Axisymmetric free-vibration problems by boundary element method. J. Appl. Mech. ASME 55 (1988) 437-442

Wei Feng; Yong-Yuang Zhang: A boundary element procedure for estimating the residual fatigue life of three-dimensional crack problem under cyclic loadings with varying amplitudes. Engineering Fracture Mechanics 29 (1988) 151-157

Zamani, N.G.: Boundary element simulation of the cathodic protection system in a prototype ship. Appl. Math. Comp. 26 (1988) 119-134

Zhang, J.D.; Atluri, S.N.: Post-buckling analysis of shallow shells by the field-boundary-element method. Int. J. Numer. Methods Eng. 26 (1988) 571-587

Index

J. Mackerle, University of Linköping, Sweden; **C. A. Brebbia,** Southampton, UK (Eds.)

The Boundary Element Reference Book

1988. IX, 382 pages. ISBN 3-540-18584-4

Contents: Books: Review of Some Fundamental Concepts and Milestone References on Boundary Elements. Boundary Element Textbooks. Special Interest Books. State of the Art Books. Proceedings. Training Material. – **Software:** Introduction. Tabular Code Presentation. Codes: Heat Transfer and Potential Problems. Fluid Flow. General Purpose. Special Purpose. Coupled BE/FE Programs. Pre- and Post-Processors. – Who's Who in Boundary Elements.

Jointly published by
Springer-Verlag Berlin Heidelberg New York London Paris Tokyo Hong Kong and Computational Mechanics Centre Publications, Southampton-Boston

C. A. Brebbia, University of Southampton, Great Britain; **J. C. F. Telles, L. C. Wrobel,** University of Rio de Janeiro, Brazil

Boundary Element Techniques

Theory and Applications in Engineering

1984. 284 figures. XIV, 464 pages. ISBN 3-540-12484-5

Contents: Approximate Methods. – Potential Problems. – Interpolation Functions. – Diffusion Problems. – Elastostatics. – Boundary Integral Formulation for Inelastic Problems. – Elastoplasticity. – Other Nonlinear Material Problems. – Plate Bending. – Wave Propagation Problems. – Vibrations. – Further Applications in Fluid Mechanics. Coupling of Boundary Elements with Other Methods. – Computer Program for Two-Dimensional Elastostatics. – Appendix A: Numerical Integration Formulas. – Appendix B: Semi-Infinite Fundamental Solutions. – Appendix C: Some Particular Expressions for Two-Dimensional Inelastic Problems. – Subject Index.

Springer-Verlag
Berlin Heidelberg New York
London Paris Tokyo Hong Kong

Springer

Topics in
Boundary Element Research

Edited by C. A. Brebbia

Volume 1

Basic Principles and Applications

1984. 144 figures, 11 tables. XIII, 256 pages.
ISBN 3-540-13097-7

Volume 2

Time-dependent and Vibration Problems

1985. 140 figures, 5 tables. XIV, 260 pages.
ISBN 3-540-13993-1

Volume 3

Computational Aspects

1987. 126 figures, 28 tables. XIV, 296 pages.
ISBN 3-540-16113-9

Volume 4

Applications in Geomechanics

1987. 87 figures. XII, 173 pages.
ISBN 3-540-17497-4

In preparation

Volume 5

Viscous Flow Applications

1989. 64 figures. Approx. 200 pages.
ISBN 3-540-50690-8

Volume 6

Electromagnetic Applications

1989. ISBN 3-540-50607-1

Volume 7

Electrical Engineering Applications

1989. ISBN 3-540-50605-5

Springer-Verlag
Berlin Heidelberg New York
London Paris Tokyo Hong Kong